LIVING WITH
THE CHANGING
CALIFORNIA COAST

D0872417

The publisher gratefully acknowledges the generous contribution to this book provided by the Gordon and Betty Moore Fund in Environmental Studies.

LIVING WITH THE CHANGING CALIFORNIA COAST

GARY GRIGGS, KIKI PATSCH, AND LAURET SAVOY

With contributions by

Reinhard Flick

Kim Fulton-Bennett

Karen Grove

Cheryl Hapke

Kenneth R. Lajoie

Charles F. Lester

Scott Mathieson

Dorothy Merritts

Robert M. Norris

Antony R. Orme

Bernard Pipkin

Derek Rust

Douglas Sherman

Robert Walker

Jerry Weber

UNIVERSITY OF CALIFORNIA PRESS
Berkeley   Los Angeles   London

University of California Press, one of the most distinguished university presses in the United States, enriches lives around the world by advancing scholarship in the humanities, social sciences, and natural sciences. Its activities are supported by the UC Press Foundation and by philanthropic contributions from individuals and institutions. For more information, visit www.ucpress.edu.

University of California Press
Berkeley and Los Angeles, California

University of California Press, Ltd.
London, England

Library of Congress Cataloging-in-Publication Data

Griggs, Gary B.
  Living with the changing California Coast / Gary Griggs, Kiki Patsch, and Lauret Savoy ; with contributions by Reinhard Flick ... [et al.].
      p. cm.
  Rev. ed. of: Living with the California coast / edited by Gary Griggs and Lauret Savoy. 1985.
  Includes bibliographical references and index.
  ISBN 0-520-24445-1 (cloth : alk. paper) — ISBN 0-520-24447-8 (pbk. : alk. paper)
  1. Shore protection—California. 2. Coasts—California. 3. Coastal zone management—California. I. Patsch, Kiki, 1977- II. Savoy, Lauret E. III. Living with the California coast. IV. Title.
  TC224.C2G75 2005
  551.45'7'09794—dc22

                          2004030728

Manufactured in the United States of America

14   13   12   11   10   09   08   07   06   05
10   9   8   7   6   5   4   3   2   1

The paper used in this publication meets the minimum requirements of ANSI/NISO Z39.48-1992 (R1997) (Permanance of Paper).

CONTENTS

PREFACE: LIVING ON THE EDGE

The coastline of California in many ways defines the state, whether the wide expanse of sand and palm trees at Santa Monica, the rugged mountains of Big Sur, or the driftwood-covered and often fog-shrouded beaches of Humboldt and Mendocino counties. The coast is an area originally inhabited by Native Americans, who thrived on its abundant food resources. It is also the part of California that was first sighted by early explorers from Spain and initially populated by them and other Europeans. The coastline continues to attract people today. Although property prices throughout the state have been gradually rising since the 1980s, prices along the coast have risen astronomically.

Most of southern California's oceanfront is heavily urbanized, with houses and apartments, condominiums and townhouses, restaurants and hotels. Today the only near-pristine coast exists in the far northern counties and along Big Sur, where the topography is too rugged to allow development and much of the land is permanently protected in parkland or reserves.

Although much of the existing oceanfront construction occurred many years ago, when coastal hazards were less clearly understood, development has continued in recent years despite the climatic assaults and storm damage of the past several decades. Property values, the desire to live on the coast, and political pressure have all contributed to this continued development and migration to the edge.

We now understand that coastal climate—including rainfall, sea-level elevation, and storm wave attack, all of which can threaten oceanfront development—oscillates over time. Cooler, La Niña–dominated cycles, with more moderate storm climate, alternate with more damaging El Niño cycles, which characterized the period between 1978 and 1998, when storm damage in the millions of dollars could occur within a few days. We know that sea level has risen and fallen throughout Earth's history in response to changing global climate. For the past 18,000 years, since the last ice age, sea level has risen as ice caps melt and glaciers retreat. Sea level continues to rise gradually today, and this rise is expected to continue for the indefinite future. The question is not whether sea level is rising, but how high it will rise, when it will reach this level, and how this rise will affect individual coastal communities and development. The natural pattern is for the shoreline to retreat in response to a higher sea level, threatening the extensive development that exists on the beaches, dunes, bluffs, and cliffs. But as a result of human intervention for 75 years or more, many miles of the coastline are now armored with seawalls, revetments, or other protective structures. Sea-level rise, future storm impacts, coastal erosion, and seawall construction are common topics of discussion and debate in coastal communities and at public hearings from one end of the state to the other.

For many people, building on the coast or living along the shoreline is the California dream, but it can be a nightmare during severe winter storms. Thirty-three homes were completely destroyed by waves in January 1983, and additional losses occurred in 1998. Before one approves new construction and invests one's life savings in that ocean view, it is important to realize that the entire coastline of the state is subject to wave attack and erosion, and anything built to protect property is going to fail eventually. Our goal in this book is to provide some perspective on the challenges and issues involved in living with the California coast. What natural processes and hazards should we be concerned about? What can we reasonably expect in the future? What are the options available to us for dealing with these threats? And what land use or permit issues do we need to be aware of?

The first edition of this book, titled *Living with the California Coast,* was published 20 years ago. Shoreline development has continued through the intervening years, and although our awareness of the hazards has increased, so has the damage inflicted by these hazards. At the same time, property values have risen and more armor has been put in place.

The first eight chapters provide the reader with some basic background on how the shoreline works, the processes and hazards that occur here, things to consider before buying or building, options in hazardous locations, and how policies and legislation influence our response. The second part of the book consists of twelve regional chapters covering the coastline from the Oregon border to Mexico. In this section, the authors and coastal geologists familiar with specific regions describe these individual areas, including what we know about their geology, hazards, and histories. Every mile of the coast is depicted in a series of coastal maps that delineate the type of coastal environment, threats to development, locations of coastal armor, and information on coastal erosion rates and past storm damage.

Many homeowners, community representatives, and local and state agency officials are looking for long-term answers and solutions. We hope this new edition provides readers with an increased appreciation of the coastline's dynamic nature and also with some perspective on how we might respond to the hazards and issues facing California's coast.

CHAPTER ONE

INTRODUCTION

A Perspective on the Coast of California

GARY GRIGGS, KIKI PATSCH, AND LAURET SAVOY

California's nearly 1,100 miles of shoreline have something to offer almost everybody: beaches to escape the heat of the city on a summer day; fertile near-shore waters that support an extensive, but threatened, sport and commercial fishing industry; private coastal property, a place to build a home for recreation or retirement; and some wild and isolated places where one can still find solitude and peace of mind. These are but a few of the shoreline's benefits. The magnetic attraction of California's shoreline is clearly seen in a single statistic: 80 percent of the state's 35 million people now live within 30 miles of the shoreline, and this number continues to increase. In the 20 years since the first edition of this book was published, the state's population has grown 40 percent, from 25 to 35 million, and many of these people have moved to coastal communities. If we divide the state's shoreline evenly among its residents, these 35 million people would each have a little less than 2 inches to enjoy. To make matters worse, only about 400 miles of the shoreline consists of beaches, so if space at the beach is what one is looking for, California residents have only about three-quarters of an inch of shoreline apiece. Unfortunately, the state's residents must share the beaches with 100 million visitors each year, which further reduces their space. California is now the nation's most populous state and ranks as the world's fifth largest economy. The coast has become increasingly popular for homes, condominiums, hotels, and restaurants.

Coastal conflicts, however, are becoming increasingly evident. They are not unique to California, but in a state known for its excesses they tend to be more widely publicized. The newspaper headline MOVIE STARS' OCEAN FRONT HOMES BATTERED BY WAVES IN MALIBU is a good illustration. Whereas the early residents of the California coast usually set their homes back a reasonable distance from bluffs or the approach of storm waves and high tides, the desire today is often to encroach as close as permits or engineering will allow, with homes even built on the beach or cantilevered out over the water (Figure 1.1) or on the side of a cliff or bluff (Figure 1.2).

The continuing growth of California's coastal population, the desirability of living on the shoreline, and the resulting astronomical prices of oceanfront property have led to political pressure and economic motivation to develop much of the state's privately held coastal land. Because of its inherent geological instability, however, the coastline is not the safest place to build. Many diverse forces and processes interact on the coast, making the coastline one of the world's most dynamic environments. Waves, tides, wind, storms, rain, and runoff combine to build up, wear down, and continually reshape the interface of land and sea. Many of California's coastal cliffs are eroding relatively quickly (a foot or so a year), but the entire 1,100 miles of shoreline is actually undergoing erosion.

If we could look back in time about 18,000 years, we would find that the climate was considerably cooler and the Earth was in the waning stages of a period of extensive glaciation. Nearly 11 million cubic miles of seawater was bound up on the continents in the form of ice caps and glaciers that covered a large portion of the Earth's surface. The removal of this seawater from the oceans led to a worldwide drop in sea level of about 350 feet. The shoreline along the coast of California at that time was 10 to 20 miles offshore to the west of its present location. As the climate warmed, the ice caps began to melt and the glaciers retreated. The meltwater flowed into the ocean, and sea level rose globally at an average rate of nearly half an inch a year, flooding all that offshore real estate, until about 5,000 years ago. From that time until the present, the rate of sea level rise has slowed, although it has continued at about a tenth of an inch annually for the past century.

Herein lies the dilemma: Much of California's and the world's people choose to build right up to the edge, even though the location of the edge constantly changes. The dimensions of oceanfront parcels have been surveyed down to a hundredth of a foot, but the Pacific Ocean is 10,000 miles wide and doesn't care much about a few feet either way at the margins.

In the not-too-distant past, the ongoing natural processes of cliff erosion and shoreline retreat either were not recognized, were not completely understood, or were ignored by most coastal builders, developers, and home buyers. Within the past 25 years, however, this problem has come into clear focus along virtually the entire coastline of California. Public and private losses during the 1978 El Niño event, the first serious coastal storms in several decades, amounted to over $50 million in 2003 dollars (for consistency, all damage or cost values will be reported in 2003 dollars). Five years later, the high tides and storm waves during the El Niño winter of 1983 inflicted over

Figure 1.1   Houses in Malibu built overhanging the beach and on an alluvial fan at the mouth of a coastal stream. Photo by Gary Griggs and Kiki Patsch.

$184 million in damage to oceanfront property. Damage was not restricted to broken windows and flooding of low-lying areas—33 oceanfront homes were totally destroyed, and dozens of businesses, park improvements, roads, and other public infrastructure were heavily damaged (Figure 1.3). In mid-January 1988 very large waves struck the southern California coastline suddenly and left $43 million in property damage. In 1997–98, another major El Niño winter seriously impacted the state's coastline; more property was lost, and more houses were damaged or destroyed (Figure 1.4).

California is not the only state with coastline erosion problems. Much of the shoreline of each coastal state erodes, and the low, sandy coastlines common along the East and Gulf coasts often erode more rapidly than California's coastline. Atlantic and Gulf coast residents also live under the threat of hurricanes and of nor'easters, which can be more damaging than hurricanes.

The conflict between coastline development and the hazards associated with it is becoming more evident for several reasons: (1) an increased migration to coastal communities and the desirability of owning oceanfront property; (2) the progressive erosion of oceanfront yards and vacant property, causing structures and utilities to be undercut or threatened; (3) the human-induced acceleration of seacliff erosion due to cliff-top construction with its associated roof, patio, driveway, and street runoff and its landscape watering;

Figure 1.2   House on the side of a bluff at Bolinas undermined by erosion during the 1983 winter. Photo by Gary Griggs and Kiki Patsch.

and (4) an era of more frequent and severe El Niño events beginning in 1978, bringing heavy rainfall, elevated sea level, and larger waves. In addition, coastal engineering projects such as groins, jetties, and breakwaters have directly or indirectly accelerated erosion rates in adjacent areas, principally by trapping sand and starving down-coast beaches.

The climate in southern California has drawn more people to this portion of the coastline than elsewhere in the state. The warm days and wide

Figure 1.3 Logs and debris carried into Rio Del Mar by high tides and storm waves during 1983 El Niño storms. Photo by Gary Griggs and Kiki Patsch.

Figure 1.4 Bluff erosion led to the collapse of portions of homes in Pacifica during the 1998 El Niño winter. Photo by Monty Hampton, U.S. Geological Survey.

summer beaches can be deceiving, however, as cliff-top and beachfront dwellers in places like Malibu, Solana Beach, and Cardiff discover during the winter months. Beach sand moves offshore, leaving the decks and plate glass windows of homes and restaurants to absorb the full force of the destructive storm waves.

The construction of seawalls and riprap is the most common response to coastline erosion in California, and a survey completed in 2001 indicates that 107 miles, or 10 percent of the entire California coast, have now been armored. At present-day costs of $1,000 to $7,500 per front-foot of armor, or $5 to $40 million per mile, protection is getting very costly, and concerns continue to be raised about spending public funds to protect private property as well as building seawalls or revetments on public beaches. These complex issues, discussed in Chapter 7, will plague California residents for years to come.

For many, living on the ocean is a dream come true. For others it has become a nightmare and an expensive lifestyle to maintain. Joe Maschutes, for example, moved into a $350,000 beachfront house in Santa Cruz in early January 1983 (at today's prices this house would probably be valued at $3.5 million); 10 days later, after a severe storm, his house slid into Monterey Bay, leaving nothing but a small pile of wreckage. An indication that we do not always learn from the past is evident in the fact that a permit was recently issued to allow construction of a new beachfront house at this same location. Events of this sort are tragic, but we can learn something from them. The Pacific Ocean is a very powerful force to reckon with, as is global sea level rise. Nearly all of our protection efforts are, in the long run, temporary, and they simply buy a little more time at great expense. Unless we are wealthy or have recourse to some government subsidy or aid, we cannot long afford to protect an oceanfront home in an area of active coastal retreat. Yet there are literally thousands of homes in such hazardous locations, and in many other oceanfront areas development continues to be planned or proposed.

Before we invest in oceanfront land, purchase a home along the shoreline, or approve new construction in a coastal area, we should have a clear sense of the history of the site, paying particular attention to what has happened there over the past 50 or 100 years. Where was the historic shoreline at this location and how fast is the shoreline retreating? It's critical to find the answers to these questions before going any further. What can we learn from California's past history of coastline development and construction, especially during the severe winters of the past 25 years? What options exist for protecting our valuable coastal property, and how effective and costly are the different methods? Is it likely that either the California Coastal Commission or the local government permitting agency will approve a proposal for a seawall? In areas not yet intensively developed, what "nonstructural" solutions, such as setbacks or buffer zones based on long-term retreat rates, can be utilized? And finally, what types of land-use planning regulations do we need to consider, and what agencies can or must we work with in our efforts to build, plan, buy, or protect our beach houses?

This book discusses the processes, problems, and issues that anyone living or planning to live directly on the coastline must understand. This book should allow readers to make well-informed decisions about the future of their particular homes or home sites and to learn something of the past, present, and future of the individual geographic areas along the state's 1,100-mile coastline. Although the shoreline may appear stable and solid, a look at the historic photographs we have assembled in this book will provide a different perspective on the permanence of the California coastline.

CHAPTER TWO

THE EVOLUTION OF THE CALIFORNIA COAST

GARY GRIGGS, KIKI PATSCH, AND LAURET SAVOY

INTRODUCTION

The coastline of California means different things to different people depending on one's regional perspective and personal observations. To a surfer in Huntington Beach, the coast is a wide, sandy beach with a perfect wave on an uncrowded sunny afternoon. To a family from San Jose trying to escape the heat of the valley on a summer day, the coast is the boardwalk at Santa Cruz, with a wide, safe beach and gentle waves. For a retired couple living on the wild Mendocino coast, a foggy winter walk on a rocky stretch of coastline backed by steep, redwood-covered mountains is their idea of the coast. The diverse landforms of coastal California reflect the state's complex geological history and the interplay of tectonic processes, geology, climate, and the sea. Huntington Beach, Santa Cruz, and Mendocino all look very different as a result of their geological histories, and each has a story to tell if we can assemble all of the pieces. This is the job of a geologist: to look at the landforms, to study the rocks underlying the coastline, and to put this in the framework of the whole state—and the entire globe for that matter—in an attempt to understand why each part of the landscape looks the way it does and to make an educated guess at how what we see today has been created. Whether a nearly horizontal, uplifted marine terrace in San Diego, a broad, sandy beach in Santa Monica, or a steep coastal cliff in Big Sur, each region of the coast has an interesting but complex history. This history—the evolution of and changes in the shoreline—allows us to appreciate how fortunate we are to live here, but it also provides perspective on the scale of geological change and what we might expect in the future. If we are investing our life savings in a house along the coast, it makes sense to understand where we are putting our money and whether this is a geologically safe long-term investment.

GEOLOGICAL AND TECTONIC SETTING

Early travelers to California quickly became aware of the diverse landscape and topography of the state and the difficulties of moving across the Sierra Nevada and the Coast Ranges, for example, particularly after traveling across the nearly featureless plains and prairies of the mid-continent. However, it wasn't until the great San Francisco earthquake of 1906 that the state's population became aware of California's dynamic nature. A large fragment of California, hundreds of miles long, moving 15 feet along the San Andreas Fault came as a bit of a surprise and, sadly, led to the nearly complete loss of San Francisco. A few years later, in 1915, Mt. Lassen, one of the southernmost of the vast chain of Cascade volcanoes that stretches through Oregon and Washington and into British Columbia, erupted and offered another glimpse of the area's geological foundation. Subsequent major earthquakes in Santa Barbara in 1927, Long Beach in 1933, Bakersfield in 1952, San Fernando in 1971 and again in 1994, and the Santa Cruz Mountains in 1989 have provided reminders nearly each decade that California is very active geologically.

The recent geological history of the coastal areas of California, as well as Oregon, Washington, and Alaska, is strikingly different from that of the Atlantic and Gulf coasts. Even a casual visitor to the coastline will notice the obvious differences between the coastal mountains and seacliffs that characterize much of California's coastal zone and the broad, low-relief coastal plain, sand dunes, and barrier islands of Virginia or North Carolina. These landforms on the Atlantic and Gulf coasts raise their own issues for human occupancy and use of the coastal zone.

The 1960s and 1970s witnessed a revolution in scientific thinking about the Earth and its history and evolution. Much of the initial evidence for the developing theories and concepts came from exploration of the ocean basins, which began in earnest in the 1950s. The discovery of a world-encircling undersea ocean ridge that was volcanically active, a system of deep ocean trenches surrounding the Pacific Ocean with associated chains of active volcanoes (the "Ring of Fire"), and a worldwide band of earthquakes that follows these unique features led to the development of the theory of global or plate tectonics in 1968. New, exciting, and somewhat controversial 35 years ago, this theory now forms part of the basic geological history that our children learn about in elementary school. Plate tectonics provide a comprehensive explanation of the origin of the large-scale features of the Earth—the mountain ranges, volcanoes, trenches, and faults—and how they all fit together.

Because of the heat inside the Earth, the interior of the planet is largely molten. This hot fluid material slowly rises, owing to its lower density, much like a pot of water heated on the stove. Where the fluid material reaches the surface, along the ocean floor for the most part, the crust cracks open, volcanoes form (in Iceland, the Azores, and the Galapagos Islands, for example), and new ocean crust is formed. The hot material beneath the surface spreads out in both directions (a process known as sea floor spreading), carrying along the Earth's crust, the ocean floor, and overlying continents as it moves.

We now recognize that the Earth's surface consists of a series of large, thin (about 60 miles in thickness), rigid plates that move around relative to each other, driven by the flow of hot fluid material beneath the crust. Much like large icebergs floating on water, these plates are in constant although slow motion (one to several inches per year). At their edges these plates may diverge or separate, collide, or slide past one another. The interactions at the edges of these huge plates lead to tectonically active regions where most of the world's earthquakes occur. In fact, it was the recognition of the global distribution of these concentrated bands of earthquakes that led seismologists to delineate the boundaries of these plates.

The coastal region of California is the place where one very large tectonic plate (the North American Plate) has collided with the Pacific Plate (to the west) for millions of years. These collisions and the resulting plate interactions have produced California's unique and dynamic landscape. For many millions of years, California was a collision zone where the thin oceanic plate to the west collided with the thicker, continental plate to the east. This process created a deep offshore trench filled with sediments eroded off the continent, a plunging zone of earthquakes (a subduction zone) where the oceanic plate was forced downward under the continent, and, further landward, a zone of volcanoes (the Cascades and older volcanoes whose roots are now exposed in the Sierra Nevada). Volcanoes, such as Mt. Shasta and Mt. Lassen, form in areas landward of trenches in response to the partial melting of crustal material as it is heated in its descent into the Earth's interior and its subsequent rise and eruption as molten lava on the Earth's surface.

About 15 million years ago, the plates underlying California and the adjacent Pacific Ocean underwent a major readjustment in response to forces we don't completely understand. The collision of the two plates—which had been responsible for the offshore trench, the Sierra Nevada mountain range, the Cascades, and the ancestral Central Valley—was terminated for the most part, and the plates began to move alongside one another as the San Andreas Fault was initiated. Over the subsequent 15 million years, that portion of California west of the San Andreas Fault from San Diego to Cape Mendocino, and including much of California's coastline (Los Angeles, Santa Barbara, Monterey, and Santa Cruz, for example), has moved northwest about an inch and a half per year, on average, relative to the rest of the

Figure 2.1 The steep terrain of the Santa Monica Mountains adjoining the coastline near Malibu. Photo by Gary Griggs and Kiki Patsch.

state to the east. Thus, California is slowly being torn apart, and the small earthquakes that occur daily are evidence of this constant grinding along the boundary between the two huge plates.

The diverse features of California, such as the Sierra Nevada, the San Andreas Fault and its associated earthquakes, the rugged coastal mountains of Mendocino and Big Sur, and the uplifted marine terraces and coastal cliffs that characterize much of California's coast, all have their origins in millions of years of large-scale tectonic processes that continue today (Figure 2.1).

The rocks exposed along the coastline and in seacliffs provide evidence of this complex geological history and the changes the landscape has undergone. Much of the low-cliffed coastline between San Diego and Pismo Beach, and from Monterey to Half Moon Bay, consists of young, weak sedimentary rocks, such as sandstones and shales, which were originally deposited in a relatively shallow offshore marine environment. These sedimentary rocks provide evidence for this area formerly being below sea level. In striking contrast, the Big Sur coast and most of the coastline north of San Francisco is dominated by steep cliffs and rocks of very different origin. These rugged areas are dominated by hard, older materials, such as granite and volcanic rocks, cherts, and a variety of rocks that were formed in an offshore trench, scraped off the deep-sea floor, and then accreted to the continent during the

Figure 2.2 The West Cliff Drive area of Santa Cruz, where the cliffs consist of very hard mudstone. (A) Photograph taken in 1890 shows a wave motor built to pump seawater to the top of the tower.

former period of plate collision. These rocks are all part of a giant tectonic jigsaw puzzle that has slowly been assembled over many years.

It would be difficult to find a coastline anywhere in the world that has had a more complex geological history than California's. It is because of the vastly differing rock types and regional geological histories that sections of the state's coast look so different. Some rocks are so hard and resistant, for example, that photographs taken 50 years ago look identical to those of

(B)

Figure 2.2 *(Continued)* (B) Photograph taken at the same location in 2003 shows little change. Photo by Gary Griggs and Kiki Patsch.

today (Figure 2.2). In other locations, however, coastal bluff materials are so soft and weak that the coast is being eroded at rates of up to 10 feet a year.

Although large-scale coastal landforms such as mountains, uplifted terraces, and seacliffs owe their relief and origin to regional tectonic processes, they have all been subsequently altered by processes such as waves, rainfall and runoff, and landslides or other mass downslope movements. In addition, sea level along the coast has changed continuously through time such that the position of the shoreline is only a temporary one. Although these changes are not rapid, it is clear from historical photographs and geological evidence that the shoreline undergoes constant change. There is no reason to believe that the processes of the past will not continue into the future, and thus will have significant impacts on our attempts to occupy the coast with "permanent" structures.

THE DISTRIBUTION OF COASTAL LANDFORMS

One logical breakdown of the diverse coastline of California is to categorize coastal areas as either (1) steep coastal mountains and seacliffs with hundreds of feet of relief (Figure 2.1); (2) uplifted marine terraces and seacliffs from 10 to several hundred feet in height (Figure 2.3); and (3) coastal lowlands with

Figure 2.3   The uplifted marine terraces and seacliffs of the coast of northern Santa Cruz County. Photo by Gary Griggs and Kiki Patsch.

beaches, sand dunes, and lagoons (Figure 2.4). The great majority (72 percent, or about 790 miles) of the California coast consists of actively eroding seacliffs that include areas (1) and (2). Of this 790 miles, 59 percent, or 650 miles, consists of lower-relief cliffs and bluffs typically eroded into marine terraces; the remaining 13 percent, or 140 miles, consists of high-relief cliffs and coastal mountains.

The high-relief, steep cliffs and coastal mountains are located predominantly in northern California from Del Norte County to Mendocino County, at the Marin Headlands just north of the Golden Gate (Figure 2.5), from Pacifica to Montara in San Mateo County, and along the Big Sur coast of Monterey (and San Luis Obispo counties. High-relief cliffed areas and headlands can be found along several areas of the southern California coastline as well; the Santa Monica Mountains (Figure 2.1) and Point Loma in San Diego County are two examples. These rugged stretches of coast typically consist of older, more resistant rock types and include the Franciscan Formation, a mélange of sedimentary and volcanic rocks, as well as granite. These rocks tend to be much harder and more resistant to erosion, and these rock types form many of the protruding headlands or points along the state's coastline. For example, along the northern California coast, Point St. George, Trinidad

Figure 2.4   The low-relief coast of the Newport Beach area, Orange County. Photo by Gary Griggs and Kiki Patsch.

Figure 2.5   The rugged coastal cliffs of the Marin Headlands north of the Golden Gate. Photo by Gary Griggs and Kiki Patsch.

Figure 2.6   Large-scale landsliding and road repair along the Big Sur coast. Photo by Gary Griggs and Kiki Patsch.

Head, and Point Delgada are all Franciscan Formation outcrops. Bodega Head, Point Reyes, Montara Point, Point Piños, and Point Cypress are all granite. Proceeding further south, Point Sur, Point San Martin, Piedras Blancas, and Point San Luis are all Franciscan Formation. These rock types erode very slowly but, in the case of some units of the Franciscan, may also be subject to large-scale landsliding or slumping, as along the Big Sur (Figure 2.6) and Sonoma County coastlines.

The coastline of California has many excellent examples of nearly horizontal marine terraces, which are characteristic features of collision coasts where uplift has taken place (Figure 2.3). These terraces, which typically resemble a flight of stairs, are commonly less than a mile in width and may ascend to elevations of several hundred feet above present sea level. Each terrace consists of a nearly horizontal or gently seaward-tilting erosional platform backed by a steep or degraded relict seacliff along its landward margin. These flat terraces were formed by wave erosion in the surf zone in the geological past, in much the same way as the intertidal platform that is visible during low tides is being eroded by breaking waves in the surf zone today.

Where multiple terraces have been preserved (Figure 2.3), evidence exists for several sea-level high stands that have been superimposed on a slowly but continuously rising coastline. Thus a rising coastline is a continuous strip chart on which relatively brief sea-level high stands, which are due to

Figure 2.7　An intensively developed marine terrace at Carlsbad in San Diego County. Photo © 2002–2004 Kenneth and Gabrielle Adelman, California Coastal Records Project, www.Californiacoastline.org.

warm climatic periods when more global ice melted and sea level rose, were successively recorded on the coastal landscape.

Low bluffs or cliffs cut into uplifted marine terraces, typically consisting of young sedimentary rocks, characterize much of the coastline of San Diego, Orange, Santa Barbara, San Luis Obispo, Santa Cruz, and San Mateo counties. Along the northern California coast, portions of the Sonoma, Mendocino, Humboldt, and Del Norte county coasts have also been eroded into marine terraces. The number of terraces exposed along the coast of California ranges from 1 in Santa Barbara to as many as 13 on the Palos Verdes peninsula of Los Angeles County. The distribution of these flat, nearly horizontal marine terraces has enabled California's intensive coastal development to take place (Figure 2.7). Ease of access and construction has facilitated development of oceanfront communities situated on these wide-spread terraces. Unfortunately, however, the weak sedimentary rocks that lent themselves to wave erosion, resulting in the formation of these terraces in the geological past, are the same materials exposed in the coastal cliffs today. These rocks are very susceptible to erosion by waves as well as by rainfall, runoff, and landslides, and thus continue to retreat landward. Rising sea level and storm waves will take their toll on these low terraces and bluffs in the future, as they have for the past 20,000 years, whether they are covered with houses or undeveloped.

CHAPTER THREE

WEATHER, CLIMATE CHANGE, SEA LEVEL, AND THE COASTLINE

GARY GRIGGS, KIKI PATSCH, AND LAURET SAVOY

California's climate varies widely with latitude along its nearly 1,100 miles of north-to-south–trending coastline. San Diego, at one extreme, is essentially a desert that receives only about 10 inches of rainfall annually and, before development and the importation of water, was characterized by an arid landscape with only sparse vegetation. At the other extreme, Eureka lies in the heavily redwood-forested northern Coast Ranges, where rainfall often exceeds 50 inches per year. The storms bringing the rainfall are often the same storms bringing the waves that break on the shoreline; they are strongly seasonal, with the highest rainfall and the largest storm waves typically concentrated in the winter months.

Winter storm waves erode beaches and attack seacliffs, threatening roads, houses, and businesses built on the coastline. In spring and summer, more gentle waves rebuild the beaches that provide recreation areas for millions. California's beaches thus undergo dramatic seasonal changes (see Chapter 4).

WEATHER AND THE CALIFORNIA COAST

Most of the serious coastline erosion occurs when major storms coincide with the highest tides. Seacliffs and bluffs are attacked directly by storm waves at these times, and the rainfall can loosen and wash away cliff materials, but also may saturate the cliff or bluff and cause landslides or other forms of mass failure. Before one builds, buys, or approves oceanfront construction, it's useful to know what happens along the shoreline in the non-summer months. What can coastal residents expect in the winter? From which direction do the storms and waves come? What is El Niño, and why is it important if we live on the coastline? If we are going to live or invest along the coast, we ought to understand the natural processes and realize

Figure 3.1   Seasonal changes in beach appearance in central California. (A) Santa Cruz—the beach in the summer. (B) Santa Cruz—the beach in the winter. Photos by Gary Griggs and Kiki Patsch.

that the shoreline in front of the house may look very different in January than it does in July (Figure 3.1).

The weather along California's coast, including storms, waves, and precipitation, is related to the changing seasonal patterns of oceanic and atmospheric circulation within and above the Pacific Ocean, as well as to the offshore and onshore topography at a particular location. Perhaps the single

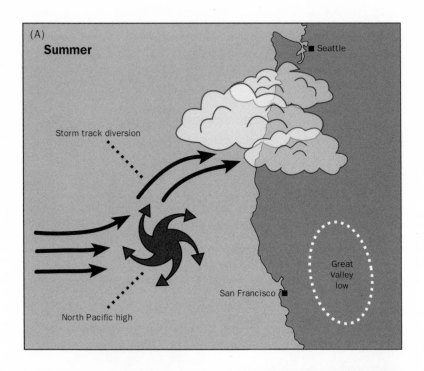

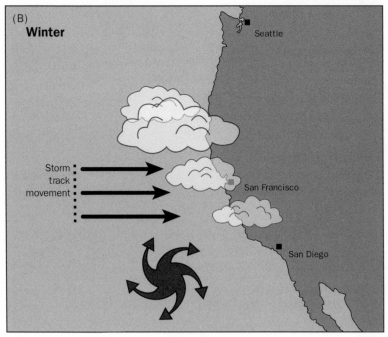

Figure 3.2 North Pacific circulation with different storm paths. (A) Summer conditions. (B) Winter conditions. Drawing by Shannon Griggs.

most important factor affecting what happens along the coast in a particular month of the year is the behavior of one large air mass, the North Pacific High. This feature forms as air that has been warmed by solar heating near the equator rises and moves northward toward the pole. As this air rises, it cools in the upper atmosphere. Some of this cool air sinks toward the ocean surface, forming a column of cool, dense air and therefore high atmospheric pressure several thousand miles north of the equator (Figure 3.2).

The North Pacific High (so named because of the high atmospheric pressure resulting from this cool descending air mass) is an active feature, moving north and south according to seasonal variations in the amount of solar energy received. During Northern Hemisphere spring and summer months, as the area of maximum heating moves north due to the tilt of the Earth on its axis, the high will migrate northward typically to about the latitude of San Francisco (38° north). In the winter months, when the Northern Hemisphere is tilted away from the sun, the area of maximum heating moves south, and in response, the high migrates southward to about the latitude of the Hawaiian Islands (20° north; Figure 3.2).

The North Pacific High varies in strength, size, and relative atmospheric pressure from year to year. It may also fluctuate in how far north or offshore it reaches, as well as how far inland it migrates over western North America. Although these variations in position and strength are unpredictable from year to year, we know that they exert a strong influence on California's coastal climate and the arrival of storms.

The summer climate of the California coast is strongly influenced by the strength and position of the North Pacific High. The high-pressure area offshore either breaks down the large storms moving east from the western Pacific or deflects them northward so that they completely miss the California coast. This is why precipitation throughout California is generally very low in the summer months and occurs only when the high is displaced far to the south or weakens enough to allow storms to move over the coast. During the winter months, however, when the high weakens and moves southward, the California coast is no longer protected, and large storms forming in the western Pacific regularly reach the coast along with rain and large waves. For this reason it is important to realize where these storms come from and how they affect the shoreline. There are three general tracks along which storms advance on the California coast.

STORMS FROM THE ALEUTIAN OR NORTH PACIFIC SOURCE AREA

The most common type of storm reaching California originates in low-pressure areas south of the Aleutian Islands and advances from the northwest down the coast of California, often bypassing the southern part of the state, before turning eastward. Possessing great energy, particularly along California's north coast, these storms may generate 20- to 30-foot waves off-shore and winds up to 40 mph. These waves usually approach from the northwest and impact both the beaches and seacliffs. A close look at the historical record of coastal storm damage in California indicates that the periods of greatest erosion and damage correspond to times when very large waves and high tides occurred simultaneously. A severe storm of this type occurred in December 1940 and January 1941, causing considerable damage along the entire coast, particularly the central and south coast. In December 1977 and January 1978, the north and central coast experienced severe storms. During the winter of 1982–83, similar storms did great damage along nearly the entire length of California's coastline, as did the storms of 1997–98.

During El Niño years, such as 1977–78, 1982–83, and 1997–98, because of the southward displacement of the North Pacific High, the storms approaching the coast of California from Aleutian and North Pacific sources are diverted further to the south and approach the shoreline from the west or southwest. As a result, those coastal areas that are normally protected from northwesterly waves by headlands and wave refraction are often severely damaged (see the following discussions of El Niño and ENSO). Northern Monterey Bay experienced major coastal storm damage during the 1977–78 and 1982–83 winters, which were both El Niño–Southern Oscillation (ENSO) events with large waves approaching from the southwest arriving simultaneously with very high tides. Most of the major storm damage in the central coastal area over the past century has taken place during El Niño years when large waves reached the coast at times when sea level was elevated.

STORMS FROM THE HAWAIIAN ISLAND SOURCE AREA

Storms originating in the open Pacific—often passing through the Hawaiian Islands area, where they pick up considerable moisture from the warm tropical waters—approach the southern California coast from the west. Such storms were responsible for extensive damage to homes, piers, and roads along the southern California coast during the winters of 1977–78 and 1979–80. During these winters, high rainfall produced coastal landslides and greatly accelerated cliff erosion.

HURRICANE-GENERATED STORMS AND SOUTHEASTERS

Hurricane-generated storms create violent winds reaching California from the south, but occur in the summer and early fall rather than during the winter months. Such storms have rarely come to the California coast in recent years, but often devastate the coast of Baja California to the south before moving eastward or westward and dissipating. If this type of storm does reach southern California, it is generally accompanied by southerly winds and huge waves that can be disastrous to south-facing coasts such as Malibu, Newport, Laguna Beach, and Long Beach. Prior to 1983, the last hurricane-generated storm to reach the southern California coast occurred in September 1939. Despite the presence of groins, jetties, breakwaters, and other coastal structures, beaches were inundated and numerous homes and structures at Long Beach and Newport were severely damaged or destroyed.

Following the warm-water ENSO event of 1982–83 in the Pacific, storms of this type have become more frequent. The Hawaiian Islands were hit by such a storm (Hurricane Iniki) in November 1982, and the most severe storm damage of the century befell the island of Kauai. On January 17 and 18, 1988, a largely unforecast "southeaster" hit the coast of southern California with peak offshore wave heights of over 30 feet. Damage to oceanfront structures and infrastructure exceeded $43 million, with the greatest damage occurring in Redondo Beach, Huntington Beach, and San Diego. Residents of the Portofino Inn, inside King Harbor in Redondo Beach, had to be rescued from the roof by helicopter.

CLIMATE CHANGE, INTERDECADAL OSCILLATIONS, AND EL NIÑO EVENTS

WHAT IS EL NIÑO?

The residents of Peru have known for at least four centuries that the intrusion of a current of warm water from the west every few years leads to a dramatic reduction in the population of anchovies in coastal waters. This anchovy decline not only severely hampers the fishing industry but also causes a decline in marine mammals and sea birds that rely on anchovies as a food source. In addition to the ecological impacts, this intrusion of warm water produces torrential rainfall accompanied by floods, mudflows, and landslides. Although the extra rainfall has a positive effect on plant growth in an otherwise dry coastal area, the large-scale negative consequences of this event over time have begun to be recognized.

In Peru, because this warm-water phenomenon often arrived around Christmas, it was given the name El Niño, Spanish for "the child," in reference to the birth of Jesus. More recently, particularly over the past 25 years, the global scale of this event and its widespread and diverse impacts have begun to be more fully appreciated. Scientists came to understand that an El Niño event is linked to an alteration of atmospheric pressure systems over the equatorial Pacific called the Southern Oscillation. This large-scale phenomenon is now more accurately called El Niño–Southern Oscillation, or ENSO, and is characterized by major, interconnected shifts in both atmospheric and oceanic circulation throughout the entire Pacific basin.

In what are thought of as "normal" times, trade winds blow from east to west, both north and south of the equator, and move warm surface waters toward the western equatorial Pacific, where a large pool of warm water accumulates. The surface currents that develop from this pool follow similar paths north and south of the equator as they flow along the edges of Asia and Australia, turn eastward, and head back across the ocean toward North and South America. Every three to seven years, however, the normal atmospheric circulation system over the Pacific Ocean breaks down, the trade winds weaken, atmospheric pressures across the Pacific reverse, and winds actually begin to blow from west to east, moving the warm pool of water gradually back toward South America and initiating an El Niño. The size of the pool of warm water is usually a good indicator of the scope of the ENSO event. Through the use of satellites and moored surface buoys, we can now accurately determine and monitor the size, temperature, and movement of this pool of warm water. It was because of this monitoring that warnings of the major 1997–98 El Niño event were circulated months in advance by NOAA and the National Weather Service. There is, however, a long time interval between the initial oceanographic observations in the spring that suggest an El Niño building in the western Pacific, and its ultimate arrival on the coast of California. Although the major 1997–98 ENSO event was predicted quite accurately, the impact wasn't really felt until January and March of 1998, almost a year after the initial predictions. As a result, many coastal residents concluded that this event simply wasn't going to take place as announced, and some editorials even proclaimed the event as El No Show.

As this warm current moves across the equatorial Pacific Ocean, it raises surface water temperatures enough to severely damage coral reefs. This has been a common occurrence in recent years, and large areas of coral have been decimated (coral bleaching) in Tahiti, the Galapagos, and many other Pacific Islands. When the warmer water reaches the shoreline near Ecuador and Peru, it moves both north and south along the coast, raising local sea level and transporting normally tropical species to more temperate waters; it also causes an increase in the number of tropical hurricanes in the eastern Pacific and typically leads to a precipitation increase along coastal South America.

ENSO EVENTS AND THE COAST OF CALIFORNIA

The effect of ENSO events on the coastline of California was not widely appreciated or fully understood until the last 25 years or so. The El Niño winters of 1977–78, 1982–83, and 1997–98 inflicted major damage on many of the developed and heavily populated portions of California's shoreline, bringing an end to the relatively benign wave and storm climate pattern of the previous three decades.

Coastal erosion and storm damage along the coast of California are maximized when several factors or processes occur simultaneously. Historical reviews focused on the central coast of California have shown that about 75 percent of the storms that caused significant erosion or structural damage between 1910 and 1995 occurred during ENSO events. ENSO-associated factors that contributed to the severe coastal damage during these events include high tides, higher-than-normal sea level, and more frequent and larger storm waves.

The ongoing natural processes of cliff retreat, storm inundation, and beach erosion were either not recognized, not appreciated, not completely understood, or simply ignored by many coastal builders, developers, realtors, and home buyers in the past. Interestingly, much of California's oceanfront development took place between the mid-1940s and the mid-1970s, a period characterized by below-average rainfall and storm frequency. It has become clear in recent years that the Pacific Ocean and the atmosphere above the Pacific undergo what are called interdecadal oscillations, or fluctuations between intervals characterized by warmer conditions and more frequent El Niño events, and intervals associated with cooler periods (known as La Niña events). It is still not completely clear what drives these large-scale, decades-long climate patterns, but careful analysis of a number of different oceanographic and meteorological parameters (sea level, water temperature, and tropical storm frequency, for example) has led to the compilation of an ENSO index, and these trends are now quite clear (Figure 3.3). Although the timing of these decade-scale changes is not predictable, it now seems clear that the Pacific basin and the coast of California in particular undergo cycles of more frequent ENSO events (1978 to 1998, for example), when storm activity, cliff erosion, and beach loss may be far more severe than during La Niña periods (the mid-1940s to 1978, for example).

THE 1982–83 ENSO EVENT

During the months of January, February, and March of 1983, eight major storms struck the coast of California. These storms produced an offshore *significant wave height* (the average height of the highest one-third of the waves) between 16 and 22 feet. The elevated sea levels and large waves damaged breakwaters, piers, park facilities, seawalls, coastal infrastructure, and public and private structures. Winter storm damage along the coast of California totaled over $184 million (in 2003 dollars). Thirty-three oceanfront homes were completely destroyed and 3,000 homes and businesses

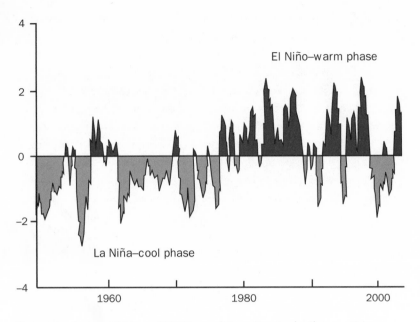

Figure 3.3 A 50-year history of El Niño conditions. Drawing by Shannon Griggs.

were damaged (Figures 3.4 and 3.5). Oceanfront public recreational facilities suffered an estimated $56 million in damage (in 2003 dollars).

Along the central coast, wave impact, flooding, seacliff erosion, and undermining of coastal structures occurred from Pacifica to Monterey Bay. Over 65 feet of bluff recession in Pacifica forced the removal of an entire row of mobile homes in a large oceanfront mobile home park (Figure 3.6). The interior of Monterey Bay provided the best example of the problems associated with the erection of permanent structures on the beach. The common wave approach along the central coast is from the northwest; thus the beaches in the northern corner of Monterey Bay tend to be protected because of the high degree of wave refraction, the process in which waves are bent and redirected when they interact with irregularities on the ocean floor, around Point Santa Cruz. A wide and nearly permanent sandy beach along the shoreline of northern Monterey Bay from New Brighton Beach to Moss Landing at the center of the bay has given homeowners on the back beach and dunes in this area a sense of security. Many of these homes and other structures were built during a long period prior to 1978 with relatively few El Niño events, when the shoreline seemed quite stable. During the first three months of 1983, however, characteristic of ENSO events, large storm waves approached the coast from the west and southwest. As a result, there was very little energy loss through bending or refraction, so that the full force of the

Figure 3.4   Debris washed into downtown Capitola in Santa Cruz County by the storm waves and high tides of the 1983 El Niño. Photo by Gary Griggs and Kiki Patsch.

Figure 3.5   A house collapsed onto the beach in Rio Del Mar as a result of beach scour undermining the support pilings during January 1983. Photo by Gary Griggs and Kiki Patsch.

Figure 3.6 Erosion of the weak bluff materials in Pacifica in 1983 led to the removal of an entire row of mobile homes. Photo by Gary Griggs and Kiki Patsch.

waves struck the coastline and the associated development and infrastructure. Private homes, seawalls and bulkheads, a state recreational vehicle campground, sewer lines, and roads were severely damaged (see Chapter 14).

THE 1978 AND 1997–98 ENSO EVENTS

Significant coastal storm damage also occurred in 1978 (Figure 3.7) and in 1997–98 (Figure 3.8), both ENSO years. Although most indices suggest that the 1997–98 ENSO disturbance was more intense than the 1982–83 event, the state's coastline suffered far more damage in the earlier event. Some important differences have been documented that explain why the 1982–83 event was so much more damaging than the 1997–98 event and why property losses were so high. During the first three months of 1983, eight major storms struck the coast. The two largest storms struck during high spring tides, so that the waves scoured away more beach sand, reached farther inland, and damaged structures higher on the shoreline. In 1997–98, the largest waves from the two biggest storms hit during lower periods in the monthly tidal cycles, significantly reducing the impact of the waves on the shoreline. Another important factor contributing to the disproportionate damage between the two winters was the higher percentage of shoreline that had

Figure 3.7 Severe beach erosion undermined decks and stairways at Pot Belly Beach in Santa Cruz County during the 1978 El Niño. Photo by Gary Griggs and Kiki Patsch.

Figure 3.8 Erosion of the weak bluffs in Pacifica, northern San Mateo County, undermined bluff-top homes during the 1988 El Niño. Photo by Monty Hampton, U.S. Geological Survey.

been armored between 1983 and 1997. Most of the areas significantly damaged in the 1982–83 winter were protected by more substantial seawalls or revetments by the time the 1997–98 ENSO event took place.

The statewide damage from the 1982–83 ENSO event was a wake-up call for coastal California and, at least in the short term, changed popular perceptions of the stability or security of oceanfront development. Development restrictions were tightened, many homes went on the market, insurance rates rose, and new seawall permits were submitted. Then it was back to business as usual, as new arrivals to the coast outbid each other for the same oceanfront homes damaged a few years before.

Any evaluation of coastal erosion rates or the long-term stability of an oceanfront property in California must factor in the ENSO cycles and the interdecadal oscillations that are now well documented. The record of coastal change in a given location for a particular interval of time—1945–75, for example, an interval characterized by relatively few and modest El Niño events—may seem to encompass a long time frame but may not be at all representative of the longer-term conditions that the site has experienced or may experience during an interval dominated by ENSO conditions.

SEA-LEVEL FLUCTUATIONS

Anyone who either lives on or visits the coast recognizes that sea level changes throughout the day as the tides go in and out. In addition to these daily variations, long-term fluctuations in the level of the ocean along the shoreline take place over hundreds and thousand of years. One of the challenges we face today is accurately predicting sea levels of the future. Rising sea level is creating significant problems for the 100 million people around the world today who live within 3 feet of sea level. Just how high the level of the ocean will rise in the decades ahead is uncertain. There is agreement among scientists, however, that the sea level has been rising for about 18,000 years and that it will continue to rise in the foreseeable future, very likely at an accelerated rate.

THE TIDES

Because the waters of the ocean move freely, they are affected by the gravitational attraction of both the sun and the moon. Over 300 years ago, Isaac Newton discovered that the moon and the sun exert a gravitational pull on the Earth, and this attraction distorts the oceans on a regular basis as the relative positions of the Earth, moon, and sun constantly change. The sun and moon provide the major forces that influence the short-term level of the ocean at any particular time during the day. Newton discovered that the amount of pull that any one body exerts on another is directly proportional

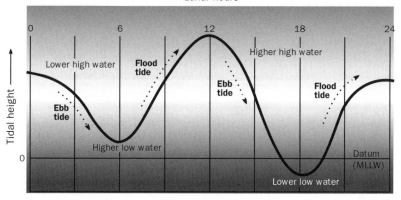

Figure 3.9 Daily pattern for a semidiurnal tide as occurs in California. Drawing by Shannon Griggs.

to their masses and inversely proportional to the distance between them. Although the moon is much smaller than the sun, it exerts about twice as much influence on the tides because it is so much closer. Because we know the orbits of the moon and sun quite well, we can predict the level of the tides many years in advance.

In addition to the gravitational attraction of the sun and moon, strong winds, atmospheric pressure differences, water temperature, and surface currents affect the surface elevation of the ocean. Thus, although tide tables provide an accurate picture of the timing and magnitude of high and low tides expected each day, sea level is affected by other factors. On January 27, 1983, for example, due to the warming of the ocean and also to atmospheric pressure differences during the largest ENSO event in many decades, sea levels were the highest ever recorded at tide gauges in San Diego, Los Angeles, and San Francisco. In San Diego and Los Angeles the recorded sea levels were about a foot above those predicted by tide tables. In San Francisco, the recorded high tide was 21 inches above the predicted level, the highest in more than 130 years of record keeping.

The coast of California experiences two high and two low tides each day, with successive highs or lows having different elevations (Figure 3.9). A complete tidal cycle (two highs and two lows) takes place every 24 hours and 50 minutes, so that the highs and lows occur a little later each day. When the sun, moon, and Earth are all aligned, their combined gravitational forces are greatest and produce the largest tidal ranges, know as spring tides. This alignment occurs twice each month, at the times of full moon and new moon, and it is at these times of the month that we observe the highest high

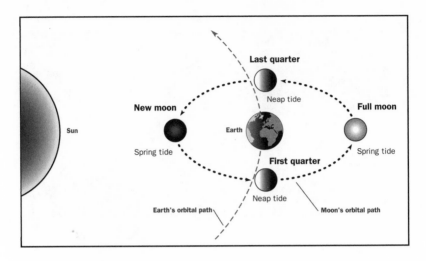

Figure 3.10 Spring and neap tides, showing the relationship between the Earth, moon, and sun at these times. Drawing by Shannon Griggs.

tides and the lowest low tides (Figure 3.10). On the other hand, when the sun and moon are at right angles to the Earth, their gravitational pulls are reduced, and we have lower tidal ranges known as neap tides (Figure 3.10).

Tidal heights, as given in tide tables, are referenced to the mean or average of the lower of the two low tides each day (designated as MLLW, or mean lower low water). In Monterey, for example, the highest tide ever recorded was 7.88 feet above MLLW and the lowest was 2.37 feet below MLLW, for a maximum elevation difference of 10.25 feet. Corresponding extremes are 11.22 feet between maximum highs and lows at San Diego and 14.08 feet at Crescent City. Tidal ranges vary geographically, however, with some mid-ocean islands experiencing only a foot or two of difference between high and low tides, and areas such as the Bay of Fundy in Nova Scotia experiencing a maximum of 56 feet between extreme high and low tides.

The tidal range at any particular location has a profound impact on how the shoreline looks and how this zone is utilized. Most of the serious coastal erosion and storm damage occurs at times of very high tides, because this is when the waves can attack the upper beach, the base of the dunes or the cliffs. The areas reached by these high tides and storm waves, the frequency of inundation, and how we choose to use or develop the back beach, dunes, or bluffs will determine the level of damage we may experience.

GLOBAL CLIMATE CHANGE AND LONG-TERM SEA-LEVEL FLUCTUATIONS

During the last ice age huge glaciers covered the mid-continent and extended down over the area now occupied by the Great Lakes, across what is now Minnesota, Illinois, and Indiana into Kansas and Nebraska. The Puget Sound area of Washington was scoured out and the area where Seattle now sits was buried under 3,000 feet of ice. A land bridge across the Bering Straits connected Russia to Alaska, and you could have walked from England to France across the exposed floor of the North Sea. Sea level was about 350 feet lower than it is today as the last ice age came to a close about 18,000 years ago.

Sea-level rise is perhaps the most obvious consequence of climate change to residents of coastal California. Although there are many editorials, debates, articles, conferences, and reports on global warming, greenhouse gases, and the impacts of human activity on global temperatures, it is clear that the Earth, oceans, and atmosphere have been warming for about 18,000 years. This isn't a new phenomenon; the history of the Earth has been one of constant climate change. The climate and temperature of the Earth is directly related to how much energy we get from the sun, and the amount of heat the Earth or any other planet receives is directly related to its distance from the sun. We now know, however, that the distance between the Earth and the sun changes over time because of irregularities in the Earth's orbit and the tilt and wobble of the Earth on its axis. As the planet moves slightly farther away, it receives less solar energy, and as a result, the Earth and its atmosphere and oceans cool somewhat. There are well-understood cycles in these orbital oscillations that span thousands of years and are associated with well-documented warming and cooling intervals that have profoundly affected the Earth's surface, its ocean basins, and its life.

During cooler periods, the water evaporated from the oceans accumulates as ice caps and glaciers on Antarctica, Greenland, and other high-latitude areas. This removal of water from the oceans leads to a global drop in sea level. The amount of sea-level drop is related to how cool the climate gets and, therefore, how much ice accumulates on the continents. During the last glacial maximum or cold period, the level of the oceans dropped globally about 350 feet, and the shoreline moved out to the edge of the continental shelf that surrounds the world's landmasses (Figure 3.11). As the Earth moved gradually closer to the sun again and began to warm, the glaciers and ice caps melted, the oceans warmed and therefore expanded, and as a result sea level began to rise. This rise was relatively rapid from 18,000 years ago to about 7,000–8,000 years ago (Figure 3.12), with an average rate of nearly half an inch per year or about 4 feet per century. Sea-level rise then slowed considerably, and for the last several thousand years or so the rate of vertical rise has been a little less than a tenth of an inch per year. Along the East Coast of the United States, for example, it has taken over a century for a one-foot rise in sea level to be recorded.

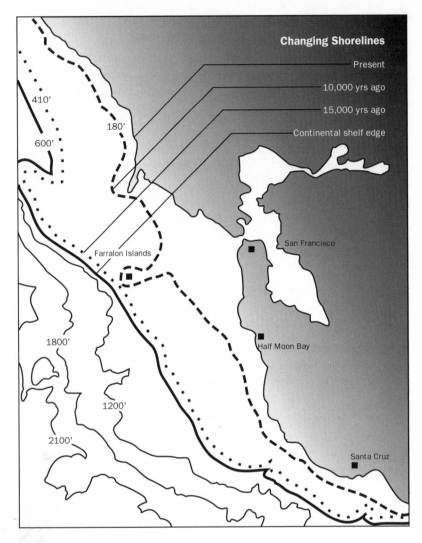

Figure 3.11   The continental shelf edge, or position of the shoreline of California ca. 18,000 years ago, 15,000 years ago, 10,000 years ago, and today. Drawing by Shannon Griggs.

Several factors determine how this gradual sea-level rise affects a particular coastal area. One important consideration is whether the adjacent landmass is stable, rising, or subsiding. Although sea level will rise and fall on a global basis depending upon how much water is in the oceans at a given time, the vertical movement of the land can compound or counteract the rise in sea level. New Orleans, Louisiana, and Venice, Italy, for example, are areas undergoing subsidence. The land in these and certain other coastal

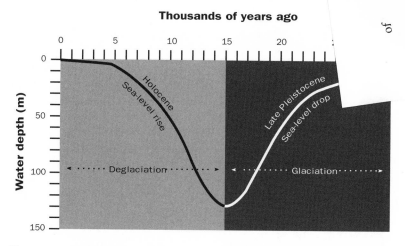

Figure 3.12   Sea-level curve for the last 20,000 years. Drawing by Shannon Griggs.

locations around the world is sinking as a result of loading of the Earth's crust by the deposition of river sediments (from the Mississippi River, for example) and groundwater withdrawal. Consequently, the relative rate of sea-level rise in New Orleans and Venice as recorded on tide gauges is much greater than the global rate because we are measuring the combined effects of local land subsidence and global sea-level rise. There are also very low-lying Pacific Island regions (the Marshall Islands, for example, or other areas in Micronesia) where a rise in sea level of several feet will lead to the disappearance of an entire nation. On the other hand, Juneau, Alaska, is actually rising more rapidly than sea level as a result of a slow rebound of the land from the removal of glacial ice, so that the tide gauge at Juneau actually shows a drop in sea level relative to the land. Residents of New Orleans are very concerned about global warming, continued sea level rise, and land loss, whereas global warming is not a significant concern in Juneau.

The other major consideration for the impact of sea-level rise in a particular location is the coastal topography or slope of the shoreline. Along low-lying, nearly flat coastal areas such as most of the South Atlantic and Gulf coasts from New Jersey to Texas, a one-foot rise in sea level may move the shoreline inland a thousand feet or more. The 1999 relocation of the Cape Hatteras Lighthouse 2,900 feet inland at a cost of $13.2 million (in 2003 dollars) was in response to a gradually rising sea level on a low-lying coastline. The lighthouse was nearly 1,500 feet from the shoreline in 1870, but 65 years later, in 1935, the shoreline had migrated about 1,375 feet closer to the lighthouse, an average rate of 21 feet per year. Thus the consequences continued rise in sea level along much of the Atlantic and Gulf coasts are very serious. Similarly, in places such as Bangladesh, where millions of people live on fertile but very low-lying deltaic land, a rise in sea level of a few feet, which typically happens during a cyclone or typhoon, can inundate thousands of acres. The disastrous tsunami that occurred in the northern Indian Ocean in December 2004 was a deadly reminder of the risks of living close to sea level. On the other hand, for California, where cliffs or low bluffs back much of the coastline, a one-foot rise in sea level may not be a major problem. There are many areas of coastal California, however—San Francisco Bay and much of the coastline of Los Angeles County (Figure 2.4), for example—where the shoreline is quite flat or where intensive development has taken place close to beach level, and where a rise in sea level of a foot or two could have significant effects. Remember, it's not just the rise in sea level but also the waves that break closer to shore that will impact coastal development and lead to inundation, damage, and destruction.

Coastal residents have built thousands of homes directly on the shoreline, in some cases cantilevered out over the beach, but we need to realize that the present position of the shoreline or seacliff is a temporary one, and simply a reflection of where sea level is today. The questions we need to answer are: How much higher will sea level rise? How long will it take until this level is reached? How is human activity influencing global warming and sea level rise, and is there anything we can do to reduce this impact? And how do we respond to a continuing sea-level rise?

GLOBAL WARMING AND FUTURE PREDICTIONS: WHAT CAN WE EXPECT IN CALIFORNIA?

Globally, 60 percent of the world's 6 billion people live within 35 miles of a shoreline and 100 million live within 3 vertical feet of sea level. Along the shoreline of California, houses regularly are sold with multi-million-dollar price tags. What many coastal residents, managers, federal agencies, and insurers need to know is what we can expect during the twenty-first century in terms of global climate and sea level. This is a complex issue and one that is difficult to find complete agreement on.

What we do know is that human activity since the Industrial Revolution in the late 1800s has significantly increased the concentration of carbon dioxide and other gases (methane, nitrous oxide, and chlorofluorocarbons) in the atmosphere. These greenhouse gases allow shortwave solar radiation to pass through the atmosphere and warm the Earth. However, much like the glass in a greenhouse or in your car windows, these gases don't allow the heat that is re-radiated by the Earth to escape into space. Just as your car heats up on a sunny day with the windows rolled up, the Earth's surface, oceans, and atmosphere have been gradually warming as the greenhouse gases build up from the combustion of fossil fuels, the clearing and burning of tropical rain forests, and certain other industrial and agricultural processes that produce nitrous oxide, methane, and chlorofluorocarbons.

Separating out the impacts of human activity from the ongoing natural cycle of global warming is a difficult problem, and many different studies have attempted to sort out these influences. Even though many models have been developed and numerous calculations have been made using different assumptions, challenges remain in assessing and quantifying the impact of the factors contributing to global warming. There is no question that the globe and oceans are gradually warming, but it may be a number of years before we can establish with certainty how much of this is due to the greenhouse effect and to what extreme sea level might rise in the near future. Most scientists recognize that rates of sea level rise are increasing and that acceleration beyond present rates is likely. We are still uncertain, however, whether sea level will rise another 9 inches or so, as experienced during the last century, or 3 feet or more, as has been predicted by some. Either way, the increase will mean the inundation of low-lying coastal land and the erosion of the shoreline, processes we should begin to anticipate and plan for.

CHAPTER FOUR

UNDERSTANDING THE SHORELINE

GARY GRIGGS, KIKI PATSCH, AND LAURET SAVOY

INTRODUCTION

California has one of the most spectacular and diverse coastlines in the country—one of high mountains plummeting to sheer, rocky cliffs; long stretches of sandy beach; and extensive marshes and wetlands. It is a state that in many ways is defined by its coast. It is little wonder that such an environment continues to draw people to the shore. Yet the shoreline is also the battleground between land and sea, a line that is constantly changing as sea level rises and falls in response to global climate change. As people continue to move to coastal communities in California, the dynamic processes operating at the shoreline have a more pronounced effect on coastal development. In addition, our greater usage and demand of coastal areas affect the stability of the shoreline itself. Anyone contemplating buying or building on the coast should be aware of the losses and lessons of the past 25 to 50 years. We need to make decisions based on the recent history of the shoreline and events to be expected in the future. We are long past the time when we could proceed with naïve or uninformed planning and decisionmaking.

The mere existence of a sheer seacliff fronting the ocean means that the coastline is actively retreating through erosion, whether slowly or rapidly. The presence of loose sand and driftwood high on the back beach indicates that this landform has been washed over by waves and high tides. Although a particular beach may appear high, wide, and stable during the summer months, it can disappear virtually overnight during a severe winter storm (Figure 3.1). Similarly, cliffs can retreat many feet during a single winter. We need to know what happens at the coastline, how frequently it happens, and how the shoreline may change during a severe storm or a typical winter, or over a number of years. No structure will last forever, but before any new construction is allowed, the Coastal Commission and most local government permitting agencies require studies be undertaken to evaluate site stability to determine whether a particular building site will survive for 50 or 100 years. It is only with an understanding of coastal hazards and processes that we can identify the restrictions that the coastal environment places on our plans, and thereby prevent costly and repeated attempts to protect or save structures built too close to the ever-changing edge.

While walking along the shoreline, whether at Solana Beach, Malibu, Santa Barbara, Stinson Beach, or countless other places along California's coast, you are likely to see sandy beaches backed by either dunes, low bluffs, or abrupt cliffs on one side, and the repeated run-up of waves onto the shore on the other. Perhaps this scene evokes, in addition to a sense of beauty and energy, a feeling of permanence or even timelessness. Such a feeling, however, can be misleading. Each feature of the coast—whether steep cliffs, sand dunes, or the beach itself—is dynamic and delicately balanced with the processes that shape the shoreline, such as the action of waves, tides, and winds, which change from day to day, month to month, and year to year. Unfortunately, millions of people and billions of dollars in structures and development now lie within a few feet of a sea level that is constantly changing; this is the overriding problem and the focus of much of this writing.

WAVES AND THE SHORELINE

A good place to start a discussion of the shoreline is with the waves, the driving force that continually shapes and reshapes the shore. Anyone who has seen storm waves batter a beach realizes the tremendous power they expend. A 10-foot wave can exert over 1,000 pounds of pressure per square foot, and wave pressures over 12,000 pounds per square foot have been recorded. Along the northern Oregon coast, just south of the mouth of the Columbia River, huge waves threw a rock weighing 135 pounds more than 140 feet above sea level, through the roof of the Tillamook lighthouse. In 1914, wave run-up extinguished the Trinidad Head Light, at an elevation of nearly 200 feet above sea level. Along Cardiff's restaurant row in San Diego County, restaurants were severely damaged during the January–February 1983 storms when large waves threw cobbles and riprap weighing up to 1,200 pounds into the buildings and onto the highway.

The waves we see breaking on the beach begin to form when wind creates friction as it blows across the ocean's surface. As the wind continues to blow, its energy is transferred to the water surface, and ripples and then small waves begin to form, just as when one blows across the surface of a cup of hot coffee. The sizes of the waves that break on the beach on a particular day depend mainly on the offshore wind characteristics: how long the wind blows, how fast it blows, and over what distance it blows across the sea surface. The longer, faster, and farther the wind blows, the larger and more energetic the resulting waves will be. Waves formed where and while the wind is blowing over the sea surface during a storm are typically irregular and choppy; this is the condition known as sea commonly experienced by ships far from shore. Driven by the wind, these waves gradually move away from a storm source area and sort themselves out into a more regular pattern known as swell. Swell waves can travel great distances from the source storm area. In fact, the waves that we see breaking on our beaches may have traveled hundreds or thousands of miles across the ocean from their stormy point of origin.

Most waves arriving at the shoreline approach the coast at an angle rather than approaching from directly offshore. Along much of the California coast, for example, many of the waves reaching the shoreline originate from storms in the North Pacific and, therefore, approach from the northwest. As the water gets shallower near the coast, the underwater portion of the wave closest to the beach "feels" the sea floor first and begins to slow down, while the seaward portion of the wave crest in deeper water continues to travel at its original speed. This results in bending or refraction of the wave fronts as they approach the shoreline (Figure 4.1). On an irregular rocky coastline, refraction causes wave energy to be concentrated at points and promontories, or over shallow areas, and dissipated in bays or over deeper areas such as submarine canyons (Figure 4.2). The bending and gradual breaking of the waves around a point or headland produce some of the best surfing spots in California, such as Malibu, Rincon, and Steamer Lane, to name a few (Figure 4.1).

As waves shoal, or reach the shallow water near the shoreline, they undergo some important changes. The portion of the wave beneath the sea surface slows down from bottom friction while the portion that we see at the sea surface continues to travel at a faster speed. As the water continues to shallow, the wave height increases (Figure 4.3). Eventually the waveform above the sea surface becomes too steep and unstable, and when the ratio of wave height to water depth—also known as wave steepness—is about 3:4, the wave will break (a 3-foot high wave will break in about 4 feet of water).

Waves are rarely completely refracted and thus usually break on the beach at a slight angle. As each wave breaks, it stirs up and momentarily suspends sand and washes it up the beach face at an angle. The return flow, or swash, moves the sand and water back down the beach face and slightly down coast. With thousands of waves breaking on the beach every day, the net result is a slowly moving current of water and beach sand that flows down coast and parallel to the shoreline, known as a longshore current (Figure 4.4). Longshore currents exist along most of California's beaches,

Figure 4.1   Waves refracting around a point (Rincon, California). Photo by A. Paul Jenkin.

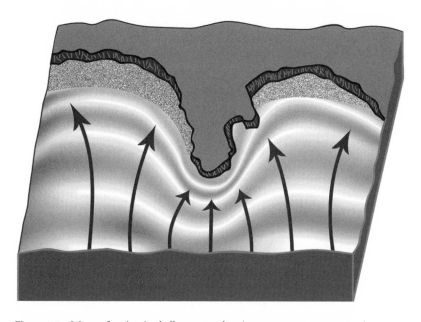

Figure 4.2   Wave refraction in shallow water showing wave energy concentration at a headland and wave energy dispersion in an embayment. Drawing by Shannon Griggs.

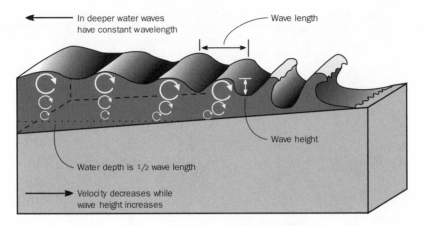

Figure 4.3 Waves undergo changes as they approach the shoreline. The height increases and the wavelength and speed decrease, but the wave period remains unchanged. Drawing by Shannon Griggs.

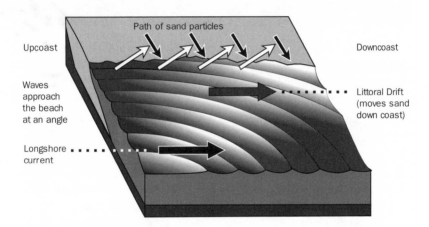

Figure 4.4 A longshore current forms along the shoreline where waves approach the beach and break at an angle. This flow of water carries sand along the shore as littoral drift. Drawing by Shannon Griggs.

are driven by waves breaking on the beach at an angle, and are very important in transporting sand along the shoreline. This alongshore transport of sand is known as littoral drift and can be thought of as a river of sand moving parallel to the coast. Hundreds of thousands of cubic yards of sand may travel as littoral drift each year along California's shoreline (Table 4.1). The amount of sand moving along the shore is related to the amount of wave energy and the angle at which waves approach the shoreline, with waves

LOCATION	ANNUAL RATE (CUBIC YARDS)	DIRECTION
Santa Cruz	300,000	East
Santa Barbara	300,000	East
Ventura	600,000–1,000,000	Southeast
Santa Monica	275,000	Southeast
Oceanside	350,000	South

breaking parallel to the beach transporting almost no sand and those breaking at a greater angle transporting more sand.

Most of the large and damaging waves that impact the coast of California arrive in the winter months and come from North Pacific storms. A south swell, or set of waves arriving from the Southern Hemisphere, is common during the summer months in southern and central California and may provide excellent surfing conditions at a time when waves are normally considerably smaller. Large Southern Hemisphere storms generate these south swells (originating during the winter months), having traveled thousands of miles from their origin, perhaps off Antarctica, New Zealand, or Australia. Waves during El Niño winters typically approach the California coast from the southwest or west, and this may drive littoral drift northward.

A TIDAL WAVE OR TSUNAMI?

Wind does not create all waves. Just as you can create small ripples by shaking a bowl of water or small waves by jumping into a swimming pool, any large disturbance or sudden movement in the ocean can also create waves. Enormous oscillations of water caused by submarine earthquakes, underwater landslides, or exploding undersea volcanoes result in waves known as tsunamis or seismic sea waves. These waves have often been incorrectly called tidal waves, but they have nothing to do with the tides. The word *tsunami* is derived from two Japanese words: *tsu*, which means harbor, and *nami*, which means wave. Some of the greatest historical impacts of tsunamis have occurred in harbors or port cities.

In the open sea, a tsunami will typically travel at a speed of 450 or 500 miles per hour, but it is almost imperceptible, with a height of only several feet. As the tsunami shoals, or reaches shallow water near the shore, however, its height increases dramatically. Waves of this type reaching elevations of over 100 feet have at various times hit the coasts of Japan, Alaska, Chile, and the West Indies. Tsunamis that reach California can originate almost any-where in the Pacific Ocean, from Alaska to Chile to Japan. Because of their great speeds, long wavelengths, and wave heights at the shoreline, tsunamis typically possess considerably more power than normal wind-driven waves. So although they might travel thousands of miles across an ocean from their source, tsunamis can cause extensive damage upon reaching the shoreline.

The impact of elevated sea levels and tsunami wave inundation is recorded along the coastline not only as damaged structures and vehicles, but also as a set of visual markers, high-water marks on buildings, and debris lines where the water deposited floating material inland. These high-water marks are now documented and mapped by geologists, and can provide the information that is needed for developing tsunami warning maps for specific coastal areas. The State Office of Emergency Services has funded work to develop tsunami risk or inundation maps for the populated areas of the California coastline.

In 1812, a tsunami caused by an earthquake in the Santa Barbara Channel generated waves reportedly 30 to 50 feet in height that pounded the shore near Refugio and Santa Barbara. This area was relatively uninhabited at the time, however, so the reliability of these measurements is somewhat uncertain.

The most recent destructive tsunami to batter California's coast accompa-nied the great 1964 Alaskan earthquake. Crescent City, on the northern California coast, was the hardest hit, being inundated by a series of waves that pushed buildings off their foundations and into other structures, and swept vehicles and buildings into the ocean. Wave run-up extended 800 to 2,000 feet inland in the commercial and residential areas of the city with water depths of up to 8 feet in city streets and 13 feet along the shoreline. The worst waves struck the waterfront area at 1:45 A.M., drowning 12 people, demolishing 150 stores, and littering the streets with huge redwood logs from a nearby sawmill. Most of the city's downtown was either damaged or totally destroyed, and rather than being rebuilt, the blocks nearest the harbor were made into a park. According to the U.S. Army Corps of Engineers, property losses approached $168 million (in 2005 dollars). Southward, damages from the same tsunami ranged from $767,000 at Noyo Harbor in Mendocino County, to almost $1.5 million at Los Angeles Harbor. The size of a tsunami and its impact can vary widely depending on the magnitude of the generating event, the nature of the offshore bathymetry, and the geometry of the shoreline and coastal topogra-phy. Because most tsunamis approaching the coast of California have come from either Alaskan or South or Central American source areas (earthquakes are generated at the trenches or subduction zones in these areas), they must pass over a significant length of continental shelf before they reach the coast-line. As a result, wave energy is reduced and damage has historically been minor to moderate compared with many other areas around the Pacific basin.

Although destructive tsunamis are not everyday events in California, they occur more often than you might think. At least 19 tsunamis have reached California over the past 200 years, and many more will arrive in the future. Historically, however, only the tsunamis of 1812 and 1964 have caused major damage. The 7.1 magnitude 1989 Loma Prieta earthquake that occurred in the Santa Cruz Mountains generated a small (about 1 foot in height) tsunami in adjacent Monterey Bay. A tsunami generated by a large slump in the head of the Monterey submarine canyon is probably a more significant concern around the margin of Monterey Bay, however. Records from geophysical surveys of the canyon head indicate that large slumps and slides have occurred frequently on the steep walls of the canyon during its development.

Recent discoveries by geologists studying preserved sediments along the coastline of northern California (in the area between Eureka and Crescent City), Oregon, and Washington have provided evidence of a large tsunami that struck this area about 300 years ago. This entire area is a subduction zone or a boundary where one large plate, the Gorda Plate, collides with the North American Plate and is forced down beneath the continent. As the Gorda Plate slowly descends, there is tremendous friction as it scrapes against the bottom of the North American Plate. Most of the time these two plates are locked, but when the accumulated stress causes the plates to unlock, and these massive plates readjust, a very large amount of energy is released and we experience large earthquakes. There is mounting evidence along the coastline of the Pacific Northwest that very large sea floor earthquakes (Richter magnitude 8 to 9) occur every several hundred years in the offshore area between Cape Mendocino and Puget Sound and produce large tsunamis. Some of the newer findings include evidence that these waves moved a considerable distance inland into estuaries and left behind beach sand deposits within the muddy organic material that would normally be deposited in these environments. In addition, during a major earthquake of this type, a large portion of sea floor and shoreline may suddenly move upward or downward. So in some coastal areas, large conifers that formerly lived a few feet above sea level were submerged and died when they came into contact with salt water. The growth rings on these old trees can be counted, and age of the trees can also be estimated using carbon-14 dating. From a number of investigations of this sort—studying the sediments left behind, the trees that died, and the ages of these materials—we have good evidence that high-magnitude earthquakes occur along the coast of the Pacific Northwest on average about every 300 to 500 years. We also have discovered recently that the last major earthquake that affected this area took place in 1700. As a result, there is an increasing level of awareness and concern about when another earthquake of this magnitude will occur, and how the associated tsunami could impact a shoreline that, although still relatively undeveloped, is home to thousands of people who weren't here in 1700.

BEACHES

Beaches are the buffer zones or shock absorbers that protect the coastline, seacliffs, or dunes from direct wave attack. Because California is still geologically quite young, and therefore very active, the presence of rising mountains and steep cliffs at some locations (Big Sur and portions of the Mendocino coast, for example), has not allowed space or time for beaches to form. At other locations, rivers or streams have cut through the coastal cliffs, or waves have eroded embayments or bays, forming small pocket beaches (Figure 4.5).

Along many areas of California's lengthy coast, beaches have formed up coast of natural headlands or artificial obstructions. The rocky headlands or points trap the sand moving down coast as littoral drift and thereby widen the up-coast beaches. Jetties, groins, and breakwaters also serve as obstructions and have either created beaches or stabilized or expanded existing beaches to the delight of beachgoers and to the dismay of down-coast property owners deprived temporarily or seasonally of the beach sand that formerly protected their homes.

At still other locations, where resistant headlands extend into the ocean, waves approaching from the northwest commonly bend around the headland and erode the weaker down-coast material. In this case, a hooked or spiral-shaped bay and a long, sandy beach form over time. Good examples of these hooked-shaped bays include Half Moon Bay, Bolinas Bay, Mission Bay, and northern Monterey Bay (Figure 4.6).

Last but not least, where the coastline relief is low and ample sand is available, long, straight, sandy beaches form, often backed by extensive dunes such as Silver Strand in San Diego; Newport, Huntington, and Redondo beaches; Zuma Beach near Malibu; East Beach in Santa Barbara; Seacliff and Sunset beaches near Santa Cruz; and Pelican Beach north of Crescent City (Figure 2.4). It is these long and typically wide stretches of beach that attract many of California's local residents and tourists and thus are important income generators for local economies.

WHERE DOES THE BEACH SAND COME FROM?

Along the California coast, most of the beach sand comes from river and stream runoff, with a lesser amount resulting from the erosion of coastal cliffs and bluffs. Measurements of sediment transport in rivers indicate that coastal streams, particularly during times of flood flow, are the major suppliers of

Figure 4.5 A typical pocket beach at the mouth of a small coastal stream south of Los Osos in San Luis Obispo County. Photo © 2002–2004 Kenneth and Gabrielle Adelman, California Coastal Records Project, www.Californiacoastline.org.

Figure 4.6 The gentle curve or hook-shaped coastline at Santa Cruz. Photo by Gary Griggs and Kiki Patsch.

Figure 4.7   The San Lorenzo River drains the Santa Cruz Mountains and discharges a large volume of sand each year, through a flood control channel, to the coastline of northern Monterey Bay. Photo by Gary Griggs and Kiki Patsch.

sand to the beaches (Figure 4.7). Along most of the California coast, estimates indicate that 80 percent to over 95 percent of the beach sand was originally stream derived. Physical and chemical weathering slowly break down rocks in the coastal mountains or watersheds into smaller fragments or into their constituent minerals or grains. Rainfall, runoff, and slope failures such as landslides begin to move the gravel, sand, silt, and clay downslope into the creeks, where it is gradually sorted out and transported downstream into the larger rivers. After many months and miles of abrasion and sorting, the smaller particles ultimately reach the shoreline while the larger boulders and cobbles are left behind in the streambeds. The sand and gravel end up on the beach; the silt and clay are carried offshore in suspension by coastal and offshore currents and are ultimately deposited on the sea floor somewhere a few miles to hundreds of miles offshore.

California is fortunate to have many relatively large rivers that deliver great volumes of sand to the coastline to nourish the beaches. The Eel, Russian, Santa Maria, Ventura, and Santa Clara rivers are all important sand suppliers. Beaches have often been observed to be much wider in the summers following winters with high rainfall because of the delivery of large amounts of sand to the local beaches by high stream flow or flood events. Sediment delivery by rivers to California's shoreline has been shown to be extremely episodic, with most of the sediment discharged during a few days of high flow each year. Additionally, sediment discharge during a single year of extreme flood conditions may exceed decades of low or normal stream flow. A recent study of major rivers in central and southern California has shown that sediment discharge during flood years such as 1964, 1969, 1983, and 1998 averages 27 times that during drier years. In 1969, over 100 million tons of sediment (about 7.5 million dumptruck loads) was flushed out of the Santa Ynez Mountains, more than during the previous 25 years combined. In a single day, December 23, 1964, the Eel River in northern California transported 57 million tons of suspended sediment (about 4.2 million dumptruck loads), which represents 18 percent of the sediment load of the river over the previous 10 years.

Coastal cliffs may also contribute sand to beaches if they consist of material that breaks down into sand-sized particles (sandstone or granite, for example). Cliffs and bluffs that are composed of silty or claylike material (shales or mudstones, for example), on the other hand, will not contribute significantly to the beach. Although the contributions of beach sand by coastal cliffs may be important locally, particularly where cliffs are rapidly eroding, this source probably accounts for no more than 10 to 15 percent of the beach sand in a given area, and usually much less.

THE BEACH—A RIVER OF SAND

Once sand arrives at the shoreline, waves and wave-induced currents provide the energy necessary to sort out sediments, push the sand up on the shoreline to form a beach, and then gradually move the sand along the coast. The direction of this alongshore movement of sand or littoral drift is determined by the dominant angle of wave approach. For example, along nearly the entire central and southern California coast, waves from the northwest drive littoral drift southward along the beaches (Figure 4.8). North of Cape Mendocino and between San Diego and the Mexican border, the flow of sand is often northward, or sand may move to the north and to the south at different times of the year depending on the dominant direction of wave approach.

It is very important to know something about the rate and direction of littoral drift prior to any human intervention in the nearshore system. The planning and design of coastal engineering structures to prevent erosion, the construction of harbors and channel entrances with their associated jetties and breakwaters, and questions of beach nourishment and long-term stability of beach material are all tied to the littoral drift system. As with the construction of a dam or reservoir on a stream, virtually anything we insert into this river of sand is going to disrupt the sand's natural flow. The number of

Figure 4.8   Littoral drift directions along the coastline of California are shown by arrows. Drawing by Kiki Patsch.

past instances where littoral drift rates were either not well understood or not considered prior to construction of large engineering structures is unfortunately quite large in California. The consequences quickly became apparent, and the resulting dredging, down-coast cliff or beach erosion (Figure 4.9), associated loss of property, and subsequent armor emplacement have cost hundreds of millions of dollars. Today, however, as a result of about a hundred years of experience and coastal engineering projects, we are aware of these systems and structures. We have a reasonably good understanding of

Figure 4.9   The construction of a breakwater at Santa Barbara in 1928 created a large sand trap. Down-coast beaches were starved of sand, which led to shoreline erosion and damage to oceanfront homes at Carpenteria.

littoral drift directions and rates for most areas, so there is no longer any reason for repeating the mistakes of the past or planning projects without a full awareness of the long-term costs of littoral drift disruption.

Average yearly littoral drift rates along California's coast show a considerable range, from about 30,000 cubic yards at Redondo Beach to nearly 1 million cubic yards at Ventura (Table 4.1). One million cubic yards is equivalent to about 100,000 dumptruck loads each year. If transported evenly through the year, this would require 275 dumptrucks full of sand moving along the beach each day, or about one every 5 minutes, 24 hours a day. Disrupting or blocking this amount of sand every year can have massive and devastating impacts (Figure 4.10). Imagine the size of a pile of sand that would be formed if a constant stream of dumptrucks dropped their loads every 5 minutes for a year, and how long it would take to move the resulting pile. This is the scenario faced by many of California's harbors.

Between 1970 and 2001, for example, over 10 million cubic yards of sand were dredged from the Santa Barbara Harbor, nearly 20 million cubic yards were dredged from the Ventura Harbor, and roughly 28 million cubic yards of sand were dredged from the Channel Islands Harbor, at a cost of tens of millions of dollars. Dredging from just these three harbors totaled over 58 million cubic yards or 5.8 million dumptruck loads of sand over this 31-year period. These are just a few of many large sand traps that exist along the coast of California, where annual dredging is a way of life.

Figure 4.10   Littoral drift has accumulated up coast of a jetty at Santa Cruz. Photo by Gary Griggs and Kiki Patsch.

WHERE DOES THE SAND GO?

We now know that beaches in most places along the coast are continually supplied with sand (although in decreasing amounts in southern California) and that longshore drift is constantly moving this material down coast. Where, then, is all of this sand going, and why aren't the beaches growing wider and wider as you move down coast? If we follow the paths of typical sand grains, we observe several possible routes by which sand may leave the beach permanently as it moves down the shoreline by wave action.

SAND DUNES   Dunes occur inland from beaches at many locations along the coast of California and can act as sinks where beach sand may be temporarily or permanently lost from the shoreline. As a beach widens and the area of dry sand on the back beach expands, inland transport by wind may occur. Generally, dunes form wherever ample dry sand is available, wind blows in a persistent onshore direction, and a low-lying or low-relief area exists landward of the beach where the sand can accumulate. If high cliffs or bluffs back the beach, dunes will have nowhere to migrate and will not develop to a significant size. If the wind direction is steady enough, sand will move, grain by grain, over the dune's surface, resulting in a downwind migration of the dune. Much of this sand is permanently lost from the beach. It has been estimated, for example, that 200,000 cubic yards of sand are blown inland each year along the 35-mile coastline from Pismo Beach to Point Arguello (Figure 4.11). Other major

Figure 4.11   Intensive public use of beach at Pismo Dunes in San Luis Obispo County. Photo © 2002–2004 Kenneth and Gabrielle Adelman, California Coastal Records Project, www.Californiacoastline.org.

California sand dunes occur just north of Crescent City, at Humboldt Bay, at Point Reyes, at the Ocean Beach area of San Francisco, in southern Monterey Bay between Marina and Monterey, at Morro Bay, at the Oxnard Plain, from Santa Monica to El Segundo, and from San Diego to the Mexican border.

Dunes serve as important coastal buffers because they are flexible barriers to storm waves and provide protection to the lower-lying back dune areas. In fact, wherever dunes can be created and stabilized, they often work better than seawalls. Dunes also maintain a large stockpile of sand that feeds the beach during severe storms or prolonged periods of wave attack. Under storm assault, the beach is first cut back, and if wave erosion continues, portions of the frontal dune may be eroded (Figure 4.12). This sand is moved offshore, where it is stored in sand bars that tend to reduce the wave energy impinging on the shoreline, because the waves will break farther offshore. As the winter storms subside, smaller spring and summer waves transport sand back onto the beach, which will ultimately be rebuilt. With time or the absence of large storms, the excess sand will be moved onshore by wind and will rebuild the dunes. The process of natural dune rebuilding may take several years, however. Dune erosion, either during storms or because of a reduction in sand supply, can be destructive to any structures built on the dunes.

Sand dunes must be seen as ephemeral or temporary landforms. Storms will recur; sandbars will shift; and dunes will erode, rebuild, and migrate. In

Figure 4.12   The high tides and storm waves of 1983 seriously eroded the shoreline at Pajaro Dunes in central Monterey Bay. Photo by Gary Griggs and Kiki Patsch.

addition, dune instability may result from human impacts in the form of construction or recreation. The primary, frontal, or fore dune is particularly prone to change, as has been discovered in recent years by owners of new condominiums and houses perched on active sand dunes. After many years of observations, landscape architect Ian McHarg suggests in his classic book *Design with Nature* that no development, recreation, or human activity of any type occur on the primary or secondary dunes, which are the least stable and contain the most fragile vegetation. This principle was widely violated along the California coast in the past, with costly consequences. Development, if it is to occur at all, should take place in the back dune area, which has the advantage of protection from winter storms. Limited cluster development might also occur in the trough between dunes, provided groundwater withdrawals will not adversely affect dune vegetation, and the dunes themselves are not breached by roads, utilities, or human trampling.

Recreational impact in sand dune areas has historically been widespread and comes in the form of pedestrian traffic and off-road vehicle use. Most foot traffic impacts come from uncontrolled crossings from the back dune area to the beach. Large numbers of crossings result in dissection of the dune field with numerous paths and the consequent development of large barren areas, the creation of blowouts, decreased dune growth, and nondevelopment of new dunes. Heavy foot traffic has been shown to decrease the amount of vegetation, which in turn accelerates dune destabilization. Off-

Figure 4.13   Sand was mined for decades directly from the beaches of southern Monterey Bay. Photo by Gary Griggs and Kiki Patsch.

road vehicle traffic, which is a common recreational practice in some dune areas, has the same effects—loss of vegetation and destabilization of the dunes, which are difficult to repair. The fragile nature of sand dunes is now more widely appreciated, and more attention is paid to proposals that would remove vegetation, lower groundwater levels, or breach the dunes themselves. Restricted and fenced pathways, boardwalks and revegetation are paying off by protecting vegetation and thereby aiding dune stabilization.

SAND MINING   In addition to the natural processes of sand loss through onshore winds, sand has historically been mined directly from some California beaches and dunes. Three major sand-mining companies removed sand directly from the beach face in southern Monterey Bay for nearly 90 years (Figure 4.13). The high-purity quartz sand was in great demand for many industrial uses, including water filtration, abrasives, and various coatings. Concern, however, as to whether the approximately 300,000 to 400,000 cubic yards of sand being removed each year exceeded the volume entering the area naturally and whether the mining was a significant factor in the regular retreat of the shoreline led to termination of beach sand mining in the 1980s (see Chapter 14). If the average volume of sand removal through direct mining each year is greater than that added by natural sources, shoreline retreat will take place. Although extensive sand and gravel mining operations still take place along streambeds in a number of locations in California, direct sand removal from beaches has been almost completely terminated.

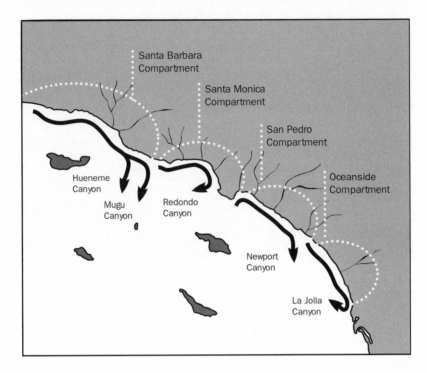

Figure 4.14   The beach compartments or littoral cells of southern California. Drawing by Shannon Griggs.

SUBMARINE CANYONS   The greatest, but also the least visible, loss of sand from California's beaches takes place through the many submarine canyons along California's continental shelf (Figure 4.14). Where these canyons extend close to shore, which they do in many places, they intercept the littoral drift and funnel it away from the beach into deep offshore basins. The canyons of southern California have been recognized for over 70 years and are the ultimate sinks for most of southern California's beach sand. Every year, for example, the Scripps and La Jolla submarine canyons at La Jolla swallow an estimated 350,000 cubic yards of sand, enough to form a beach 10 feet deep, 90 feet wide, and 2 miles long. The Mugu Submarine Canyon, near Ventura, is even more impressive as a sand sink; it siphons off over a million cubic yards of sand each year on average, enough to build a beach 125 feet wide, 10 feet deep, and 6 miles long.

Monterey Submarine Canyon, which bisects Monterey Bay and extends almost into the shoreline at Moss Landing, is one of the world's largest submarine canyons—over 6,000 feet deep and of sufficient size to hold the Grand Canyon of the Colorado River (Figure 4.15). Every year, virtually all of the

Figure 4.15   Monterey Submarine Canyon, which extends almost to the beach, drains off the littoral sand moving south along the shoreline. Courtesy of Friends of Monterey Bay.

nearly 300,000 cubic yards of sand that is transported down the coast of northern Monterey Bay, some from as far away as the entrance to San Francisco Bay, is carried offshore into deep water by this vast underwater conveyor belt.

Once sand starts moving into one of these canyons, it is lost permanently to the beach. Many observations and years of research indicate that in the steeper canyon heads this sand may simply flow downslope, grain by grain, until slopes decrease, and that transport to many miles offshore is achieved by submarine mud flows known as turbidity currents. Turbidity currents are large masses of sand, mud, and water driven by their greater density relative to seawater that are capable of flowing many miles down submarine canyons over very low slopes. Ultimately, these former beach sediments are deposited on the sea floor as deep-sea fans, much like alluvial fans at the foot of a mountain range. With their final resting places at depths of 10,000 to 15,000 feet below sea level, these sands are in effect permanently lost to the beach. Although proposals have been made to dam these offshore canyons as a way to trap this sand so that it can be pumped back onto the beaches, it would be far easier to halt or slow the flow of sand on the beach or at the shoreline than to do so offshore in a canyon head in deep water.

LITTORAL CELLS OR BEACH COMPARTMENTS

Many years of study of the coastline of southern California led to the recognition of a series of distinct beach compartments or littoral cells. These are self-contained segments of the coast characterized by distinct sources of sand, longshore transport or littoral drift of the sand, and a sink or area where the sand is lost from the system (Figure 4.14). It was the discovery of the many offshore submarine canyons and the realization that sand moving down coast along the beaches was flowing into these canyons that led scientists to this concept. Down drift of many of the canyons, the beaches narrow or disappear and rocky headlands or points typically mark the end of each compartment. Continuing down drift, streams again begin to deliver sand to the shoreline, beaches appear, and as more sand enters the cell, the beaches begin to widen. Some distance down drift another submarine canyon intersects the shoreline, the beach sand is conveyed offshore, and the end of another compartment or cell is reached. In an idealized littoral cell or beach compartment, there is little to no longshore sand transport between cells.

Along the southern California coast, between Santa Barbara and San Diego, four major littoral cells have been recognized: the Santa Barbara Cell, the Santa Monica Cell, the San Pedro Cell, and the Oceanside Cell (Figure 4.14). Each is characterized by individual sand sources (primarily streams and cliff erosion), down-coast littoral drift, and a sink or sinks (coastal dunes and submarine canyons). Other littoral cells have been recognized along the entire length of the California coast, although the boundaries and the total volumes of sand moving through these cells are still being measured and documented.

Although we don't yet completely understand the boundaries of each of California's beach compartments, the volumes of sand provided by each source, the amount lost to each sink, or the volume moving alongshore annually as littoral drift at various locations within the cell, this information is required prior to consideration of any large-scale intervention in this coastal sand transport system. Today, the greatest concern to the state's coastal residents, planners, regulators, and coastal engineers is how to deal with the threats from the reduction of beach sand supply and seacliff erosion. Numerous conferences, workshops, hearings, and local government and citizen's groups focus on issues that are intimately connected to the budget of sand in any individual cell and how that budget has been altered by human activity such as beach nourishment and replenishment, sand supply reduction, bluff armoring, and dredging and dredge disposal.

ARE THERE THREATS TO OUR BEACHES?

Many human activities significantly influence the sand supply that feeds California's beaches. Since the 1950s many of California's rivers and coastal streams have been dammed for water supply, flood control, and recreation. Debris basins have been built throughout southern California to protect development from mud or debris flows from the steep slopes of the San Bernardino, San Gabriel, Santa Monica, and Santa Susana mountains. These basins trap coarse sediment while allowing the passage of water and fine sediment. In addition, armoring the seacliffs, whether through riprap or seawalls, effectively cuts off the supply of sand that would have been added from the eroding cliffs.

DAMS, RESERVOIRS, DEBRIS BASINS, AND SAND REMOVAL
Dams along coastal streams that effectively impound water also trap the sand destined for the coastal beaches. Thus, the benefits of flood control or increased water supply and recreation have been countered by the gradual reduction of sand input to the coastline. Dams reduce sediment supply to beaches in two ways: by trapping sediment behind the dams and by reducing the peak river flows or floods that, under natural conditions, would have transported sand downstream to the shoreline. More than 1,400 dams over 25 feet high or impounding more than 50 acre-feet of water have been constructed across California, with 539 of these dams located in the coastal watersheds that drain directly into the Pacific Ocean (59 dams are in watershed

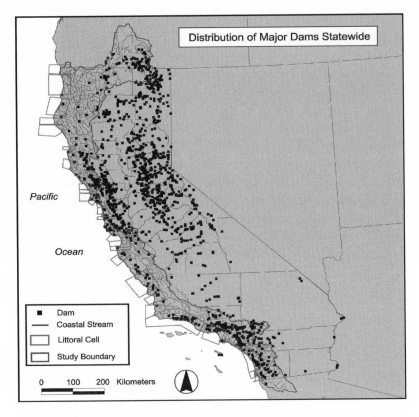

Figure 4.16   The distribution of large dams in California. Map by Cope Willis, from California Beach Restoration Study, State Coastal Conservancy.

areas that extend into Oregon and Mexico). Since the construction of the first coastal dam in California in 1866, an average of 3.5 dams per year have been built, and today 480 large dams reduce the flow of sediment to the coast (Figure 4.16). These dams affect more than 38 percent of California's total watershed area and impact sand contributions from over 16,000 square miles. The combined effect of these coastal dams has been to reduce the average annual sediment supply by more than 25 percent to California's 20 major littoral cells. In Southern California (Point Conception to San Diego), where demand is the greatest for recreational beaches, sediment supply has been reduced by over 50 percent to half of the littoral cells. A close look at 14 of the largest dams in southern California indicates that they have been impounding about 1.33 million cubic yards of beach sand on average annually for over 50 years, which means that 66 million cubic yards of sand (6.6 million dumptruck loads) have been withheld from the beaches and lie beneath the waters of the reservoirs. The sand trapped each year by these 14 large dams alone is enough to build a beach 10 feet deep, 100 feet wide, and 7 miles long!

With the construction of dams, the sand load of the Santa Maria River has been reduced by an estimated 68 percent; the Ventura River by 53 percent; the Los Angeles, San Gabriel, and Santa Ana rivers by 67 percent; and the San Dieguito River by 79 percent. Throughout southern California, 311 water supply lakes and flood control reservoirs trap beach materials. In addition, the construction industry has historically mined over 20 million tons of sand and gravel from beaches, dunes, and riverbeds each year. Although mining of the beaches and dunes no longer takes place, sand and gravel is still removed in large volumes from seasonally dry streambeds and alluvial floodplains in southern California by the construction and aggregate industries. In Los Angeles County alone, about 21 million cubic yards of sand and gravel have been extracted from the floodplains. Each cubic yard of sand extracted from a streambed or alluvial plain or fan represents a potential reduction in sand supply to the downstream beaches. There are also 115 debris basins in Los Angeles County alone that trap the coarse sediments coming out of the steep mountain drainages. Over 20 million cubic yards of material, much of it sand, have been removed from these basins and deposited in inland disposal sites; this removal also represents a significant reduction in sand delivery to the area's beaches.

The northern California sand supply is considerably less altered. Major dams are few, and sand and gravel mining is more limited. The sand supply of the Klamath River has been reduced by 37 percent, the Russian River by 17 percent, the Salinas River by 33 percent, and the Carmel River by 59 percent because of the construction of dams. Although dams have been built on some of the other coastal rivers and streams, none of these major rivers have had their sand supply reduced by more than 10 percent.

Fortunately for much of southern California, artificial nourishment associated with a number of large coastal construction projects essentially kept pace with sediment losses from dam construction over much of the twentieth century. Several new marinas were dredged out of coastal wetlands (Marina del Rey and Oceanside Harbor, for example), and many large construction projects were undertaken in areas occupied by coastal dunes (the Hyperion Sewage Treatment Plant, the El Segundo Oil Refinery, and the Los Angeles International Airport, for example). Between 1940 and 1960, over 130 million cubic yards of sand were placed on the region's beaches by these large construction projects near the coast, so that the beaches remained wide and healthy.

By the late 1960s, however, harbor and marina construction and the associated beach nourishment activities were terminated. In some areas, the nourishment projects built beaches that were broader than those previously maintained by natural sand supply. In other areas, nourishment from construction balanced the losses from dam construction. Today, the nourishment from construction projects is over, but the sand impoundment behind the many dams continues. With the damaging ENSO events and severe storms of 1982–83, 1988, and 1997–98 and the apparent narrowing of beaches in southern California, there has been considerable discussion and several proposals put forward to remove dams that are filled with sediment and no longer serve any useful purpose. Although many dams impound large volumes of sediment and have the potential to be removed, two dams in particular—the Matilija Dam on the Ventura River and the Rindge Dam on Malibu Creek—have been targeted for study and possible removal.

Matilija Dam is located on Matilija Creek, a tributary to the Ventura River, about 18 miles inland from the coast. The Ventura County Flood Control District built the dam in 1946 for water supply and flood control. The dam is a concrete arch structure about 200 feet high that originally held about 2,000 acre-feet of water. When surveyed in 1999, however, the reservoir contained about 6 million cubic yards of sediment, and about 93 percent of the storage capacity had been lost. In 1999, partly because of concerns about the effects of sand supply reduction along the shoreline, Ventura County proposed the removal of the dam and requested that the U.S. Bureau of Reclamation undertake an appraisal study of the removal. Although the dam could be taken down and sediment removed through incremental notching (Figure 4.17), the reliance on periodic high stream flows to transport the sediment to the coast has the potential to impact downstream properties, water intakes, and stream and estuary flora and fauna. On the other hand, dam removal could restore a remnant steelhead run in the Ventura River. One possible solution that would avoid many of the potential impacts is a slurry pipeline that would transport the sand and fine-grained sediment to the coast in a controlled manner.

Rindge Dam on Malibu Creek is about 2 miles inland from the coast near the city of Malibu. A private landowner built this concrete structure in 1926 for private water supply. The dam is about 100 feet high and originally stored about 600 acre-feet of water. By the late 1950s sediment had completely filled the reservoir, which now stores about 800,000 to 1.6 million cubic yards of sediment, about half of which is believed to be suitable for beach replenishment. Malibu Creek has historically been an important sediment source for the Malibu coastline, and the California Department of Fish and Game initiated a study on the feasibility of the removal of Rindge Dam in 1992 as part of an effort to restore a steelhead run. Both the U.S. Bureau of Reclamation and the U.S. Army Corps of Engineers have subse-

Figure 4.17 The Matilija Dam on the Ventura River, which is now filled with sediment. Photo by A. Paul Jenkin.

quently studied the feasibility of removing the dam. They concluded that there are shoreline benefits to dam removal and sediment delivery to the coast, but, as with the Matilija Dam or any other dam removal, potential habitat impacts must be resolved or mitigated. The most recent action is a work plan for a reconnaissance study by the Corps of Engineers to determine the most feasible and environmentally sound way of removing the dam and transporting the stored sediment to the coast.

ARMORING THE SEACLIFFS   Protective barriers such as riprap and seawalls that are built to halt or control seacliff erosion can also reduce sand supply to the beaches from cliff erosion. Studies were recently completed that documented the significance of cliff erosion to the beach sand supply for two lengthy sections of the California coast: the 48-mile-long segment from Dana Point to La Jolla (the Oceanside littoral cell, which includes southern Orange County and northern San Diego County) and the 144-mile segment from the Santa Maria River to Point Mugu (the Santa Barbara littoral cell, which includes Santa Barbara County and nearly all of Ventura County). An important concern that precipitated this study was the increasing amount of coastal cliff armoring (seawalls and riprap) that had been approved and constructed in recent years and the perceived impacts of this armor on beach sand supply through the reduction of cliff erosion.

In the Santa Barbara and Ventura county area, although coastal cliffs are widespread and are eroding, they supply less than 1 percent of the total beach sand to the Santa Barbara littoral cell. Most of the cliffs of this area consist of fine-grained shale with little to no sand, and thus the eroding cliffs contribute very little to the beaches. In addition, this region also has four major rivers (the Santa Maria, Santa Ynez, Ventura, and Santa Clara) and several smaller streams whose collective sand production is very large, over 2 million cubic yards a year on average, overshadowing the minor amount of sand supplied from cliff erosion.

The Dana Point to La Jolla area (Oceanside littoral cell) is somewhat different. The cliffs contain more sandy material and the streams are much smaller, but there are numerous gullies and short drainages that do produce considerable sand. Nonetheless, cliff erosion in this area provides only about 11 percent of all of the sand on the beaches of the region. Thus, although shoreline armoring in both of these cells is of concern for many reasons, sand supply reduction due to seawalls and riprap, particularly in the Santa Barbara cell, does not significantly affect the supply of sand to the beaches of these areas.

LITTORAL DRIFT INTERRUPTIONS

Much as a dam on a river traps the sand moving downstream, many large coastal engineering structures have had similar impacts on the transport of littoral sand down coast along the shoreline. The impoundment of sand behind engineering structures is not a unique issue to California. There are, however, many groins, jetties, and breakwaters along the length of the state's coastline that have intentionally or unintentionally disrupted the natural littoral drift and trapped large volumes of beach sand with some very clear impacts.

GROINS   Groins are relatively short structures designed to trap the longshore drift of sand and have been successfully used in California and elsewhere to create, widen, or stabilize beaches. Many of California's beaches exist because of natural littoral drift barriers such as headlands, and many others owe their existence to artificial barriers such as groins, jetties, and breakwaters. Groins mimic natural headlands and, with appropriate planning and at appropriate locations, can be used to hold in place more of the sand on California's beaches, thereby increasing shoreline protection and improving recreational areas with far less maintenance, at less cost, and with less negative environmental impact compared with armoring or artificial nourishment.

Groins are usually built perpendicular to the shoreline and may be several hundred feet or more in length. They can be built of rock, concrete,

Figure 4.18 A groin field has created and stabilized a wide beach at Ventura. Photo by Gary Griggs and Kiki Patsch.

timber, steel, or even sand-filled geotextile bags. They can be permeable or impermeable and can extend above or below beach level. Because a groin forms a partial trap or barrier to littoral drift, sand tends to accumulate on the updrift side of the structure. The amount of sand trapped and the size of the beach formed depend on the height, length, and permeability of the groin. Short, low, permeable groins, for example, will allow more sand to move past the structure than higher, longer, impermeable groins.

Although single groins can be constructed to stabilize or form a local beach, they are often built in groups (called groin fields) in order to form or stabilize a beach along an extended stretch of coastline (Figure 4.18). The predictable impact of a single groin, or of a groin field when the individual groins are placed too far apart, is not only deposition updrift from each groin, but also erosion down coast. Depending on the nature and use of the beach or coastline in these down-coast areas, a certain amount of liability may be expected by those constructing groins. The most sensible approach is to completely fill or charge the area up coast from each groin so that there will be no net sand removal from the natural system. Because of the barriers they form and their possible down-coast effects, groins should not be built unless properly planned for the particular site and until the spacing and effects on adjacent or down-coast beaches have been thoroughly considered.

Although groins can be quite effective in widening or stabilizing an existing beach, they do not provide total protection from wave action. For

Figure 4.19   (A) Winter beach erosion at Capitola due to wave attack from the west. (B) Wide summer beach at Capitola. Photos by Gary Griggs and Kiki Patsch.

example, when large storm waves approach the coast directly or head on, rather than from an up-coast direction, the beach, which has been widened because of the groin, may be eroded and not provide the desired protection from wave attack. This has happened repeatedly at the small resort town of Capitola near Santa Cruz. A single groin was constructed in 1969 to trap sand in order to form a permanent beach. As long as the waves come from the northwest, the protective beach remains. During the large El Niño winters that have occurred over the past 25 years, however, major storms attacked the coast from the west and southwest. The 200-foot-wide sand beach has been eroded and the waves have overtopped a low seawall, battered the beachfront restaurants, and flooded the city streets with debris (Figure 4.19A). In the spring and summer, as waves from the northwest begin to dominate, littoral drift is again trapped to form a wide beach (Figure 4.19B). The important thing to remember is that groins trap sand but do not provide complete protection in themselves, and that there are predictable down-coast effects.

JETTIES AND BREAKWATERS   Jetties and breakwaters have a totally different purpose from groins, although they often produce some of the same effects. Jetties are usually built of very large rocks, or large interlocking concrete modules (Figure 4.20) and are built in pairs, perpendicular to the shoreline, to protect or stabilize a harbor entrance channel or river mouth against wave action, and thereby provide safe navigation for boats entering or leaving the harbor. The jetties at Humboldt Bay, Santa Cruz, Moss Landing, Ventura, and Channel Islands harbors are good examples. An additional goal is to control or reduce shoaling of a channel by littoral drift. Jetties, because of their length, initially impose a total littoral barrier to sand transport. Accretion or sand build-up will occur up drift or up coast from the jetties where there is a dominant littoral drift direction. Beach narrowing and cliff or bluff erosion will occur down coast as the beaches are starved, unless sand bypassing or dredging is initiated. In this respect, jetties have the same effects on littoral drift as groins, except that jetties, being typically much longer, impose an even greater barrier to sand transport. The major difference is that groins are constructed to trap sand and widen a beach, and jetties are not built with this intent. The California coast has many examples of sand impoundment due to jetty construction that has required expensive annual or biannual dredging (Figure 4.21 and Table 4.1). Although many jetties were built in the past as a result of political pressure for the creation of a harbor to serve local interests, without regard for the impacts of construction, planning for any harbor today where littoral drift rates are high must be preceded by a thorough investigation of the coastal processes, the possible impacts of any structure, and methods for mitigating those impacts. Because of the now well-understood impacts of jetty construction on littoral transport, and the long-term costs of dredging or bypassing large volumes of sand, it is unlikely that the California coast will ever see the construction of additional harbors or marinas.

The jetty construction at the Santa Cruz Small Craft Harbor, completed in 1965, provides a good record of the direct and indirect effects of coastline

Figure 4.20 Jetties at the Channel Islands Harbor along the Ventura County coast. Photo by Gary Griggs and Kiki Patsch.

Figure 4.21 Dredge in operation at the Santa Cruz Small Craft Harbor. Photo by Gary Griggs and Kiki Patsch.

alterations of this sort (Figure 4.22). The combined effects of this project include (1) the formation of a wide, permanent up-coast beach that protects a previously eroding stretch of urbanized cliffs and provides an important recreational beach; (2) shoaling of the harbor entrance channel with average annual dredging of 200,000 to 250,000 cubic yards at a cost of about $500,000; (3) increased erosion rates for unprotected and already eroding

Figure 4.22 (A) Pre-harbor coastline of the Seabright Beach area of Santa Cruz. (B) Post-harbor coastline of the same area.

Figure 4.23   Large breakwater at Half Moon Bay in San Mateo County. Photo by Gary Griggs and Kiki Patsch.

down-coast bluffs following harbor construction, requiring armor emplacement; and (4) the loss of a public beach at the town of Capitola, farther down coast, over the next several years. The harbor jetties have thus had a significant and immediate impact on private and public property up coast and down coast. Regular dredging now puts the littoral sand trapped by the up-coast jetty on the down-coast beaches, so that the system has more or less returned to equilibrium.

Breakwaters are large rock and/or concrete structures built to protect harbors or anchorages from wave action. Crescent City, Half Moon Bay, Monterey, Port San Luis, Santa Barbara, Redondo, and Long Beach harbors are all protected and formed by large breakwaters (Figure 4.23). The Ventura and Channel Islands harbors combine jetties and breakwaters (Figure 4.24). Regardless of the placement and orientation of a breakwater, it will affect the adjacent coastline in several ways. The primary and intended effect is to greatly reduce or nearly eliminate wave action within the protected area behind the breakwater. This reduction of wave energy, however, usually produces the secondary effect of interrupting littoral drift because the energy is no longer available to move the sand through the harbor area.

At Santa Barbara, an initially detached breakwater built in the late 1920s proved almost immediately to be an effective sand trap (see Chapter 17). Sand moving down drift first filled a large embayment west of the breakwater. A college stadium, a large parking area, and a city park were subse-

Figure 4.24   Ventura Harbor with jetties and breakwater in 1967. Photo by A. Paul Jenkin.

quently built on this sand fill, which extended the beach seaward more than 300 yards from the original shoreline. As the sand continued its path down coast it swung around the tip of the breakwater and began to settle out in the protected waters of the harbor, now cut off from wave energy. The breakwater was extended a few years later and the harbor has been dredged continuously since 1935, removing an average of about 300,000 cubic yards per year of sand at a current cost of about $750,000 annually. In the early years of sand accumulation against the breakwater and prior to the initiation of regular dredging, beach erosion followed by bluff erosion proceeded many miles down coast and eventually damaged a number of oceanfront homes. Artificially filling the area up coast from the breakwater and then initiating sand bypassing at the end of the breakwater, although costly, could have mitigated the down-coast erosion damage.

The situations at Santa Barbara and Santa Cruz are not unique but provide well-documented examples of direct and indirect impacts of major coastal engineering structures that were not anticipated. Although jetties and breakwaters are designed to provide protected harbors, in many locations they have had significant impacts on littoral drift and coastal erosion. Thus, homes or other structures built on eroding seacliffs or on the beach have often been the indirect victims of these large public projects.

The other direct cost associated with many of these harbors, which have intercepted large quantities of littoral drift, is the annual dredging of sand.

This is a process that in some cases is done year round, in some cases annually, and in some cases biannually. The total cost of dredging California's coastal harbors is now many millions of dollars annually. This will go on forever as long as we have littoral drift, and the cost will continue to increase.

INCREASING THE SAND ON CALIFORNIA'S BEACHES—BEACH NOURISHMENT

Beach nourishment is the artificial placement of sand on a beach and is intended either to replace sand that has been prevented from naturally reaching the beach (because of dam construction, for example) or to widen a beach that is naturally narrow. Widening a beach produces two primary benefits: increasing the area available for recreation and providing a greater buffer to the coastline from storm wave attack.

There are essentially two different types of beach nourishment. The first type, opportunistic beach nourishment, has been practiced in California for many years and involves dredging sand up drift of a harbor breakwater or jetty or from the entrance channel of a harbor and placing it on the down-drift beach. Examples of harbors utilizing this type of nourishment are Santa Cruz, Santa Barbara, Ventura, and Oceanside. No new sand enters the system, but some sand dredging or transport system is used to pump the sand onto the down-coast beaches. These projects simply move sand across or under a barrier.

The second type of nourishment involves transporting sand from some off-site or distant source, either an offshore or onshore location, to the beach. Whereas East Coast states such as Florida, New York, and New Jersey obtain millions of dollars in federal funds to carry out this type of beach nourishment, this is not the case in California. The lack of funding for beach nourishment is in part due to the fact that California's coastlines are very different from the East Coast, and most of its beaches have historically been wide and healthy. The state's rivers provided large quantities of beach sand, eroding bluffs added to this budget, and interruptions in the sand flow were few. In addition, particularly along the Santa Monica to Redondo Beach shoreline, several large coastal construction projects that took place in the coastal dunes added huge quantities of sand to the beaches as a by-product of construction. This has gradually changed, however with the damming of rivers and the concreting of river channels, the construction of debris basins to collect sediments, the use of breakwaters and jetties to trap large quantities of sand, and since the late 1970s, a substantial increase in wave and storm damage combined with a slowly rising sea level. There are also more revetments and seawalls on beaches that cut off the sand supply from eroding bluffs in many locations. Beaches in some places are or are perceived to be narrower due to this drastic reduction in sand supply. There is also a desire, by both oceanfront property owners and southern California communities whose economies are dependent on tourists, to restore or widen their beaches.

On the surface, beach nourishment sounds like a wonderful compromise between armoring the shoreline and retreating. Few would argue that to the degree that we can widen our beaches, we all benefit. Some important questions and environmental issues need to be addressed, however, before we embark on any significant beach nourishment program or project.

The first issue to be resolved is where the sand comes from and how much it will cost. An essential requirement, whether the sand originates from onshore or offshore, is that it be compatible with the beach to be nourished. If the sand is too fine-grained, it will be washed away by wave action. If it is too coarse, it may remain on the beach but be less desirable to beach users.

Although there are on-land sources such as existing sand and gravel quarries, debris basins, and reservoirs now full of sand, these require excavation, washing or cleaning in some cases, and transport by truck and, as a result, are expensive. At 2003 prices of about $23 per cubic yard to deliver clean sand from a nearby dune area to the shoreline in southern Monterey Bay, it would cost $5.5 million for a proposed nourishment project of 240,000 cubic yards to protect a 1,000-foot-long section of eroding shoreline with a beach 100 feet wide. It would require 24,000 10-cubic-yard dumptruck loads to deliver this sand, or a truck every 5 minutes, 8 hours a day, 5 days a week, for an entire year. These trucks would clearly create some significant roadway and beach problems and would essentially render the beach unusable for a year.

Offshore sources also exist, although there are increasing concerns in certain areas with the sea floor disturbance associated with dredging up large quantities of sand. Along 200 miles of the central California coast, it is doubtful that sand dredging for beach nourishment would be allowed within the 5,300 square miles of the Monterey Bay National Marine Sanctuary. Although there are some offshore sand resources of the appropriate grain size, costs are considerable and replenishment will be an ongoing need because of California's high littoral drift rates. The first large artificial nourishment project in California was completed in the spring of 2001 in San Diego County. This effort took a number of years to plan, coordinate, and fund, and it involved regional, state, and federal agencies. Two million cubic yards of sand were dredged from six offshore sites and pumped onto 12 San Diego County beaches at a cost of $17.5 million, or $8.75 per cubic yard. Beach surveys indicated that within a year most of the sand had moved alongshore or offshore.

Artificial nourishment is thus costly, and it is not clear where the money to maintain the effort in perpetuity will come from. In California, where typical littoral drift rates are on the order of 300,000 cubic yards a year, the life spans of nourished beaches at most sites can be expected to be relatively short. Artificial beach nourishment can provide short-term benefits, but at significant costs, and we need a permanent and sustainable solution.

WHAT YOU SHOULD REMEMBER ABOUT BEACHES

Important things to remember about beaches include: (1) Beaches provide important recreational areas for residents and visitors alike, (2) they provide a buffer to wave attack of the coastline, (3) they are ephemeral and undergo regular and sometimes dramatic seasonal changes, and (4) many human activities can affect beaches.

As a dynamic and fragile feature, a beach may grow, shrink, alter its shape, or even disappear in a single storm. During the winter, large, steep, closely spaced waves scour away beach sand and move it offshore. When the weather calms in the spring, waves that are less steep and more widely spaced normally push the sand back onshore and rebuild the wide summer beach. This is a natural seasonal process by which the beach strategically adapts to high winter storm waves and low summer swell. Remember that the beach and the ocean are in a dynamic equilibrium, so that when one changes, the other must adjust. So if a house is built on a wide beach during the summer, it should be no surprise to the owner to find the ocean in the living room during a winter storm. In addition to the seasonal on- and offshore movement of sand, beach sand moves alongshore in most locations through littoral drift.

The development of California has been accompanied by large-scale changes to the landscape and major alterations or disruptions to natural systems in order to provide for human "needs." Dams and debris basins have been built, rivers and creeks have been paved or concreted over, sand has been quarried for construction, and harbors and other large coastal engineering structures have been built along the shoreline. Each of these actions has altered the natural flow of sand to or along the state's beaches, particularly in central and southern California, where the recreational demand on beaches is greatest. Although there are no simple solutions, we now recognize the problems and disruptions we have created and have begun to evaluate various responses and solutions. The approaches we develop today need to be sustainable and to rely as much as possible on natural processes.

CHAPTER FIVE

THE EROSION OF THE COASTLINE

GARY GRIGGS, KIKI PATSCH, AND LAURET SAVOY

INTRODUCTION

If you have ever watched storm waves batter the coastline in the winter months, you appreciate the tremendous power exerted by the ocean. Even during calm weather, small waves constantly wash the sand and gravel across the rocky intertidal zone—wetting, drying, and gradually weakening the rocks that make up the cliff and carrying off the bits and pieces that break loose. Although this day-to-day activity takes its toll on the cliffs, it is generally the winter storm waves at times of high tides that lead to most cliff retreat. In part, this is because the winter waves are larger and have more energy. In addition, prolonged winter rains weaken the bluffs, making them more susceptible to failure. Another important factor is the reduction in width or even the total disappearance of protective beaches during the winter. With this buffer zone of sand gone, the waves attack the cliffs directly.

During severe winters we often see television news reports or read newspaper stories about houses or roads being undermined or collapsing into the ocean. It is relatively easy to understand that the gradual retreat of the shoreline is a logical consequence of continuing sea-level rise due to global warming. As discussed in Chapter 3, the Earth has been warming for the last 18,000 years. Glaciers and ice caps have melted, adding water to the oceans. In addition, warm water is less dense and takes up slightly more volume than the same amount of cold water, so warming of the ocean raises sea level. The consequence of this sea-level rise has been a worldwide flooding of the offshore continental shelf and the ongoing landward migration or retreat of the shoreline. Sea-level rise is not a significant issue in uninhabited coastal areas, but for the developed coastlines of the world, the inland migration of the shoreline has an enormous impact. Today there are about 100 million people living within 3 feet of sea level, and the best projections indicate that about 3 feet of sea-level rise will take place by 2100. Although this may not be of much concern to inland residents, oceanfront property owners have some significant issues to deal with. Those agencies that permit construction or insure coastal properties need the best information available on the likely scenarios for the decades ahead. How high might sea level rise by 2025 or 2050, for example, and where might the position of the shoreline be as a result?

Shoreline erosion means different things depending on where you live in the United States or even in California. For the residents of the Gulf or south Atlantic coasts, where the shoreline consists of a low-relief coastal plain fronted by offshore barrier islands, coastal erosion occurs in the form of landward barrier island migration. Because the barrier islands consist only of loose sand, erosion or shoreline migration can take place at a relatively rapid rate, 11 feet per year in the case of Cape Hatteras, on the Outer Banks of North Carolina.

HOW DOES THE COASTLINE ERODE?

Along the shoreline of California, coastal erosion can take several different forms. The beach erodes seasonally, but this is a reversible process and the beach sand lost each winter is generally replaced by the next summer (Figure 3.1). As discussed in the previous chapter, where a wide sandy beach has historically fronted the coast, but sand supplies have been significantly reduced through the construction of dams or large coastal engineering structures, the beach itself can be permanently eroded or narrowed. There is concern that some southern California beaches may be undergoing long-term erosion due to sand supply reduction. The beach may be rebuilt and recover, however, if sand supplies or sand flow is restored.

To most residents of California, coastal erosion is the actual retreat of the seacliff or coastal bluff. This process is distinct from beach erosion and is not recoverable, at least within our lifetimes or under natural processes. The rate at which a seacliff erodes depends primarily on its exposure to wave attack, the height and energy of the waves that typically reach the area, and, very important, the strength and physical properties of the materials making up the cliff. The hardness or degree of consolidation or cementation of the cliff rock and the presence of internal weaknesses such as fractures, joints, or faults all directly affect the resistance of the rock to wave impact or the degree of chemical weathering of the rock by seawater.

Cliff erosion takes place through several different processes, all gradually taking their toll on private or public oceanfront property. Hydraulic impact is perhaps the most important and is simply the direct impact of a breaking

Figure 5.1   Coastal cliff failure along shore-parallel jointing in sedimentary rock near Capitola. Photo by Gary Griggs and Kiki Patsch.

wave on the seacliff. Where rocks are unconsolidated or jointed, waves can dislodge large and small fragments or blocks, leading to gradual failure or retreat of the cliff. The sand and gravel and even cobbles that the waves wash back and forth across the shoreline become important abrasion tools. Just like sandpaper, the constant grinding of the rocks against the bedrock or, under large-wave conditions, the launching of gravel and cobbles against the cliffs also contribute to the retreat. The alternating cycles of wetting and drying, and heating and cooling, that take place within rocks in the inter-tidal zone, as well as the chemical breakdown of the rocks by seawater, are also important processes. The bedrock exposed in cliffs and along the shore-line is broken down by a combination of physical disaggregation and chem-ical reaction. All of these processes act in concert to either weaken the cliff-forming materials or break out and remove fragments.

Human impacts, such as street or storm drain runoff, septic tank leach-ing, and landscape watering, and the alteration of normal drainage away from natural gullies so that more water passes through or across the cliffs, also contribute to accelerated seacliff or bluff retreat. Each of these processes can erode the cliff directly or lead to weakening of the cliff material followed by slumping or sliding (Figure 5.1). Although native vegetation usually acts to stabilize seacliffs, at some locations nonnative or ornamental vegetation has been planted in an attempt to stabilize the seacliff material but has had

Figure 5.2   Irregular erosion along joints in resistant granitic rock near Carmel,
Monterey County. Photo © 2002–2004 Kenneth and Gabrielle Adelman, California
Coastal Records Project, www.Californiacoastline.org.

the opposite effect. Excessive watering may weaken the cliff-top materials.
The roots of some trees may penetrate into fractures or weak zones in the
rocks and act as wedges to pry away large blocks. In addition, ground cover,
such as ice plant, often hangs down over the cliff with so much weight that
it uproots and carries down soil and loose cliff-top material with it.

Coastlines consisting of hard crystalline rock, such as the granite of the
Monterey Peninsula, Point Lobos, or the outer end of Point Reyes, usually
erode very slowly. Within these rocky zones, however, erosion rates may still
vary somewhat. Waves attack the weaker zones, the fractures and joints, to
form inlets and coves (Figure 5.2). Good examples of the influence of rock
resistance variability on wave attack occur along the coasts of Mendocino
and Big Sur. The geological variability typified by the Franciscan
Formation, which is a chaotic mixture of very hard rocks (such as cherts,
which resist wave attack), and softer mudstones and shales (which are more
easily eroded) has produced a very irregular coastline. The more resistant
rock remains in the form of points, headlands, or offshore sea stacks,
whereas the weaker rocks erode to form embayments (Figure 5.3). In contrast, erosion can be uniformly rapid where the coastline is made up of relatively soft, weak sedimentary rock such as sandstone or shale, or even
unconsolidated materials. In these zones, the cliffs often retreat in a more
linear fashion, producing relatively straight coasts (Figure 5.4).

Figure 5.3 Nonlinear erosion of heterogeneous rocks of the Franciscan Formation at Gleason's Beach in Sonoma County. Photo © 2002–2004 Kenneth and Gabrielle Adelman, California Coastal Records Project, www.Californiacoastline.org.

Figure 5.4 A linear coastline eroded in unconsolidated terrace deposits south of Half Moon Bay. Photo by Gary Griggs and Kiki Patsch.

Figure 5.5   Erosion of weak sedimentary rocks in the Isla Vista area of Santa Barbara County has undercut and threatens cliff-top structures. Photo by Gary Griggs and Kiki Patsch.

Coastal cliff retreat is one of the more dramatic processes taking place along the coastline. Undermined foundations, dangling decks and stairways, and structures that have collapsed onto the beach are harsh reminders of the ongoing retreat of the California coastline as sea level continues to rise (Figure 5.5). Many coastal communities have lost entire oceanfront streets through continuous erosion over the years (Figure 5.6).

The retreat or erosion of seacliffs observed along much of California's coast is due not only to waves gnawing at the cliff base, but also to landsliding, slumping, and rockfalls originating higher on the cliff (Figure 5.7). For example, the combination of jointing patterns, percolation of groundwater from street runoff and yard watering, wedging by tree roots, and undercutting by waves has led to massive rockfalls along the seacliffs flanking northern Monterey Bay (Figure 5.8). These rockfalls commonly occur after intense storms, when rainfall and runoff have been high and the surf heavy. During March 1983 a slab 100 feet long and 6 to 12 feet wide collapsed at Capitola and

Figure 5.6   Severe wave attack combined with high tides ultimately led to the loss of an entire row of oceanfront houses in Pacifica in the winter of 1998. Photo by Monty Hampton, U.S. Geological Survey.

Figure 5.7   Weak coastal bluffs near Jalama, north of Point Conception in Santa Barbara County, continue to fail from terrestrial processes. Photo by Gary Griggs and Kiki Patsch.

fell to the beach below, removing part of a city street. Cliff failure several years earlier at this same location led to the relocation of a two-story house to a safer site two blocks inland. During the 1989 Loma Prieta earthquake, which had a magnitude of 7.0 on the Richter scale, additional cliff failure occurred and six cliff-top apartments had to be demolished in the same area. Seismic shaking during this major earthquake triggered rockfalls and bluff failures from Marin County to Monterey Bay, a distance of over 100 miles (Figure 5.9).

Figure 5.8 Failure of the cliffs in the Depot Hill of Capitola continues to take place along a set of joints that are oriented parallel to the coastline. Photo by Gary Griggs and Kiki Patsch.

Landslides or slope failures can be much greater than the Capitola example, as illustrated by the Portuguese Bend landslide in the Palos Verdes Hills along the Los Angeles County coastline. In this massive failure, which covered about half a square mile, movement began in 1956 and continued for a number of years. More than 200 homes were destroyed or damaged in this event, and property damage (in 2003 dollars) amounted to about $67 million.

Large coastal landslides are frequent along the Big Sur coast due to weaknesses of the rocks exposed, the very steep slopes, and high rainfall (Figure 2.6). Many of these slides extend from the ridgetops to the shoreline and also involve California Highway 1, which is a challenge to keep open during years of heavy rainfall. Devil's Slide, just north of Half Moon Bay and south of Pacifica, presents another example of the challenges of maintaining a major state highway through an active coastal landslide. Some large landslides may not be easily visible from land but can be clearly seen from offshore along the central California coast (Figure 5.10). Prolonged rainfall tends to initiate or reactivate these slides. They present problems for any type of land use because landslides of this scale are very difficult and expensive to stabilize or eliminate. Removal and repair costs to the state of California for one large Big Sur landslide in the mid-1980s were nearly

Figure 5.9   The 1989 magnitude 7.0 Loma Prieta earthquake generated widespread coastal bluff failure, including this massive slide in Pacifica. Photo by Gary Griggs and Kiki Patsch.

Figure 5.10   A large coastal slump at Moss Beach, in San Mateo County. Photo by Gary Griggs and Kiki Patsch.

$12 million (in 2003 dollars), and there have been dozens of these along this stretch of coastline (see Chapter 15).

HOW FAST IS THE COASTLINE ERODING?

Variations in the physical forces that produce cliff erosion or retreat (wave exposure, rainfall and runoff, earthquakes, etc.), as well as in rock strength or the resistance of seacliff materials to failure, have resulted in wide-ranging erosion rates along California's 1,100 miles of coastline. The shape of the coast we see today at a particular location reflects the ability of the materials making up the coastline to resist the forces that are acting to break down the cliffs. The hardest materials stand out as points or promontories, such as Point Loma, Dana Point, Point Fermin, Palos Verdes Point, Point Dume, Point Conception, Point Arguello, Point Sal, and Point San Luis. In contrast, where the materials exposed along the coastline are weak or where the topography is very low, we see embayments, such as Mission Bay, San Pedro Bay, Santa Monica Bay, Estero Bay, and Monterey Bay, to name a few.

One of the things we have learned from studying coastal erosion in California is that coastal retreat varies widely in time and space. Because of vast differences in the erosional resistance of granite on the one hand and unconsolidated alluvial materials on the other, local coastal erosion rates may vary significantly. At some sites cliff erosion has been negligible for the 75–100 years of reliable historic records, simply because the rocks are so hard and resistant to wave attack. At other locations the average rate of retreat may be as high as 5 to 10 feet per year. At Point Año Nuevo, in southern San Mateo County, the erosion rate of the low bluffs has averaged about 9 feet per year for the last 300 years, one of the highest natural rates along the state's entire coastline. The low bluffs at this point are only several feet above sea level, so that storm waves can easily reach and erode the material. In general, the sedimentary rocks such as sandstones, siltstones, mudstones, and shales that form much of California's coastline have retreated at long-term average rates of a few inches to perhaps a foot per year, or more in some places.

Coastal erosion also tends to be an episodic process, with much of the long-term failure or retreat taking place within a few severe storm events every 5 or 10 years. For example, erosion in the resistant sandstones and siltstones at Sunset Cliffs, San Diego, had averaged about a half an inch per year for the 75-year period prior to 1973, and 75 percent of the area studied had undergone no significant erosion during this period. However, in 1976, the U.S. Army Corps of Engineers reported that at the foot of Del Mar Street, Sunset Cliffs, the toe of the bluff had retreated landward 38 feet and the bluff top 40 feet between 1962 and 1976, for an average rate of about 2.7 feet per year.

The arrival of large storm waves at times of high tides and elevated sea level, which occur during major El Niño events, can produce severe coastal erosion in areas formerly thought to be relatively stable. Chapter 3 described the intense 1982–83 El Niño event and its impact on the coast of California. Recent research on changing climate and weather conditions has identified periods lasting several decades when El Niño events were much more severe (Figure 3.3). Many areas of the coast have undergone significant erosion and storm damage during these periods. The most recent cycle of more intense El Niño events began in 1978 and has been characterized by increased rates of cliff and bluff retreat as well as beach erosion. This post-1978 period follows the era when most of California's coastal development took place, which was a period of less intense El Niños, when coastal or beach erosion was not generally perceived as a major hazard or limitation to development.

Many of the old piers, breakwaters, and other structures that were damaged during the storms of early 1983 had survived earlier severe weather, evidence that this winter's damage was extreme by any measure. The 1983 losses resulted from the repeated coupling of elevated sea levels, high tides, storms, and large waves. The rapid erosion that occurs during major storms of this sort can greatly alter coastal cliffs and beaches that have survived many years of more moderate winters and can change our perceptions of what is stable and safe and what is not.

As an example, one section of seacliff in the city of Santa Cruz eroded about 25 feet in the interval from 1931 to 1982. During the January 1983 storms, waves removed about 46 feet of cliff top. This single storm increased the "long-term" average erosion rate from about 6 inches per year to over 16 inches per year! Such observations suggest that average erosion rates be used with great caution and that every effort be made to research the historical changes from maps and aerial photographs as far back in time as possible.

MEASURING COASTAL EROSION RATES

Coastal or shoreline erosion rates are most often determined using historic data on the former position of the shoreline, such as old survey maps or historical aerial photographs, and comparing the former position or line with the present location of the shoreline. City and country assessors, surveyors, and public works and planning departments usually have very accurate historical surveys of developed or improved oceanfront parcels. If the length of an oceanfront lot or the distance from the street to the edge of the seacliff has been recorded on early maps of known dates, an accurate present-day measurement of the corresponding distance and a comparison of the two measures may be all that is necessary to determine the erosion rate.

Dividing the amount of land lost by the number of elapsed years between surveys will produce an average annual erosion rate for the particular site, usually listed in feet per year or inches per year. The longer the available record from old maps and photographs, the more accurate the results will be, simply because of the episodic nature of coastal erosion.

At the town of Bolinas, for example, north of San Francisco in Marin County, comparing the location of the cliff edge during a 1984 survey with that depicted in a 1927 subdivision map indicates that at one particular location near Duxbury Point, the coastline retreated about 120 feet during this 57-year period. Dividing the 120 feet by 57 years gives an average cliff erosion rate of 2.1 feet per year—a high rate, since many coastal lots may not even be 120 feet deep.

However, years of observation, particularly during severe winters such as 1983 and 1997–98, have shown that erosion is usually episodic and irregular. Although the average rate of retreat of the cliffs above Capitola in northern Monterey Bay is about one foot per year, large slabs may collapse overnight, suddenly moving the cliff edge back 5 to 10 feet, followed by little change for a number of years. Short-term cliff erosion rates will often be much different from long-term (30–50 year) averages. This is an extremely important concept for anyone contemplating the purchase, construction, or remodeling of oceanfront property, or anyone involved in approving permits for such construction. A thorough and professional analysis of long-term coastal erosion or change should be required prior to the approval of any oceanfront construction or the purchase of such property. Most coastal properties look very different in the summer, when the FOR SALE signs are out, than they do in the middle of a winter storm. Aerial photographs are often one of the most readily available sources of data for determining historic changes in the shape of the coastline and, therefore, rates of coastline or cliff retreat. Local university or college libraries, the U.S. Geological Survey, the Army Corps of Engineers, the Soil Conservation Service, and the U.S. Forest Service, as well as local or regional government agencies, all have aerial photographs on file and/or contract to have such photographs taken on a regular basis. The oldest aerial photographs available for the California coast were shot in the late 1920s and thus provide an approximately 80-year record of coastal change.

Aerial photographs vary in scale, resolution, and quality and can be difficult to work with unless you have considerable experience. In order to determine the actual ground distance from some feature to the edge of the seacliff in order to see how this distance has changed over time, you need to know the scale of the photograph and the position of the cliff edge from a reference feature such as a street or building. Scales must be determined from ground measurements, and it is a good idea to hire a geologist or engineering geologist with experience in aerial photo interpretation to carry out an investigation of this sort. The accuracy and value of the erosion rates thus determined depend on the observer's experience in working with aerial photographs; the quality, distortion, and scale of the photographs; one's familiarity with the area; and the time period covered by the photographs. In recent years much more sophisticated methods for working with aerial photographs have been developed as a result of GPS (global positioning system) technology as well as improved computer and GIS (geographical information system) software capabilities.

Although "average" erosion rates for some of the more populated southern and central California coastal areas have been determined, this is not the case for most of California. A Federal Emergency Management Agency (FEMA)–funded study, completed in 1999, documented long-term coastal cliff erosion rates for nearly all of San Diego and Santa Cruz counties. This is the most comprehensive study of cliff erosion to date along the coast of California.

The regional section of this book contains the most comprehensive data available to the authors on cliff erosion rates in California, although there are many locations where rates simply have not been determined or are not readily available. Determining cliff erosion rates is a time-consuming, laborious process and requires a long-term aerial photograph or map base. It is evident that the cliffs and bluffs are undergoing active erosion along virtually the entire coast of California, but in many areas the long-term rates have not been measured or published.

The erosion rates presented in the following chapters are based on actual measurements and observations using readily available historical maps and photographs and should be considered reasonably representative of the adjacent areas, although these rates are not intended to be used on a site-specific basis. Local variations in the characteristics of the rock constituting the cliffs can produce wide ranges in erosion rates in adjacent areas. If the rates given in the vicinity of a particular parcel are high and/or a hazard designation has been given to that particular stretch of coast, the prospective property owner would be well advised to look further into any foreseeable problems or the potential for future cliff erosion.

CHAPTER SIX

BUILDING OR BUYING ON THE COAST

GARY GRIGGS, KIKI PATSCH, AND LAURET SAVOY

INTRODUCTION

Anyone who has ever stood on the edge of a coastal bluff during a winter storm doesn't need to be convinced that the coastline is a dynamic environment. Unfortunately, however, most coastal properties and homes are bought and sold during the warm, calm days of summer, when there are no obvious threats to these parcels. Coastal change occurs both over short time intervals (for example, the impact from a single storm or winter) and over longer intervals (for example, the progressive erosion of an unstable bluff area over years or decades). Both types of change can affect an individual property or building, and both should be seriously evaluated prior to investing one's life savings. Although many people fall in love with that special oceanfront house or lot during those long, sunny summer days when the beach is wide and sandy, any potential buyer should wait until the winter storms arrive before making a final decision, or at least make some inquiries to local residents to find out what happens under more severe conditions. Talk to the neighbors, the people who have lived nearby for many years, to see what they typically experience during the winter months. Where does the high tide reach, and what types of changes or damage have they seen in their years of observation? That wide, protective beach can disappear quickly during a major storm, and before you know it your concrete patio and redwood deck are undercut by waves (Figure 6.1). Many oceanfront residents have discovered too late that sliding glass doors and half-inch-thick plywood siding are no match for the large driftwood logs or cobbles thrown about by the surf crashing through their front yards (Figure 6.2).

Two areas of major concern that are addressed in this chapter are the coastal environments that are potentially hazardous for construction and

Figure 6.1   A collapsed deck that has been undercut by wave erosion of the bluff. Photo by Gary Griggs and Kiki Patsch.

Figure 6.2   A row of houses built on a low fill terrace constructed above the beach received major damage from waves overtopping the riprap during the 1983 El Niño event. Photo by Gary Griggs and Kiki Patsch.

the structures, either existing or proposed, that we build in these environments. Considering both the hazards that affect coastal areas and the very high cost of oceanfront property, anyone contemplating such an investment is strongly advised to hire a professional geologist or engineer with experience in the coastal zone to evaluate the stability of the property and/or its structures. For any proposed new construction along the California coast, several local government hurdles must be negotiated before a project is approved. During this process, issues of coastal erosion rates, geological stability, and hazard exposure should be completely evaluated. As a result of these findings, specific conditions such as a building envelope and cliff setbacks will probably be attached to any permit approval (see Chapter 8). On some sites building simply may not be possible or acceptable.

Along the California coast there are three particular physical or geomorphic environments in which widespread development has taken place, regardless of the potential hazards to structures. These environments include (1) the beach itself, (2) sand dunes, and (3) eroding cliff tops. Processes, concerns, and examples of problem areas in each of these settings will be treated individually.

BEACH CONSTRUCTION

Throughout many areas of coastal California, houses have been built too close to the shoreline, directly on the beach. Homes have been built on concrete slabs or on pilings above the beach sand and driftwood (Figure 6.3). These beachfront areas are similar in many ways to the flat floodplains adjacent to creeks and rivers. Just as floodwaters during major rainstorms overflow the stream banks and wash over the floodplains, large waves combined with high tides periodically sweep completely across the beach and up to the flanking seacliff or dunes. This inundation may not happen every winter, but it's important to realize how the sand beneath your house and the driftwood and other storm debris in your front yard got there. It may not have occurred last winter, but the evidence is a reminder that it has happened before and will happen again.

Much of the over $50 million of damage in 1978 and the $184 million of damage in the 1983 storms (in 2003 dollars) occurred when waves attacked beach-level or back-beach homes in places such as Stinson Beach, Rio Del Mar, Aptos, Solimar, Faria, Malibu, Del Mar, Oceanside, and Imperial Beach. A look at old newspapers indicates that the same housing developments are regularly damaged or threatened. Both 1978 and 1983 were major El Niño years, and as a result, the entire state coast-

Figure 6.3   The pile of debris on the left side of the photograph is all that remains of a house built on the beach on a concrete slab along the shoreline of northern Monterey Bay that was undermined during 1983 El Niño wave attack. Photo by Gary Griggs and Kiki Patsch.

line experienced extremely high tides (7–8.5 feet). Higher-than-predicted tides are typical during a large El Niño, when storm surge of a foot or more combines with the increased sea surface elevation that accompanies the warmer waters. During the 1983 El Niño winter, 12 to 15 major storms with very large waves (over 10 feet in height) reached the California coast during the months of January, February, and March. Seven of these storms arrived during very high tides. As a result, much of the beach sand was removed by waves during the first day or two of these storms. With this natural buffer or shock absorber gone, the wave energy from subsequent storms attacked the coastline with greater force, increasing the impacts on coastal structures and property (Figure 6.2). As a result, the 1983 wave attack and damage was perhaps the most severe in half a century.

In contrast, although the 1997–98 El Niño was predicted to be as large and potentially damaging as the 1982–83 event, the largest waves in 1998 arrived during lower tides and coincided with lower wind velocities. Less damage was recorded to beachfront homes because a higher percentage of

Figure 6.4   A house on shallow pilings collapsed following heavy wave attack and beach scour in Rio Del Mar, 1983. Photo by Gary Griggs and Kiki Patsch.

the shoreline had been armored by 1997–98. The addition of substantial sea-walls and revetments in the intervening 15 years protected most of the areas significantly damaged in 1982–83.

Damage in 1983 included undermining of shallow pilings or piers so that homes on the beach collapsed onto the sand (Figure 6.4). Structures on low pilings were also uplifted by waves at high tide and smashed through the pilings as they fell. In addition, waves overtopped low, protective seawalls or other barriers and either damaged or destroyed the home fronts facing the

Figure 6.5   Ocean Front Walk on South Mission Beach, San Diego, flooded by storm waves and high tides, January 18, 1988. Photo by Dana Fisher. Reproduced by permission of the San Diego Union-Tribune.

sea (Figure 6.2). In some cases structural damage of this sort led to house collapse. Nonstructural damage, such as the loss of decks, stairways, patios, yards, and landscaping, was widespread (Figure 6.1).

Strong winds and extremely large waves resulting from an unforecast southeaster struck Pacific Coast beaches from Baja California to San Francisco on January 17 and 18, 1988. The measured wave heights from this exceptional storm exceeded anything ever recorded along the southern California coast. The Redondo Beach–King Harbor area was attacked and severely damaged by large waves that found a route through the normally sheltering offshore Channel Islands. Wave energy within the harbor was severe enough to destroy the front of the Portofino Hotel, and 54 guests had to be evacuated by helicopters as waves washed a barge into the bottom floor. Other structures severely damaged included the yacht club, two restaurants, and many shops located on Horseshoe Pier. Beaches from Santa Barbara to San Diego lost 75 to 150 feet of beach width during the two-day storm (Figure 6.5). Total damage (in 2003 dollars) reached $42 million.

There are some logical questions to ask now that we have seen the storm damage to beachfront property of even the most recent years. Can we recognize such inundation- or wave –impact–prone areas in advance and avoid them altogether, or can something be done to protect existing structures?

The answer to the first question is a clear yes. As discussed in Chapter 5, such beach-level and back-beach areas covered with sand and driftwood are easily recognized. Some may even be partially vegetated at times. All, however, are at the mercy of the waves and tides, some more often than others. A wide beach between a house or potential house site and the ocean should not be accepted as a permanent feature or evidence of long-term security from wave attack. History tells us the beach won't always be there. A brief investigation or inquiry into the storm history of a particular area can provide more specific information that will help in assessing risk. All real estate agents are now required to complete a disclosure form that states whether or not a house or property is located in a recognized and mapped floodplain, fault zone, or other hazardous area. The history of storm inundation and damage along the shoreline may not be designated on any map, however, and therefore won't necessarily be disclosed. A simple measurement of the lot or house elevation compared with information on the maximum expected tidal heights, storm surges, and storm wave heights will provide a clearer sense of the risk of a particular site. The damage to beachfront homes throughout the California coastal zone during the winters of 1978, 1983, and 1988, for example, is clear testimony either that these risks were not adequately evaluated or understood, or that these hazards were disregarded in the design, planning, and/or construction process.

There are some engineering solutions for protecting homes already constructed on beaches, at least for the short term. These include reinforced or elevated foundations, pilings and support structures, as well as seawalls or revetments that can help protect buildings from direct wave attack. Coastal protection is covered in detail in Chapter 7, so our discussion here will be somewhat brief.

If one is to build at all on the beach, it only makes sense to design a foundation and structure with minimal exposure to wave action. A timber piling or concrete caisson foundation that extends into the subsurface or beach well below any potential depth of wave scour is the most sensible starting point (Figure 6.6). These pilings should be sufficiently large and adequately cross-braced, and the structure itself elevated safely above the maximum inundation or wave impact level. Shallow piers can collapse from scouring of the supporting sand. Conventional concrete slab or spread footing foundations are at beach level or at the level of direct wave impact, and therefore are easily reached by winter waves. The value of elevating homes on deep pilings or concrete piers above all potential wave impact was clearly revealed at Stinson Beach in Marin County during the 1983 storms. Houses that faced the sea but were elevated in this manner were virtually untouched, whereas homes built on conventional foundations farther landward were seriously damaged.

Figure 6.6 Multi-million-dollar home built on concrete caissons on the beach along the northern Monterey Bay shoreline. Photo by Jeremy Lezin, courtesy of David Mann.

No doubt there have been countless situations where homes were built without the benefit of adequate engineering or where engineers used sub-optimal design parameters. Common problems with older beach homes include underestimating the depth of wave scour such that pilings lose support and begin to tilt, or underestimating wave run-up so that homes are built too close to sea level.

The emplacement of armor or the construction of a protective wall is another common approach to dealing with the problems inherent in living on the beach (Figure 6.7). Homeowners do not always understand that any coastal protection structure provides only a certain degree of protection, which is provided at considerable cost, and that any structure has a finite life span (see Chapter 7). The questions that have to be answered through a discussion between the homeowner and engineer include: (1) What are the options for

Figure 6.7 The construction of a layered revetment showing the underlying filter cloth, smaller core stone, and larger cap stone. Photo by Gary Griggs and Kiki Patsch.

Figure 6.8 Failure of a concrete panel seawall in Rio Del Mar in 1983. Photo by Gary Griggs and Kiki Patsch.

protecting the house or property? (2) What will it cost to build and maintain the protection? (3) How long will the protection last? Once some engineering options have been developed, the next step is to gain approval for the project, which normally involves a more complicated interaction with the permitting agency. The primary issues to be resolved are the potential impacts of the structure on the coastline (discussed in Chapter 7).

One should not assume simply because there is a seawall or revetment fronting a beachfront home that the home is protected forever from storm waves. During the winter of 1983 many protective structures were over-topped or completely destroyed by the severe high tide and wave conditions, much to the surprise of homeowners (Figure 6.8). Other structures, how-ever, served their purpose reasonably well. The point is that the presence of some protective structure in itself does not guarantee complete protection. If you live behind such a structure, do not maintain a false sense of security. Be prepared to evacuate your home in the face of a storm that may overtop or destroy the seawall and flood your house.

CONSTRUCTION ON DUNES

Because the California coastline is geologically young and tectonically active, much of the coast is characterized by coastal mountains, steep cliffs, and low bluffs, occasionally broken by stream valleys or low coastal plains. As a result, sand dunes—which need a dominant wind direction, a source of sand, and a flat, low-lying area to form on—are somewhat restricted along California's coast. Low-relief areas where dune formation is possible back only 28 percent, or 300 miles, of the state's 1,100-mile coastline. The more geologically stable, low-relief East Coast of the United States, by con-trast, has far more extensive sand dune development.

The primary dunes, or those closest to the ocean and tied to the beach for their sand supply, must be understood as constantly changing landforms. Although there may be older dunes farther inland that have been stabilized by vegetation and no longer depend on the beach for fresh sand, the ocean-front dunes change regularly. Storm waves cut away at the base of the dunes during the winter months, and summer wind and swell bring the sand back onto the beach to help rebuild the dunes. During severe winters the most seaward dunes may be severely eroded or breached altogether.

Dune stability is easily affected by human intervention, as has been common along stretches of the New Jersey coast. Houses were built on conventional foun-dations atop beachfront dunes, destroying the grasses and other stabilizing veg-etation. These dunes were breached to provide vehicle and human access to the beach, groundwater was withdrawn for landscape watering, and large areas were paved over for streets and parking. In March 1962 a violent nor'easter attacked

Figure 6.9   Aerial photographs of the Pajaro Dunes area of central Monterey Bay (A) in 1931 and (B) in 1989. Notice the active and nonvegetated dunes extending up to the oval racetrack in the 1931 photograph, and also the nonvegetated area of wave overwash at the mouth of the river in the 1989 photograph. These areas have been developed with houses and condominiums that have subsequently been threatened by wave attack. Photo in (B) by Gary Griggs and Kiki Patsch.

the entire Atlantic Coast with 60-mile-per-hour winds combined with high tides and waves 30 to 40 feet high. In three days, 2,400 homes were destroyed or damaged beyond repair, and 8,300 homes were partially damaged. Foundation exposure was the most common problem. Those houses that sat high on the dunes, with the best views, found the sand swept out from under them until their foundations or pilings collapsed or were damaged beyond repair.

Figure 6.10   Emergency rock emplaced at Pajaro Dunes to protect houses from further dune erosion. Photo by Gary Griggs and Kiki Patsch.

One of the most extensive areas of home development on oceanfront dunes in California is central Monterey Bay. Although the developed dune area is relatively small (a total of about 2 miles of oceanfront) compared with East Coast dune development, the problems and hazards are identical. Where individual homes, townhouses, and condominiums were built on conventional foundations and directly on the frontal or primary dune, periodic wave erosion has threatened and partially undermined the buildings. For example, a careful look at the old aerial photographs and historical records at the Pajaro Dunes development, located immediately north of the mouth of the Pajaro River, indicates that one large area of low dunes now occupied by 57 condominiums was completely washed over around 1930 (Figure 6.9).

Since construction began at the Pajaro Dunes development in 1969, severe storms have threatened major portions of this frontal dune development during three different winters, all El Niño years. Up to 40 feet of dune erosion occurred during the winter of 1983 at this location. A number of structures were threatened, foundations were undermined, and about $3.7 million (in 2003 dollars) worth of emergency rock was brought in and quickly piled against the eroding dune face to protect the structures (Figure 6.10).

Although the recent history of the site was evident on the aerial photographs, the dynamic nature of the dune environment was given little if any consideration in the planning of this development. The dunes fronting the beach at any location along the California coast should be seen as active, moving landforms with no guaranteed long-term stability. An experienced geologist can obtain historical aerial photographs for virtually any coastal site in California, determine the magnitude of shoreline change that has transpired over the past 50 or 60 years, and provide you with either some sense of security or reason for concern.

DUNE BUILDING AND STABILIZATION

For centuries, shoreline inhabitants around the world have searched for ways to protect themselves from wave attack and erosion. Although many different engineering solutions have been tried, with varying degrees of success, sand dunes are one of the most effective natural buffers to wave attack. When the sand supply is high, and weather, topography, and winds are favorable, dunes build up a large reservoir of sand. When wave and tidal conditions combine to attack the shoreline, the dunes can be eroded and will renourish the fronting beach. Preserving and respecting the dunes for the role they play is a lesson northern Europeans learned long ago. Queen Elizabeth I established a law prohibiting the removal of marram grass (often called European beach grass) during spring to avoid destabilizing the dunes. The Dutch, however, are the best-known users of this natural dune defense from the sea. They have known for centuries that a large and stable reservoir of dune sand is an inexpensive seawall to build and maintain.

The quickest way to build a dune, besides using a bulldozer, is to erect picket sand fences (similar to snow fences) perpendicular to the dune-building winds. The fence slows the wind, and much of the sand bouncing or rolling along the dune surface drops out at the base of the fence. There are two common fence styles, the old-fashioned slat fence and newer fabric fences. In the most efficient spacing of slats, the open area is slightly less than the width of each slat. Although fabric fences are easier to install than the more traditional slat fence, they may cost up to twice as much; they tend to sag under heavy sand loads, reducing the effective height; and in populated areas, fabric fences appear to be more susceptible to vandalism.

Sand fences are used extensively and are usually quite successful. Yet they cease to trap additional sand as soon as sand accumulation reaches the top of the fence. At this point sand simply shifts from one side of the fence to the other with changes in wind direction. To trap more blowing sand, another fence must be built either upwind, downwind, or on top of the old fence, or in some combination. An alternative measure is just to establish plantings of beach grass or other suitable native vegetation. For optimal protection, a uniform beach grass cover should be maintained, which may require some initial watering.

Methods that use sand fencing and plants in combination have many advantages. First, the fence can be put in place at any time during the year and is fully effective as soon as it is erected. Plants, on the other hand, may be planted only in late winter or early spring in many locations and do not become effective sand traps until they develop later in the season. Second, wind velocities may be too great to allow for the establishment of vegetation. Fences significantly slow the wind, preventing scour and providing time for full below- and above-ground plant development. Plants are the most practical long-term stabilization measures. They can be less expensive and more durable than alternative measures, especially if they are natives, and also are aesthetically appealing and self-repairing.

ERODING CLIFFS OR BLUFFS

Thousands of existing homes and undeveloped lots sit atop eroding seacliffs along California's coast, far more than in either dune or beachfront settings. Coastal cliffs throughout much of southern and parts of central California between Monterey and San Francisco have been intensively developed with single-family homes, condominiums, apartments, hotels, motels, and restaurants (Figure 6.11). Although the extent of urbanization is far less along the central coast, these oceanfront cliffs have been heavily developed locally. Major portions of the bluff or cliff edges of Santa Cruz and San Mateo counties, for example, are covered with homes, and for good reason: Almost anyone would love to live or vacation right on the coast, with an unobstructed ocean view.

In 1971 the U.S. Army Corps of Engineers published a summary of California's coastal erosion problems and reported that 86 percent, or 950 miles, of California's 1,100 miles of coastline was eroding. We now recognize, however, that a continuing rise in sea level and wave attack take their toll on the entire state's coastline. It's all migrating landward, but at somewhat different rates depending on the interaction of the physical forces acting to wear down the cliffs (rainfall, runoff, wave impact, tidal range, slope failure, etc.) and the strength of the materials that make up the cliffs. Coastal geologists have documented cliff retreat of up to 5 or 10 feet per year, but we also have evidence of areas where there has been no significant or detectable change in the past 50–75 years (Figure 2.2). Where average cliff retreat rates are a foot or more annually, there are some serious problems facing any

Figure 6.11　Intensively developed bluffs in Solana Beach in San Diego County. Photo © 2002–2004 Kenneth and Gabrielle Adelman, California Coastal Records Project, www.Californiacoastline.org.

Figure 6.12　Small-scale bluff failure in the Aptos Seascape area of southern Santa Cruz County resulting from seismic shaking during the Loma Prieta earthquake. Photo by Gary Griggs and Kiki Patsch.

home or property owner. On the other hand, where retreat rates are very low, homes or parcels may be secure for decades to come, depending, however, in large part on sea-level rise rates and future storm or wave climate, as well as other severe events, such as large earthquakes.

A professional evaluation of long-term (at least 50 years) cliff erosion rates using available historical aerial photographs and maps is strongly recommended to anyone seriously considering the purchase of oceanfront property. Local government planning departments or the California Coastal Commission require this type of information or investigation prior to any major land-use changes or construction proposals. Staff who work with these coastal erosion issues on a regular basis should also be able to recommend qualified geologists who will undertake the work necessary to make a careful evaluation of cliff erosion rates. The maps and references included in the specific geographical sections of this book will enable you to make a preliminary assessment of potential problems in a particular area. In general, we know most about erosion problems in those areas where intense development has taken place—for example, much of the coast of Santa Barbara, Ventura, Los Angeles, Orange, and San Diego counties in the south; and San Mateo, Santa Cruz, and Monterey counties along the central coast. Knowledge of coastal erosion rates along much of the area north of San Francisco and for other, more isolated and less developed areas is much more limited.

In 1999 FEMA published a report on preliminary coastal erosion investigations for each coastal state to evaluate the rate of shoreline retreat (which in California is usually determined as cliff retreat) along the nation's coastlines. In each state a single county was selected for detailed, mile-by-mile study and long-term (approximately 50–60 years) shoreline retreat rates. Because of the length and diversity of California's coast, two counties, San Diego and Santa Cruz, were selected. In addition to historical coastal erosion rates, the position of the shoreline 60 years in the future was projected based on past erosion rates. The FEMA study also determined the number and value of structures at risk in the 60-year erosion hazard zone and used this as a basis for evaluating whether the federal flood insurance program might be expanded to include coastal losses. In addition to these studies and the rates included in this book, there are other scattered sources of data on average annual cliff erosion rates that local geologists or city and county government planners may be able to provide. In general, however, many of the studies are site specific, completed by local consultants and, unfortunately, not easily accessible.

Coastal cliffs may fail as small fragments (Figure 6.12); as large, often joint-bounded bocks (Figure 5.8); or in the form of massive slides or slumps

Figure 6.13 (A) Bluff-top home at Moss Beach, San Mateo County, threatened by continuing erosion. This house was subsequently demolished. (B) Moss Beach with house now demolished. Photos by Gary Griggs and Kiki Patsch.

(Figure 2.6). Evidence of recent cliff failure is often easily recognized as fresh or nonvegetated scars on the cliffs themselves, or as blocks or debris at the base of the cliff. This type of evidence should alert prospective property owners to potential erosion and failure problems.

Other general observations or evidence can be useful as well. Locations where the cliffs consist of layered sedimentary rock, such as shale, and the bedding or layering is tilted down toward the beach are notoriously unstable, and large slabs will frequently break loose and fall to the beach below. Undermined, subsiding, or cracked roads, sidewalks, or patios near the cliff edge more than likely reflect recent or progressive failure in the underlying materials. The age or maturity of the vegetation on the cliffs may also provide evidence of stability or recent history. Although there are many rocky cliffs that simply do not support vegetation, nonvegetated cliffs or bluffs or those with only very young or immature vegetation may have suffered recent failure or erosion. Older trees, on the other hand, suggest at least some extended period of stability. It is also important to look not only at the cliff in front of your prospective house or property, but also at the area to either side, which is probably experiencing the same erosional processes and may provide additional evidence for stability or instability.

Many questions need to be resolved prior to the purchase of a home or future home site:

• What is the long-term cliff retreat rate?

• For an existing house, how close is it to the cliff edge, and if the long-term average cliff retreat rate persists into the immediate future, how close to the edge will you be in 5, 10, or 20 years?

• If you are considering buying an undeveloped cliff-top parcel, how deep or long is the property, and how many years of setback or security exists given the calculated erosion rate?

• What if a major storm occurs and 10 to 20 feet fail overnight?

• Is there evidence for recent failure or instability? What do the neighbors have to say?

These questions should be evaluated and your fears laid to rest before you purchase any oceanfront house or property. Some California cliff-top areas underlain by hard and resistant rock such as granite and/or protected by a wide, sandy beach have experienced little measurable erosion in 50–75 years of photographic record. Other sites, however, are underlain by weak rock or sediments, are regularly attacked by winter waves, and are failing on a regular basis to the point that homes have collapsed, require relocation, or are now threatened (Figure 6.13). The fronting seacliff should not be the only point of concern, because at some locations large slides, slumps, or earth flows may extend for hundreds or thousands of feet inland (Figure 5.9) and may be as hazardous to property and structures as failure of the cliff face itself. Many specific areas with erosion problems or hot spots will be discussed in the following area-specific chapters.

CHAPTER SEVEN

RESPONDING TO COASTAL HAZARDS

GARY GRIGGS, KIKI PATSCH, AND LAURET SAVOY

INTRODUCTION

Once a home or other structure has been built in a location that is prone to wave damage or coastal flooding, several options exist. One is simply to take your chances—do nothing and hope that while you own the property nothing happens. Depending on the particular location, elevation, setback from the shoreline, and the storm and El Niño climate, this may work for a while. The cost is nothing until a major storm finally occurs with a very high tide, and then either a rapid emergency response is required or everything may be lost or seriously damaged in a matter of hours. Emergency responses are usually costly and temporary, however.

Pajaro Dunes, an expensive development of vacation homes, townhouses, and condominiums built along a mile of oceanfront sand dunes in central Monterey Bay, is a case in point. Some dune erosion occurred during the first year after development began in 1969. A few old automobile bodies were brought in and placed at the base of the dunes to halt the localized erosion. Nine years later, during the winter of 1978, which was the precursor to a 20-year period characterized by more severe El Niño events and associated storm damage along the California coast, storm waves attacked and eroded the dunes at several locations and came to within 10 to 15 feet of several groups of homes. Consultants were hired to evaluate the erosion threat and recommend possible solutions. By the time their work was completed, however, summer had come, the beach had been rebuilt, and most of the absentee property owners had forgotten about the threats to their investments from the previous winter. They decided to do nothing. Five years later, in January 1983, during one of the most severe El Niño events in 50 years, the combination of elevated sea level and high tide, storm surge, and very large waves took its toll, as up to 40 feet of the dunes were eroded (Figure 4.12). This left a near-vertical scarp 15 to 18 feet high that came right to the foundations of many of the homes; at least one house was in imminent danger of collapse and had to be shored up. Emergency riprap was placed along the entire seaward frontage of the development at a cost of more than $1 million (Figure 6.10). The placement of riprap was followed within a year by the construction of a permanent revetment with an additional cost of nearly $3 million.

As discussed in Chapter 6, had the project planners, engineers, and their consultants looked at the previous storm history, including historical aerial photos of the area, it would have been evident that the frontal dunes had been periodically eroded and then slowly rebuilt, and that the sand spit where more than 50 condominiums had been built was completely overwashed by waves several decades earlier (Figure 6.9). The erosion problems and the need for costly coastal protection could have been avoided altogether had the site's prior history been thoroughly studied. An adequate setback from the front edge of the dunes that would have accommodated erosion and accretion of the dune edge, as well as avoidance of the low sand spit, for example, may have provided a safe development area for the lifetime of the structures. In this and in many other coastal locations, however, development plans have historically been driven by economics rather than environmental or hazard constraints. The closer you can build to the shoreline and the more houses or condominiums you can build on the parcel, the higher the profit that can be generated. The quality of planning has changed significantly over the past two decades, particularly with the increasing awareness of decadal shifts in climate and thus coastal storm frequency and intensity.

CONTROLLING OR REDUCING COASTAL EROSION

The most common approach to coastal erosion problems where public or private structures or utilities are at stake has, until very recently, been some type of erosion control structure or armor. An astonishing 110 miles, or 10 percent, of the coastline of California has been armored as of the year 2000. Armoring varies considerably in cost, size, effectiveness, and life span. At one extreme, large pieces of broken concrete, asphalt, and even car bodies were simply dumped at the base of a cliff in the not-too-distant past in an attempt to slow wave erosion (Figure 7.1). In Sand City, along the coastline of southern Monterey Bay, concrete trucks routinely dumped any extra concrete after a delivery down the edge of the eroding dune face in an attempt to reduce dune retreat (Figure 7.2). Many examples of these unplanned and unpermitted efforts are still apparent today.

Figure 7.1   Concrete rubble was widely used in attempts to slow cliff erosion in the 1950s and 1960s along the coast of California. Photo by Gary Griggs and Kiki Patsch.

Figure 7.2   Concrete has been poured on the sandy bluffs at Sand City, Monterey County, in an attempt to slow bluff retreat. Photo by Gary Griggs and Kiki Patsch.

Figure 7.3 The O'Shaughnessy seawall was a very expensive, carefully engineered seawall built along Ocean Beach in San Francisco in 1928 that still provides coastal protection today. Photo by Gary Griggs and Kiki Patsch.

At the other end of the spectrum are massive, carefully engineered concrete seawalls such as the O'Shaughnessy seawall along Ocean Beach in San Francisco, which was built in 1929 and is still structurally sound (Figure 7.3). The purpose of these structures is essentially to reduce, minimize, or halt coastal retreat and thereby protect threatened structures and property from wave attack. Today, a permit is required for virtually any coastal protection structure, although under emergency conditions permits may be obtained relatively quickly. Either the city or the county in which the property is located is normally the primary permit-granting agency, although in many instances California Coastal Commission approval is required as well.

For any coastal armor permit application submitted today, a set of engineered plans for the proposed structure is necessary as is a strong rationale for the project. Additionally, an evaluation of the physical conditions that the structure would have to withstand, including expected wave heights, maximum sea levels, and depth of scour, is necessary. Proposals for coastal protection structures designed to protect a large area or entire development usually are also required to have a thorough environmental impact report completed by an independent consultant. This report must consider alter-

Figure 7.4 An older timber and concrete seawall failed during an El Niño event in 1978 in Rio Del Mar, Santa Cruz County. Fifty-gallon oil drums filled with sand were unsuccessfully used as temporary protection. Photo by Gary Griggs and Kiki Patsch.

natives and evaluate all of the environmental impacts of the proposed project as well as each of the alternatives. There are many examples of coastal protection structures that were built without an adequate knowledge of the forces and conditions to be experienced in the surf zone, and that subsequently failed (Figure 7.4). Wave overtopping, undermining, outflanking, and battering by debris are all potential concerns for any coastal protection structure.

The following discussion addresses the reduction or control of marine erosion caused by wave impact on coastal dunes and cliffs. Structures built on cliffs, however, must also contend with erosion or cliff failure due to terrestrial processes. These include erosion or gullying from uncontrolled surface runoff, or cliff failure from slumping, landsliding, or rockfalls. Countless examples from one end of the state's coastline to the other indicate that excess water, whether surface or subsurface (from landscape watering, or roof or street runoff, for example), is often the catalyst for failure. Large earthquakes, although occurring less frequently than major rainstorms, are also capable of generating widespread cliff failure (Figure 5.9).

Controlling on-site or off-site runoff at cliff-top locations will reduce the potential for erosion or failure of the slope, and is an important and relatively inexpensive step any homeowner can take. In particular, the runoff from rain gutters and downspouts, patios, driveways, and other large, impermeable surfaces should be collected and routed inland either well away from the cliff edge and house foundation or, if necessary and feasible, carried to the base of the cliff in an enclosed pipe. Uncontrolled runoff and drainpipe discharge on the bluff face are common causes of failure. Few landslides or slumps occur during the dry summer months, simply because the excess water is not normally present. It is in the winter months, particularly during major storms, that the cliffs become saturated to the point of failure.

RIPRAP

Where protective beaches are narrow, seasonal, or nonexistent, and property on the beach or cliff is eroding or endangered, the emplacement of large rocks, or riprap, has been a common solution. In fact, riprap protects more of California's coastal property (roughly 65 miles) than any other type of armor. Riprap has been used on a widespread basis both at the base of retreating cliffs and in front of structures built on the beach or sand dunes. The rock is stacked against the cliff, dune, or beach with the intent of absorbing most of the wave impact. Although quite costly, as are all coastal protection measures, riprap has been one of the most common protective measures used by individual property owners, in part because it doesn't require any complicated engineering or construction methods. It can be used to protect a single structure and can be placed where access is limited.

Observations of many areas where rock or riprap has been emplaced indicate that, for optimal results, the individual rocks must be large enough (usually 3 to 8 tons) to remain stable under the most extreme wave conditions that may be experienced at the particular location. The rock must also be durable so that it does not break up or disaggregate readily. The availability of durable rock of the size needed may place constraints on usable rock and at some locations may add considerably to the cost of the project if rock must be brought in from a great distance. Typical costs along the central California coast in 2003 were $80 to $120 per ton for riprap in place. Normally at least 10 to 25 tons of rock per linear foot of property are required, which amounts to $800 to over $3,000 per foot, depending on the dimensions of the structure. Riprap protection for a property 60 feet wide would cost on the order of $48,000 to $180,000 (in 2003 dollars). These costs will vary somewhat from place to place as a function of rock availability, required rock volume, access, and foundation materials, but are very high in any case.

From a careful study of the history of a number of different armor projects, including some that have performed poorly and others that have lasted for many years, certain design considerations have been shown to be critical to the success of a particular structure: (1) The riprap must be large and heavy enough to remain stable under storm wave conditions. (2) The riprap must be stacked high enough that waves do not overtop it and damage the structures being protected. (3) The rock must be stacked at a stable slope (2:1, horizontal to vertical, is recommended). (4) The rock must be placed in such a way (either on bedrock or deep enough in the sand) that it will not be susceptible to undermining or scour.

Riprap has had limited effectiveness or a short life span at many locations in the past because these design criteria have not been met. One reason for this is that riprap is often emplaced under emergency conditions in the midst of a storm. The rock has simply been dumped or stacked quickly on the sand in front of threatened structures at low tide, and no real design principles are followed. These "emergency" walls may need to be rebuilt later, and in some cases have had to be removed altogether. Placing rock during the winter months, when the beach has been cut back and the underlying bedrock may be exposed as a solid base or foundation, will produce a more stable structure. If riprap is placed directly on the beach or berm, wave scour will eventually remove or gradually liquefy the underlying sand. The removal or destabilization of the supporting beach sand followed by collapse or settlement is the most common mode of riprap failure. Protection is then either lost or greatly reduced and the riprap must be rebuilt. As a result, poorly planned and installed riprap will often need to be rebuilt or supplemented with additional rock after 5 to 10 years of wave attack. Undermining appears to be a particular problem where no bedrock exists beneath the beach sand for support. The initial construction costs, therefore, may be only the beginning.

At Aptos-Seascape in northern Monterey Bay, a group of expensive homes was built on an elevated fill terrace directly on the beach in 1968. Riprap about 15 feet high was placed on sand along the entire ocean front, and 21 homes were built on the terrace with virtually no setback from the top of the riprap. There were no problems for 15 years. However, during the damaging 1983 El Niño storms, large waves combined with several days of 6.6-foot high tides and elevated sea levels overtopped the riprap and washed through most of these homes, breaking out windows, doors, and walls (Figure 7.5). Continued wave action scoured the sand from the beach and from beneath the riprap, so that approximately 350 feet of the rock settled and collapsed onto the beach.

Another potential problem with riprap and virtually all other coastal armoring solutions is discontinuous placement or construction. One

Figure 7.5 El Niño–driven waves overtop the riprap at Via Gaviota, northern Monterey Bay, in 1983. (See Figure 6.2 for damage from the wave overtopping.) Photo by Gary Griggs and Kiki Patsch.

property owner may have a barrier constructed, whereas his or her neighbor does not. With time, erosion will continue on the unprotected area and, depending upon the end configuration of the armor, wave energy may even be reflected onto adjacent unprotected property. As a result, erosion of the cliff or dune may proceed around the structure and outflank the protected property. This issue is a difficult one for permitting agencies such as the Coastal Commission, in that the approval of a single seawall or structure will invariably lead to a request for similar protection for adjacent properties.

REVETMENTS

Although the terms *riprap* and *revetment* are often used interchangeably, a revetment is a more carefully engineered and constructed rock structure. Instead of simply stacking large rocks on the beach at the base of a cliff, a layered structure is built that is designed to minimize scour beneath the structure, limit wave reflection, prevent overtopping, and absorb the wave energy (Figure 6.7). A revetment is usually underlain by very durable and permeable filter cloth. In addition, a buttress of very large, often 5- to 8-ton rocks is embedded in the sand or keyed into the underlying bedrock to provide support for the rock structure. Smaller core stones, typically of a few hundred pounds to several tons, are then placed on the filter cloth and stacked between the cliff or dune and the large buttress stones. Finally, the largest rocks, called capstones, which are typically 3 to 8 tons apiece, are placed on the surface of the revetment. The size, shape, and objectives of a revetment are identical to those of riprap. A revetment simply has the

Figure 7.6   Revetment construction in Pacifica, San Mateo County, following wave erosion of the bluffs the previous winter. Photo by Gary Griggs and Kiki Patsch.

added advantage of the graded rock and filter cloth, which reduce scour and make the structure less susceptible to undermining and subsequent settlement or collapse. The other design considerations for riprap hold for revetments as well. The rock must be durable, large, and stacked at a low enough slope to remain stable under extreme wave conditions. The rocks cannot be stacked too steeply, and the structure needs to be high enough to prevent overtopping.

Following the collapse of many older sections of riprap during the 1983 winter along the central coast, engineered rock revetments were built to replace these structures. It was anticipated that the design improvements in the revetments would provide for a much longer life span and therefore a higher degree of protection. Many thousands of feet of revetments were built in 1983–84 along the central coast, but two deserve mention: those built along the Esplanade in Pacifica (Figure 7.6) and at Pajaro Dunes in central Monterey Bay (Figure 7.7). Both structures utilized filter cloth, large buttress rocks, layered stones, and large capstones, and both were built on deep sand beaches with no near-surface bedrock foundation. During the next major damaging El Niño winter in 1997–98, both engineered revetments experienced some settlement, with rocks scattered across the beach (Figure 7.8). At the Pacifica site, despite emergency efforts to restack the rock, heavy equipment became stranded on the beach and

Figure 7.7 Pajaro Dunes revetment construction following 1983 dune erosion. Photo by Gary Griggs and Kiki Patsch.

Figure 7.8 Collapse and settlement of the engineered revetment at Pajaro Dunes (see Figure 7.7) during the 1998 El Niño. Photo by Gary Griggs and Kiki Patsch.

Figure 7.9   Failure of the Pacifica revetment during the 1998 El Niño winter. Photo by
Monty Hampton, U.S. Geological Survey.

seriously damaged, and continuing bluff failure ultimately led to the loss
of an entire row of oceanfront homes (Figures 7.9 and 5.6). The Pacifica
revetment was a complete loss.

SEAWALLS AND BULKHEADS

Seawalls have been used widely in many developed areas along the California
coast with varying degrees of success. By 2000, approximately 44 miles of the
state's coast had been armored with some type of seawall. The terms *seawall*
and *bulkhead* are often used interchangeably. These structures are built parallel to the shoreline using concrete, wood, steel, or a mixture of these materials and have a wide variety of shapes or configurations. The exposed face of
a seawall may be vertical, concave, stepped, or some combination thereof.
Each design has its own advantages and disadvantages. A seawall's primary
purpose is to protect seacliffs, buildings, roads, or other improvements from
direct wave attack. In many cases they also serve as retaining walls to keep
weak cliffs or fill from slumping or failing.

    For years there has been a general perception that a revetment, because
it is more permeable than a seawall, should absorb more wave energy and
therefore have less impact on the fronting beach or adjacent coastline. Field
observations, particularly under large wave conditions, however, indicate
that revetments do not absorb all of the wave energy that reaches them. In
fact, much of this wave energy is actually reflected. During the 1983 El Niño
winter, major damage occurred at Aptos-Seascape, where a revetment

became a ramp for large waves that caused heavy damage to oceanfront homes (Figure 7.5). A more significant difference between seawalls and revetments is the amount of beach area they cover. Typical revetments extend seaward from the base of the cliff or dune a distance of 1.5 to 2 times their height, so a 20- to 25-foot-high revetment will cover 30 to 50 feet of beach. Most seawalls, in contrast, are very narrow—a few to perhaps five feet—and thus take up relatively little beach area.

EFFECTIVENESS OF SEAWALLS

Observations of many different types of seawalls and how they have performed over time and under severe wave attack provide a great deal of practical information that isn't always available in textbooks or manuals. Several processes have been responsible for most seawall damage or failures of the past, including overtopping, undermining, outflanking, and wave or debris battering or impact. Those involved in designing, approving, or constructing seawalls will benefit from a careful examination of the lessons of the past as they consider armor for the future.

For some protective structures the past history of failure has been disregarded and the same structures rebuilt again and again. A piling and timber bulkhead protecting Seacliff Beach State Park in northern Monterey Bay has been destroyed or seriously damaged 10 times in the past 75 years. The typical design used at Seacliff consists of 12-inch-diameter wooden pilings set about 8 feet apart and tied together with 3" × 12" timbers that are simply nailed in from the back (landward) side. In general, timber and wood have not performed well along the coast of California, in part because of the deterioration of the wood over time with exposure to seawater and salt air, and in part because of the impacts of large logs on the wooden pilings and planks. The first timber bulkhead of this type was built at Seacliff in 1927 and was destroyed the following winter; it was rebuilt in 1940 and destroyed the next winter (Figure 7.10). Similar structures were again destroyed in 1978 and in 1980. An identical structure was rebuilt and completed in late 1982 at a cost of over $3 million, and was projected to last 20 years. Six weeks after completion, the El Niño storm waves of late January 1983 destroyed 700 feet of the new 2,600-foot-long structure and inflicted about $1.5 million in damage (all in 2005 dollars; Figure 7.11).

The failure pattern of the piling and timber bulkhead at Seacliff State Beach was essentially identical each time. Large logs and other floating debris that had been brought down by high flows on the local streams battered the front of the wall. Simultaneously, waves and logs overtopped the structure and began to break through the asphalt parking lot behind the bulkhead. As the timbers were loosened and the asphalt torn up, the under-

Figure 7.10   The remains of a series of timber and steel bulkheads are visible at low tides in the winter months along the beach at Seacliff State Beach. Eight different sea-walls have been built and destroyed at this site over the past 70 years. Photo by Gary Griggs and Kiki Patsch.

lying fill began to wash out, even with filter cloth behind the bulkhead. Once the integrity of the structure was lost and the support of the back fill was removed, the waves easily battered and removed the timbers, leaving the pilings behind. In high-energy environments, timber and piling bulkheads constructed in this fashion have not proven effective over the long term. They are very susceptible to battering by heavy logs and trees, especially

Figure 7.11   Destruction of a portion of the new Seacliff State Beach timber bulkhead during the 1983 winter due to the impact of logs and debris. Photo by Gary Griggs and Kiki Patsch.

where the timber or lagging is too thin (2 to 3 inches) and simply nailed to the backside of the pilings.

A much more resistant timber seawall was constructed along an adjacent beach to protect houses, a road, and utility lines. In this case, steel H-piles were used instead of wooden pilings, and the timber lagging was 6" × 12". With the timber held in the slots of the H-beams (Figure 7.12), it cannot be battered easily from either side, as is the case with wood pilings.

The possibility of wave overtopping must be considered in the design of any seawall, and there is often a compromise between cost and loss of view and what past or expected wave conditions dictate. Seawalls and bulkheads must be capable of withstanding the pressure exerted on the face of the structure by waves, logs, and other debris, but also be able to deal with the pressures exerted by water and fill behind the wall. Rainwater, groundwater, runoff, or water from wave overtopping will percolate into the fill material behind a wall and can build up sufficient pressure to cause failure. Drain or weep holes at the base of a seawall should be installed to relieve this water pressure.

Seawalls have also failed as a result of poorly constructed drainage systems. In one case, in Rio Del Mar, as beach sand levels dropped during a

Figure 7.12 Although the concrete stairway was detached from the seawall, this wall consisting of steel H-piles and 6" × 12" timber has done very well for nearly 20 years. Photo by Gary Griggs and Kiki Patsch.

period of large storm waves, the 4-inch-diameter drainage holes were exposed (Figure 7.13). As wave energy at high tide impacted the wall, it forced seawater through the weep holes, washing out the sand fill that had been placed behind the very thin (8-inch) concrete seawall. With much of the supporting fill gone, the wave impact quickly began to crack the concrete panels and damage the wall, which had been completed only a few months before (Figure 7.14). A simple solution to this problem would have been to use rock fill behind the wall, or coarse gravel with permeable caps over the weep holes so that water could drain but the gravel could not be washed out.

Another common seawall problem is scour or undermining. Six to 8 vertical feet of beach sand can be removed during a major storm, reducing support and allowing the structure to settle, lean, or collapse (Figure 7.15). Typically engineers and contractors try to set the base of a structure below the maximum expected scour depth, which is not always easy to determine. Excavation on the beach with a high water table and between the high tides is challenging, however, and revetments and seawalls have been undermined and then collapsed or settled because the base of the structure was not deep enough.

Figure 7.13 Severe winter scour lowered the sand level in front of this new concrete panel wall on Beach Drive in Rio Del Mar, leading to wave surging through drainage holes and removing the sand support behind the wall. Photo by Gary Griggs and Kiki Patsch.

Figure 7.14 Same concrete seawall as in Figure 7.13 after sand had been washed from behind the wall and wave impact had battered and cracked it. Photo by Gary Griggs and Kiki Patsch.

Figure 7.15   Beach scour undermined the base of this seawall at Moss Landing and caused it to settle and fail. Photo by Gary Griggs and Kiki Patsch.

Another problem with seawalls results from different types of walls being built next to one another. This hazardous arrangement is understandable in that different structures may have been built by adjacent property owners with different budgets and different engineers and at different times. If these are older structures, they probably aren't tied together structurally and they will undoubtedly provide differing degrees of resistance to wave attack. Once a single section of seawall or the weakest link fails, wave surge can remove the sand behind this segment and begin to erode the sand or fill from behind the adjacent sections of wall. Because many seawalls or bulkheads are really just thin barriers that rely on the mass of the fill behind the wall to resist wave impact, when the fill is gone many structures cannot resist the continued wave impact. Once the fill is removed, an entire stretch of adjacent seawall segments may simply unzip, regardless of the attributes of the individual structures. Where protective structures such as seawalls or bulkheads are being planned in front of adjacent properties, homeowners are strongly advised to develop a uniform and structurally coherent approach rather than a haphazard and irregular one. If a solution of this sort is well planned for the particular site conditions, the structure will be more economic and will also survive longer. It's important to keep in mind for any proposed seawall, however, that most protection structures are very expensive and have a limited life span. There is no such thing as permanent protection, and there is no guarantee today that a seawall permit will be approved.

COMBINED APPROACHES AND OTHER STRUCTURES

Erosion control measures may also be used in combination, either to complement one another or to accomplish what a single measure might not do alone. A common beachfront approach is to utilize riprap, for example, in front of a bulkhead or seawall (Figure 7.16). The riprap will absorb some of the wave energy while the bulkhead serves as a retaining wall in front of beachfront fill and buildings. Although a combined structure that is high enough to avoid most wave overtopping and constructed to withstand wave attack is probably going to be more costly, it may provide better protection and support than either structure individually.

Over the years, coastal homeowners, an occasional maverick engineer, and Army Corps of Engineers demonstration projects have experimented with several other, often low-budget devices in an attempt to reduce erosion and provide protection for particular problem areas. Many of these structures have been placed and/or tested in a low-energy lake, bay, or estuary that is in no way representative of the physical forces that any structure would need to withstand on an exposed high-energy coast such as California. Although some of these methods may provide short-term protection, it can probably be assumed that if any of these approaches had been highly successful and superior to past methods, they would have been utilized quickly on a widespread basis. This is not to say that the existing, most commonly used methods are fail-safe, or that we should not try to develop better ways to cope with coastal erosion, but rather that many novel approaches have been tried and all appear to have their limitations. The best evidence for how well a particular structure may fare at a specific location is to investigate the history, life span, and success of other protection structures in the immediate area. History is usually the best teacher, although we need to keep in mind that some structures have survived for a number of years under modest wave conditions (during a period with few major El Niño events, for example), only to be damaged or destroyed by a fatal combination of elevated sea level, high tide, and large waves. This is exactly what happened in the winter of 1982–83 along many areas of the California coast. The San Pedro Breakwater, built between 1899 and 1912, withstood numerous storms in its 75-year life; however, waves in February and March of 1983 overtopped the breakwater, displacing 10- to 20-ton granite blocks and causing $7.3 million (in 2005 dollars) in damage.

Figure 7.16 Combined use of concrete and riprap for bluff protection. Arrow points to gunite-covered bluff above seawall. Photo by Gary Griggs and Kiki Patsch.

CLIFF STABILIZATION AND ARTIFICIAL ROCK WALLS

At several coastal locations where steep cliffs undergo rapid erosion, the entire cliff has been sprayed with gunite or shotcrete for protection (Figure 7.17). This approach has, however, usually been of limited effectiveness. If adequate drainage is not provided for water seeping through the cliff, excess water pressure can develop behind this thin barrier, which will ultimately lead to collapse. Erosion will still occur at the margins or base of the concrete, gradually or rapidly undermining it or removing its support over time. Again, an isolated approach of this sort on a single parcel has its limitations. Gunite is also typically unattractive and unpopular with the beach-going public and permitting agencies.

In recent years, a significant improvement on the gunite walls has been developed that is increasingly being proposed and used. The gunite wall is more appropriately referred to as cliff or bluff stabilization rather than a seawall. The objective is to harden the cliff so it is more resistant to erosion, but not to create a large, imposing structure. The approach, also known as artificial rock or a soil-nail wall, has been used to infill caves, notches, or

Figure 7.17　Bluffs at Carlsbad in San Diego County have been covered with gunite or shotcrete for protection. Photo © 2002–2004 Kenneth and Gabrielle Adelman, California Coastal Records Project, www.Californiacoastline.org.

voids in the cliffs of San Diego County (Figure 7.18) along the coastline of Pebble Beach on the Monterey Peninsula, and can also be used to strengthen an entire section of eroding cliff. A series of steel anchors or tiebacks are drilled or embedded into the area of bluff or cliff to be protected and used to support a mesh of reinforcing steel. The steel mesh is shaped to mimic the existing rock or topography of the cliff, in the way a swimming pool is formed. This structure is then coated with about a foot of gunite or concrete applied with a high-pressure hose, again exactly as in the construction of a swimming pool. A second layer, about a foot thick, is then applied over the first and is both colored and sculpted to match the existing rock as closely as possible (Figures 7.18 and 7.19), much like the artificial rock that one would find in animal enclosures in a zoo. The objective is to harden a weak or eroded section of cliff, thereby slowing erosion, or to fill a void or cave that might lead to collapse of the overlying material, but without the visual impact of a large seawall and without the loss of beach that a revetment would produce. Concrete can be mixed to be extremely durable, even in a saltwater environment, and the reinforcing steel can be epoxy coated so that it too resists saltwater attack. Where these soil-nail walls have been built to date, it is often difficult for the casual observer to notice them; but how closely they blend into the surrounding cliffs depends on the skill of the construction crew. Some of the newest examples of this approach have

Figure 7.18   The lower portion of this cliff in San Diego County has been armored with reinforced concrete, tied back to the cliff, and colored and textured to look like the existing cliff rock. Photo by Gary Griggs and Kiki Patsch.

involved making a mold of the adjacent rock and then using this to repro-duce the texture and color in concrete on the area to be protected (Figure 7.20). This approach does eliminate some of the negative impacts of revet-ments and seawalls and seems to be a significant improvement over many of the large structures built in the past.

Before embarking on a coastal protection project of any scale or cost, it is critical to question the effectiveness and probable life span of the structure being proposed. The owner should seek out a coastal engineer with a successful track record because coastal engineering is a highly spe-cialized profession. It is clear from the failure of countless shoreline structures that most or many of the structures built in the past have been

Figure 7.19 A soil-nail wall, or tied-back reinforced gunite wall, sculpted to match the adjacent cliff on Cowells Beach in Santa Cruz. The darker-colored material extending from the left side of the cliff to just right of center is the soil-nail wall. White streaks at the base of the wall are from water draining from the cliff through weep holes or drainpipes. Photo by Gary Griggs and Kiki Patsch.

Figure 7.20 Concrete was colored and textured to match the adjacent rock in order to minimize the visual impact in the protection of these low bluffs in Pebble Beach on the Monterey Peninsula. (A) Two sections of bluff in the background are both textured and colored concrete. Photo by Lesley Ewing, California Coastal Commission.

Figure 7.20 (*Continued*) (B) The middle section of the photo is concrete. Photos by Lesley Ewing, California Coastal Commission.

inadequate to withstand the forces in the surf zone over the long term. It also should be emphasized that there is really no such thing as low-cost shore protection in a high-energy environment. All evidence indicates that every structure has a finite life span, with the most important influence probably being the uncertain occurrence of the next major storm, with simultaneous large waves and an elevated sea level at high tide. It is important to consider the cost of rebuilding or repairing the structure you are considering and to be aware of the level of protection you are buying.

EFFECTS OF COASTAL PROTECTION STRUCTURES

In contrast to the oceanfront homeowner's concern for the cost, life span, or effectiveness of a coastal protection structure, considerable opposition has arisen in recent years around proposals for new seawalls or revetments because of the perceived direct or indirect impacts of these structures. Many of the concerns, including aesthetic or visual impacts, restrictions on beach access, reduction of sand supply from previously eroding cliffs, and the loss of the beach beneath the riprap or seawall, revolve around the issue of the degree to which private property owners should be allowed to impact public beaches as they attempt to protect their own property. In the case of government-funded projects, the issue becomes how much taxpayer money

Figure 7.21 Almost all protective structures produce a visual impact. Photo by Gary Griggs and Kiki Patsch.

should be spent in efforts to stabilize the position of an otherwise eroding coastline.

The continued widespread reliance on protective structures and the controversies they generate result in part from the limited, ambiguous, and often conflicting policies set forth in the California Coastal Act of 1976. The reason for part of this uncertainty is that at the time the Coastal Act and the interpretive guidelines were written, the state was in the midst of a 30-year period of modest storm and wave climate; coastal erosion and El Niño events were not major concerns. Over the past several decades, however, concern has increased regarding the direct impacts of seawalls or revetments on beaches. As the percentage of armored coastline has increased in recent years, the general public has become more aware of the visual and access impacts of shoreline armoring. As of 2000, some sort of armoring structure "protected" almost 30 percent of the coastline of Santa Cruz County, and 34 percent of the entire 234 miles of the combined coastlines of Ventura, Los Angeles, Orange, and San Diego counties. For anyone who frequents the beach, it is not difficult to notice the increasing amount of rock and concrete along the shoreline. In the 21 years between 1971 and 1992, the extent of armoring along California's coast increased 400 percent. Visual impacts are perhaps those noticed by most people.

Figure 7.22 Many seawalls—depending upon their height, length, and configuration, as well as their location on the beach profile—will restrict either vertical access to the beach or horizontal access along the shoreline at high tide. Photo by Gary Griggs and Kiki Patsch.

Although these impacts are a somewhat subjective issue, there are numerous existing structures that many would say detract from the natural experience they desire when walking along the shoreline (Figure 7.21). At the other extreme are the newer cliff stabilization projects designed to mimic the natural rock that makes up the local cliffs and, where well done, are difficult to recognize.

Many older large seawalls and revetments limit or restrict beach access (Figure 7.22); this is an area where the California Coastal Commission has been extremely vigilant. There are ways in which access can be built into seawalls, although this is far more difficult with large revetments. Beach

Figure 7.23   A concrete panel wall has been used to protect the weaker terrace deposits at Lighthouse Point in Santa Cruz, rather than using riprap that would have extended to beach level and covered the beach (see Figure 7.24). Photo by Gary Griggs and Kiki Patsch.

access will remain a concern and must be resolved in individual coastal armoring proposals.

Where erosion of coastal cliffs provides a significant portion of the region's beach sand, armoring the bluff and halting or slowing erosion will reduce the sand supply. The perceived cumulative impact of seawalls on sand supply reduction led the California Coastal Commission to develop a policy in 1997 requiring a sand mitigation fee to be paid by property owners requesting permission to armor their particular bluffs. The amount of sand that would be removed or prevented from ending up in the beach or littoral system annually as a consequence of the construction of a seawall can be calculated if one knows the length of cliff to be armored, the height of the cliff, the average annual rate of cliff retreat, and the percentage of the material making up the cliff that is of beach sand size. Along the California coast, however, rivers and streams supply most of the sand for the state's beaches. A recent study, completed in 2002, of the Santa Barbara beach compartment (which extends from the Santa Maria River to Point Mugu) indicates that throughout the entire 144 miles of this littoral cell, erosion of the cliffs contributes less than 1 percent of the total beach sand supply. In the Oceanside littoral cell, which extends from Dana Point to La Jolla, about 11 percent of the total beach sand along this region is supplied by cliff erosion. Reduction

Figure 7.24   The placement of riprap will cover a large area of beach. Photo by Gary Griggs and Kiki Patsch.

of beach sand supply through armor emplacement is therefore a potential impact, but the significance of this impact varies geographically depending on how important the erosion of the cliffs is in contributing sand to the beaches of the particular cell.

Simply placing a protective structure on the beach will, depending on its size and shape, cover a given amount of beach sand (known as placement loss). If this structure is a vertical, relatively narrow seawall, the amount of beach lost is very small. Some walls are built only on the upper portions of cliffs, where they protect the more erodible marine terrace sediments and therefore don't impact the beach directly (Figure 7.23). On the other hand, if a revetment is built on the beach, due to the height normally required to prevent wave overtopping and the slope necessary to develop a stable configuration, the wall may extend 20 to 50 feet seaward, in some cases completely covering the usable beach (Figure 7.24). There are trade-offs in using riprap to protect eroding cliffs. The ongoing erosion, which may threaten private homes or public infrastructure, can be temporarily reduced or controlled using riprap, but the public beach is going to suffer a loss; the amount of loss is a function of the size and footprint of the protective structure built.

An additional seawall impact is termed passive erosion. Wherever a hard structure is built along a shoreline undergoing long-term net erosion, the shoreline will eventually migrate landward behind the structure. The effect of this migration will be the gradual loss of beach in front of the seawall or

Figure 7.25   Passive erosion at Stillwell Hall in central Monterey Bay. A beach still exists to either side of the riprap where there is no armor, but the beach in front of the riprap has been lost. Photo by Gary Griggs and Kiki Patsch.

revetment as the water deepens and the shoreline moves landward (Figure 7.25). This has also been called the peninsula effect because the armored area becomes a peninsula over time. Passive erosion has been a very contentious issue along the Atlantic Coast, where offshore barrier islands are migrating landward as sea level rises, except where the coastline has been fixed with seawalls. Passive erosion has also been recently documented in Oregon, Washington, and in the Hawaiian Islands. Although homes or other buildings may be temporarily saved, the public beach is gradually lost. The process of passive erosion is a result of fixing the position of the shoreline on an otherwise eroding stretch of coast, and is independent of the type of seawall constructed. As the amount of coastal armoring increases in California, passive erosion and the resulting loss of beach through placement of these protective structures are going to become increasingly significant issues.

The potential for a seawall or revetment to cause or accelerate beach erosion has also been the source of considerable controversy over the past two decades. The general perception is that building an impermeable seawall along a shoreline will lead to wave reflection and therefore beach erosion in front of the seawall. A similar concern involves reflection from the ends or sides of walls; it is believed that the construction of a solitary structure will immediately increase the erosion on adjacent properties. Although these processes make intuitive sense, an 8-year-long study completed in 1996 that monitored the impacts of several different types of seawalls and revetments on adjacent beaches in northern Monterey Bay indicated that although there were local moderate-scale winter impacts, there were no permanent effects on the beaches or cliffs studied. This study took place, however, along a stretch of coastline that has a high littoral drift rate and is not undergoing any net long-term erosion but does undergo expected seasonal beach changes. While riprap revetments have often been judged by permitting agencies to be more permeable and therefore expected to have less impact on beaches than "impermeable" seawalls, this has not been supported by field observations and surveying. Thus, the issue of the direct effects of seawalls on wave reflection, and therefore beach erosion, needs to be evaluated in each specific location. Similarly, if a structure is to be built directly against the base of an eroding bluff, the design or treatment of the end sections needs to be carefully considered so that wave energy will not be reflected onto the adjacent properties, thereby increasing erosion in those areas.

Coastlines are receiving more national attention as their economic and recreational importance continues to grow and as the issue of coastal erosion or retreat begins to affect more public and private property and development. The terms *beach erosion* and *coastal erosion* are often used interchangeably in these discussions and in requests for government assistance and protection permits. There is an important distinction between these terms, however. Beach erosion refers to the permanent or long-term loss of beach sand or reduction in beach width. Beach erosion can be reduced or reversed if the natural supply of sand is returned or replenished. Coastal erosion, on the other hand, is the retreat of the cliff or bluff edge and is an irreversible, unidirectional process, at least within our lifetimes. Statistics on beach usage and tourist dollars have often been used to justify "coastal protection." Seawalls and revetments are designed and built to protect property and structures on dunes, bluffs, or cliffs and not to protect public beaches. To our knowledge, a seawall has never been built to protect or save a beach.

RETREAT OR RELOCATION OF A STRUCTURE

An alternative response to a threatened house or structure on an eroding beach or cliff is to relocate or remove the structure. There are many sites along the coast of California where there are no other reasonable or acceptable alternatives. If the threatened property is large enough to accommodate the structure being moved back far enough to significantly extend its

Figure 7.26 Ongoing failure of loose bluffs at Pacifica is beginning to threaten a large apartment complex. Photo by Gary Griggs and Kiki Patsch.

lifetime, this should definitely be considered. The size, condition, and physical setting of the dwelling are critical considerations that need to be assessed by a professional house mover and possibly a structural engineer. During the winter of 1983, as a result of 30 to 50 feet of rapid bluff retreat at Pacifica, a three-story building was moved away from its oceanfront site, as were 23 mobile homes (Figure 3.6). An adjacent row of bluff-top homes was also threatened by cliff erosion during the 1983 winter, but the decision was made to construct a revetment at the base of the highly erodible bluff (Figure 7.6). In the intervening years, specifically during the severe El Niño event of 1997–98, the revetment settled and collapsed, and failed to protect the bluffs from waves. Houses were undermined (Figure 3.8) and began to collapse, and before the winter was over, they were demolished (Figure 5.6). Here relocation would have been a better solution. A large apartment complex sits on the same eroding bluff a few thousand feet up coast and faces identical problems but without the option of relocation (Figure 7.26).

The cost of moving a house may be far less than that of a major coastal protection structure or the cost of repair if nothing is done. In addition, there is no guarantee that a permit for a seawall or other type of protective structure will be approved. If your parcel is not large enough to relocate the structure on-site, another parcel will obviously have to be purchased. Relocation may still be more economical in the long run, however, depending on the magnitude of the erosion problem and the costs of providing long-term erosion protection. Typical moving costs for a moderate-sized structure may be in the range of $24,000 to $32,000 (based on a 2,000-square-foot house and 2003 moving costs of $12–16/sq. ft.).

Relocation likely has not been seriously considered by most threatened oceanfront property owners simply because of their desire to preserve their home and setting at any cost. The increasing frequency and magnitude of El Niño events and coastal damage since 1978, as well as a clearer appreciation of global warming and continuing sea-level rise, have altered our perceptions about coastline stability and may lead more oceanfront property owners as well as permitting agencies to give relocation and retreat a higher priority in the future. The initial and recurring costs of providing protection under these conditions may eventually exceed the value of a structure or the cost of its relocation. It is important, therefore, to look at the real costs, and at the advantages and disadvantages of all responses to coastal erosion.

THE FUTURE

Virtually every coastal community experiences the same dilemma: increasing coastal populations, existing and proposed oceanfront development, rising sea level, and a recent history of expensive natural disasters affecting the coastal zone, with more predicted in the future. Whether it's nor'easters in New England, hurricanes on the south Atlantic Coast, or El Niño events on the Pacific Coast, we know that losses to oceanfront development will continue to increase as coastal development intensifies and property values rise. Although there are no easy answers, the need exists for long-term, balanced policies. We are going to have to look at sustainable solutions, which may well be different from the approaches that have guided us for so long.

CHAPTER EIGHT

AN OVERVIEW OF CALIFORNIA'S COASTAL HAZARDS POLICY

CHARLES F. LESTER

INTRODUCTION

The California Coastal Act is California's primary coastal hazards law. This law establishes two key policies for shoreline development. First, it requires that *new* development avoid coastal hazards if possible. Second, it specifically allows shoreline protection structures, such as seawalls and rock revetments, to be built for *existing* development that is threatened by coastal erosion, but only if there is no other reasonable way to protect the development. These policies reflect a basic objective to minimize the construction of shoreline protection structures because of their negative impacts on the coastal environment, which include blocking public access to the beach, loss of beach area, degrading scenic views, and preventing the erosion of sediments from the bluffs or cliffs that helps to maintain California's beaches.

Although the Coastal Act is straightforward in concept, applying its policies to development proposals has been challenging. Difficulties range from technical issues, such as methods for quantifying erosion rates and risks, to more basic human challenges, such as rational planning and regulation in a policy area characterized by emergency response. The private property along California's coast is also some of the most valuable in the world, which heightens the potential for political conflict when new shoreline developments are being considered. Coastal hazards policy involves high stakes, and nothing will provoke a clash between public and private perspectives like a proposal to build a new seawall.

CALIFORNIA COASTAL HAZARDS
POLICY FRAMEWORK

THE CALIFORNIA COASTAL ACT

The California Coastal Act of 1976 requires statewide planning and regulation for development in hazardous areas, including strict regulation of proposed shoreline protection structures, such as seawalls and revetments. Although the California Coastal Commission is the primary regulatory actor in this policy arena, local governments play an important role through the development of local coastal land use plans and ordinances.

Hazard Avoidance for New Development

The Coastal Act requires that new development minimize risks from coastal hazards. Section 30253 states in part:

New development shall:

(1) Minimize risks to life and property in areas of high geologic, flood, and fire hazard.

(2) Assure stability and structural integrity, and neither create nor contribute significantly to erosion, geologic instability, or destruction of the site or surrounding area or in any way require the construction of protective devices that would substantially alter natural landforms along bluffs and cliffs.

This policy specifically requires that new development not be premised on the construction of a shoreline protective device such as a seawall. Thus, section 30253 makes property owners assume the risks of developing along the coast by requiring that new development be located and designed to be safe without artificial means of protection from the forces of the ocean. This requirement is an explicit effort to stop the proliferation of seawalls, revetments, and other shoreline structures that cumulatively degrade the coastline.

Protecting Existing Development

The Coastal Act also sets standards for when and how to protect existing development from coastal hazards. Section 30235 states in part:

Revetments, breakwaters, groins, harbor channels, seawalls, cliff retaining walls, and other such construction that alters natural shoreline processes shall be permitted when required to serve coastal-dependent uses or to protect existing structures or public beaches in danger from erosion, and when designed to eliminate or mitigate adverse impacts on local shoreline sand supply.

Accordingly, shoreline protection structures such as seawalls or revetments shall be approved if an existing development is threatened by erosion, if the structure is the necessary response, and if the impacts to the local shoreline sand supply are eliminated or mitigated. For example, the law would not allow a seawall to be built for a threatened development if the development could be easily relocated out of harm's way.

Impact Mitigation

In addition to regulating proposed shoreline structures, the Coastal Act requires that any resource impacts of a new shoreline structure be fully mitigated. Section 30235 specifically addresses mitigation for sand supply impacts, but other Coastal Act policies protect public beach access, scenic viewsheds, natural shoreline processes, coastal and marine habitats, and shoreline recreational activities, such as surfing. For example, a proposed sloping revetment that covers up the public beach (Figure 7.24) may have to be redesigned as a vertical seawall to minimize beach encroachment. Visual impacts may have to be addressed through changes in the texture and color of materials used in constructing a seawall (Figure 7.19).

Local Coastal Programs

The Coastal Act requires that local governments adopt local coastal programs (LCPs). The act recognizes that local governments remain on the frontline of coastal resource protection, and that they are the central actors in local land use planning and regulation. LCPs must include policies and ordinances that reflect the Coastal Act requirements, and the Coastal Commission is responsible for reviewing and certifying LCPs as consistent with the Coastal Act. Once an LCP is certified, much of the Coastal Commission's responsibility to regulate new coastal development in that city or county through a permitting process is delegated back to the local government.

An LCP will typically identify minimum cliff- or bluff-top setbacks and procedures to ensure that new development is not located in a hazardous area. For example, the Santa Cruz County LCP has detailed criteria for development on coastal bluffs and beaches, including requirements that all development demonstrate the stability of the site, in its "pre-development condition," for a minimum of 100 years as determined by either a geological hazards assessment or a full geological report. Development must be set back from the bluff edge a minimum of 25 feet. LCPs should also include standards for approving new shoreline structures, and the Santa Cruz County LCP has detailed rules that limit construction of shoreline protection structures to situations where there is a "significant threat" to an existing structure or where adjacent parcels are already similarly protected. Consistent with Coastal Act section 30235, the LCP also requires that all applications for shoreline protection include a comprehensive analysis of alternatives to a shoreline protection structure, including an evaluation of relocation or partial removal of the threatened structure.

Finally, LCPs should include requirements for impact avoidance and mitigation. For example, the Santa Cruz County LCP specifically requires that shoreline protection structures not "reduce or restrict public beach access, adversely affect shoreline processes and sand supply, adversely impact recreational resources, increase erosion on adjacent property, create a significant visual intrusion, or cause harmful impacts to wildlife or fish habitat, archaeological or paleontologic resources." The LCP further requires that shoreline protection structures use building materials that blend with the color of natural materials in the area and meet approved engineering standards, that construction impacts be minimized, and that applications include a permanent monitoring and maintenance program for the structure.

LCPs are a critical component of the California policy framework for shoreline development. They provide valuable local guidance for applying the general requirements of the Coastal Act, particularly for locating and designing new development. In practice, though, most shoreline protection structure proposals are reviewed by the Coastal Commission because they are often located in the commission's retained original permit jurisdiction, which includes tidelands between the mean high and low tides. In addition, the Coastal Commission has appellate authority over locally issued coastal development permits along the shoreline.

OTHER FEDERAL AND STATE POLICY

Many other federal and state governmental agencies may also be involved in the management of development in hazardous coastal areas and in the regulation of proposed shoreline protection structures. Because of the complexity of the regulatory environment, close coordination and cooperation among public agencies and private project proponents is important to effective implementation of coastal erosion policy.

FEDERAL AGENCIES AND PROGRAMS

One of the more significant federal programs influencing California's coastal hazards policy is the National Flood Insurance Program (NFIP), implemented by the Federal Emergency Management Agency (FEMA). This program provides federally backed flood insurance for property in hazardous areas, if the community in which the property is located has adopted a comprehensive flood management program designed to minimize flood risks through land use planning and regulation. Federal law sets specific standards that must be met in these community programs, such as ensuring that the lowest floor of all new structures is elevated to or above the base flood elevation as estimated for a 100-year storm event. In practice, these standards are often reflected in LCPs as requirements for locating and designing new development in hazardous areas. Because the NFIP provides significant financial incentives to private development interests through federally subsidized insurance, the program directly influences development trends. Without the availability of hazard insurance, the value of coastal property would no doubt be reduced, which might in turn reverse the trend toward increased property investment in hazardous areas.

The U.S. Army Corps of Engineers is another important federal agency in coastal management. The Corps is potentially involved as an applicant in the actual construction of shoreline structure projects, and also has regulatory responsibilities for projects that affect U.S. navigable waters. In addition to the Army Corps, the federal Marine Sanctuary Program within the National Oceanic and Atmospheric Administration (NOAA) is involved in the regulation of shoreline structures, such as along the Central Coast of California, where the Monterey Bay National Marine Sanctuary has authority over projects that may alter the seabed of the sanctuary. The MBNMS works closely with the California Coastal Commission to integrate its concerns into the permitting process of the commission. Finally, depending on the impacts of specific projects, other federal agencies that may be involved in coastal erosion regulation include the U.S. Fish and Wildlife Service and the National Marine Fisheries Service, both of which protect endangered and threatened species found in the shoreline environment.

STATE AGENCIES AND PROGRAMS

In addition to the California Coastal Commission, several other state agencies play an important oversight role for development along the shoreline. The California State Lands Commission (SLC) may be involved if a proposed structure would be placed on state lands, which are generally considered to be those tide and submerged lands between the ordinary high water mark, or mean high tide, and 3 nautical miles out to sea. When a project is proposed or approved on state lands, the proponent must get SLC approval and may be required to enter into a lease for the ongoing use of state lands. Usually, however, the SLC does not require a lease, and it may not require a permit, given the uncertainty in establishing the boundary between private and state lands.

Other state agencies that may be involved in shoreline structure proposals, depending on the potential impacts of a project, include the Regional

Water Quality Control Boards and the Department of Fish and Game. The California Department of Boating and Waterways may also play an important role in the review, planning, design, funding, and construction of shoreline protective works funded by the state.

IMPLEMENTATION ISSUES

The Coastal Commission and local governments have permitted thousands of new developments along the California shoreline since 1976, the vast majority of which have been single-family homes. They also have approved many new shoreline structures to protect existing development. For example, a 1995 commission study found that over 2 linear miles of new coastal armoring had been approved in Santa Cruz County between 1978 and 1993. Along the Malibu shoreline, nearly 3 miles of new armoring was approved between 1978 and 1996. Many more repair and maintenance projects have been approved for existing seawalls and revetments. The Coastal Commission infrequently denies a proposed development, and the majority of its effort is focused on crafting conditions for approval that either modify project designs or require other mitigations to address the impacts of the proposal. In Malibu, for example, 85 percent of the proposed shoreline armoring projects were approved between 1978 and 1996.

The high approval rate for shoreline structures is not surprising, given that the Coastal Act says that shoreline protection structures "shall be permitted" when necessary to protect existing development in danger from erosion. Still, few are enamored of the proliferation of seawalls and revetments and their significant impacts on the public beach resources and natural landforms of California's coast. The approval of new development in shoreline locations, or the consideration of a new shoreline protective device, is not taken lightly by the commission or the public. At the same time, California's coastal real estate is some of the most desirable property in the world, and substantial investments have been made in residential and commercial developments. Most of the coastal erosion policy challenges faced by the commission and local governments are rooted in this basic tension between the desire of the public to maintain and protect the natural resources of the coast, and the desire of private property interests to protect their homes and investments.

GEOLOGICAL UNCERTAINTY: EVALUATING COASTAL EROSION RISKS

Risk assessment lies at the heart of Coastal Act policies to locate new development out of harm's way, and to limit the construction of new shoreline protection structures. LCPs typically require that new development be set back far enough from the bluff edge to ensure its safety for some identified "economic life," usually from 50 to 100 years. To do this, geological reports are needed to identify the erosion rates and risks at specific coastal locations. Similarly, the decision to approve a shoreline structure turns first and foremost on a finding that existing development is "in danger" from erosion. Unfortunately, geological risk assessments are subject to significant uncertainty and interpretation, making it difficult to guarantee that new development will indeed be safe without the need for shoreline protection over its life.

The case of the Cliffs Hotel in the City of Pismo Beach is a good example of this problem (see CCC Appeal A-3-PSB-98-049). In 1983 the Coastal Commission approved a permit to build the Cliffs Hotel on a 75-foot-high cliff. Based on the applicant's geological analysis, which established the erosion rate at the site at 3 inches per year, the hotel was required to be set back 100 feet from the cliff edge, which the commission found to be more than adequate to provide the hotel with 100 years of stability. The Commission further required that the cliff-top area and beach below be dedicated for public access, and prohibited any future development in this setback area except for public access improvements. As with much of the shoreline development that the commission had approved, the agency also required that the Cliffs Hotel property owners acknowledge and assume the high risks of developing on the cliff-top location. The hotel was subsequently built and opened for business (Figure 8.1).

As early as 1996, however, the Cliffs Hotel owners sought approval of a revetment from the City of Pismo Beach under the certified LCP. The stated purpose of the revetment was to protect a sewage-holding tank that had been built in the protected easement area, contrary to the original approval for the hotel. After the City approved the proposed revetment, the project was appealed to and denied by the Coastal Commission as inconsistent with both the LCP and the prior restrictions placed on the hotel development. Soon after this project denial, however, the City of Pismo Beach approved an emergency permit for the Cliffs Hotel to construct a revetment in anticipation of the 1997–98 El Niño, and rock was placed on the beach below the hotel without commission review. The required follow-up permit for this emergency revetment approval was eventually appealed to and heard by the Coastal Commission, which again denied a permit to construct the revetment. Ultimately, the commission prevailed in litigation and was able to compel the removal of the revetment and restoration of the beach below the hotel.

Although there were a number of difficult issues with the proposed Cliffs Hotel revetment project, including the fact that the rock had been placed in

Figure 8.1   The Cliffs Hotel in the city of Pismo Beach. Photo © 2002–2004 Kenneth and Gabrielle Adelman, California Coastal Records Project, www.Californiacoastline.org.

the previously deed-restricted setback areas, one of the main issues that the commission addressed was whether the Cliffs Hotel was actually at risk from cliff erosion. The applicants asserted that the hotel was indeed at risk, based on an evaluation of new geological data and an estimated erosion rate of 4 feet per year, as opposed to the 3 inches per year estimated at the time of the original approval of the hotel. The commission evaluated the geological data somewhat differently, noting that the applicant's latest erosion rates were based on an episodic loss of 6 feet of bluff averaged over an 18-month period. Naturally, this led to an exaggerated erosion rate. In addition to this variability in estimates, the commission noted that even with acceptance of the revised erosion rate, the hotel would not be directly threatened for another 19 years, leading the commission to conclude that there was no imminent "danger" to the hotel.

The commission's review of the Cliffs Hotel and proposed revetment project illustrated the variability that is often encountered in geological calculations or estimates of bluff retreat rates. Table 8.1, adapted from the commission's Cliffs Hotel findings, summarizes the various estimates that had been made through the Cliffs Hotel project history. This type of variability is not uncommon in the world of geological consulting, and much of it is due to technical difficulties in estimating erosion rates. Problem areas include the scale, resolution, and time span of the stereo aerial photographs

TABLE 8.1   **CLIFFS HOTEL RETREAT RATE ESTIMATES**

GEOTECHNICAL REPORT	ESTIMATED RETREAT RATE	SOONEST THE HOTEL WOULD BE UNDERCUT
Cliffs Hotel original geotechnical report supporting 100-foot setback	3 inches per year	312 years
1998 consultant report and 1979 draft EIR report for the area	12 inches per year	78 years
Cliffs Hotel geotechnical report for A-3-PSB-96-100 seawall project (denied 12/96)	13 inches per year	72 years
Long-term documented erosion since 1955 at the Cliffs Hotel site	14 inches per year	67 years
Cliffs Hotel emergency permit geotechnical report adjusted for seasonal accuracy	2.1 feet per year	37 years
Cliffs Hotel emergency permit geotechnical report in support of city revetment approval	4 feet per year	19 years

used; the skill and experience of the interpreter; the methods used to measure the position of the cliff or bluff edge; and the complexities of local geological conditions and erosional processes. For example, the likelihood of
large episodic cliff collapse events, which can happen instantly in the case of
a rockfall or extend over the course of a single severe storm season, compli-
cates the calculation of long-term average cliff retreat.

Apart from technical challenges, the uncertainties inherent in calculating
cliff or bluff erosion rates has meant that geological analyses are vulnerable
to interpretation by the consultants and geotechnical experts who are typically involved with shoreline projects. This vulnerability works against the
policy objective of ensuring that development is restricted to safe locations
in order to avoid the need for shoreline protection.

In another commission permit appeal, a new cliff-top home was
approved by the City of Pismo Beach in 1997, with a setback based on the
consultant's erosion rate estimate of 3 inches per year. Virtually before the
paint was dry on the new house, however, the homeowner was pursuing a
seawall to protect the house based on new geotechnical evaluations arguing
that the erosion rate at the site was closer to 2 feet per year (based on an

analysis of episodic erosion events and a "re-averaging" of actual cliff retreat over time). A seawall was eventually approved by the commission to protect the now "existing" house, a mere six years after the house had been approved (CCC Appeal A-3-PSB-02-016).

In recognition of the problem of geological uncertainty in the determination of erosion rates, and the potential abuses of the geological or geotechnical evaluation process, the commission has begun to impose a more rigorous version of the "assumption of risk" condition than it has historically imposed on coastal developments. This condition requires that landowners agree to a permit condition and associated property deed restriction that prohibits future shoreline protection structures for the development being approved. Although it remains to be seen how this restriction will fare in future legal challenges to its application, the approach effectively requires the landowner to assume the risks of coastal development by agreeing to abide by the initial geological determination that a proposed building site will be "safe" for a stated period of time, without the need for a seawall or revetment.

WHAT IS AN EXISTING STRUCTURE?

Since 1976 the Coastal Commission has had to decide what constitutes an "existing structure" for purposes of Coastal Act section 30235. Early in its history the commission drew a clear distinction between primary structures, such as residences or commercial buildings, and secondary or accessory structures, such as a bluff-top gazebo or a storage building. This was done to make clear that seawalls and revetments would not be approved to protect just any structure, but only those that represented a significant and primary use of land, such as a private home.

A different challenge has arisen in the consideration of existing developments that are not clearly "structures." Some private property owners feel that shoreline protection should be available for the backyards that sit between their homes and the cliff edge. Questions also have been raised about whether a golf course is a structure for the purposes of section 30235, and whether trails and other "nonstructural" public access features, such as bluff-top public open space, should be considered structures worthy of shoreline protection.

The biggest question concerning existing structures, though, is whether development that was built after 1976 should be considered to be "existing," and thus eligible for shoreline protection under section 30235. A strong case can be made that Coastal Act sections 30253 and 30235 should be read together as establishing a firm date—January 1, 1977—after which new development is not eligible for shoreline protection. Development that was in existence as of the date that the Coastal Act became law, it is argued, should be considered "existing development;" whereas development approved and constructed after this date would necessarily have had to have been found consistent with section 30253, and thus should not have been premised on the need for a future shoreline protective device. The commission generally has not implemented the Coastal Act in this way, however, and in some cases has approved shoreline protection for development that was approved after 1976, under the theory that it was "existing" development at that time and thus is protected under section 30235. This weaker interpretation of the Coastal Act has worked against the policy objective of limiting the approval of new shoreline structures.

THE ECONOMIC LIFE OF STRUCTURES AND "PLANNED RETREAT"

Most LCPs rely on minimum cliff- or bluff-top setbacks to ensure that new development is located in a "safe" place. The typical approach is to estimate the erosion or "retreat" rate at the development site and then calculate the distance that would guarantee the safety of the structure for a given period of time. This time is usually set somewhere between 50 and 100 years, which represents the likely economic life of the project. In theory, this ensures that no new shoreline structures will be built at the location of the new development for the life of the project. And once the project reaches the end of its economic life (and is presumably removed), a new project would also have to be set back safely and not be allowed to have a shoreline structure. Overall, this approach should support a system of "rolling" setbacks or "planned retreat" from California's eroding coastlines, at least in those locations that were undeveloped when the Coastal Act was adopted.

In practice, planned retreat has proven very difficult to implement. In addition to the already discussed difficulty of accurately projecting cliff or bluff retreat rates, which may lead to inadequate initial setbacks, a more fundamental problem with this approach is the assumption that structures have an economic life in the first place. The Coastal Commission has only been in existence since the 1970s, so it does not have any experience with whether structures approved under the "set back for economic life" approach will actually be required to be removed at the end of the originally assumed economic life. The commission does have experience, however, with redevelopment trends in the coastal zone, which suggest that aging structures do not really die so much as metamorphose into "new and improved" structures in the same place. Thus, an increasing challenge along the coastline is how to regulate the *redevelopment* of buildings that, under traditional planning and zoning concepts, might be considered "nonconforming" because they are not set back sufficiently from the bluff edge, or because they have shoreline structures that predate enactment of the Coastal Act or an LCP.

This problem is particularly vexing for those interested in pursuing "planned retreat" as a strategy in eroding urbanized areas. Although many LCPs have ordinances that require nonconforming structures to meet current regulatory standards when they are being redeveloped, the requirement is usually triggered only beyond some threshold level of redevelopment. For example, reconstruction of a building close to a bluff edge in the same location may be allowed as long as no more than 50 percent of the exterior or interior walls are altered or reconstructed in the process. Similarly, some minimal expansion in building size might be allowed without the owner having to bring the entire building into conformance. Thus, it is not uncommon to see nonconforming structures essentially be redeveloped, through progressive changes that incrementally comply with the zoning codes. Over time, an old structure is rebuilt, and nonconforming bluff setbacks or old seawalls remain unchanged. Ultimately, preexisting urban development patterns ensure that existing shoreline armoring will remain, whereas new shoreline structures will be needed for development that effectively never retreats to a safe setback distance. It is particularly difficult to implement required setbacks and "no seawall" provisions in urban areas that are substantially built out and already protected by armoring. It also seems unreasonable to some to deny the last few unprotected properties in an urban area the same protection enjoyed by the rest of the neighborhood. The commission has recognized this situation in the past through implementation of a "string-line" setback method—essentially allowing infill development to be located at a distance from the bluff consistent with adjacent existing development setbacks, regardless of the hazards.

Notwithstanding the difficulties of implementing coastal erosion policies in urban areas, there is renewed interest in planned retreat in California. In 2003 the California Resources Agency released a draft policy for coastal erosion that emphasized the need to avoid new or modified development in hazardous areas, as well as the need to relocate or even eliminate threatened existing coastal development (where feasible). Unfortunately, due to the state budget crisis of 2003–4, this draft policy was never finalized or implemented. Local communities such as Solana Beach are debating new rules to implement planned retreat. They are struggling with the question of whether a planned retreat policy would result in an unconstitutional taking of private property. Some argue that prohibiting substantial redevelopment in hazardous locations is a taking of private property because it would interfere with a reasonable investment-backed expectation to redevelop the property. Others argue that property owners with an existing use of land are not entitled to more development, particularly if it means the degradation of public resources—the beach, for example—by the often associated seawall or revetment. This legal issue is a challenging one.

PLANNING THROUGH EMERGENCY RESPONSE

A significant amount of permit activity for shoreline structures occurs under emergency or extreme winter conditions. Studies of coastal development trends in the Monterey Bay region illustrate how most of the new and expanded riprap on beaches was placed during significant storm years, particularly the El Niños of 1978–79 and 1982–83. Commission data clearly shows spikes in permit activity associated with these periods as well as the 1997–98 El Niño. Emergency situations typically do not allow for adequate project review to confirm consistency with the Coastal Act or LCPs, and most emergency shoreline projects involve the placement of riprap on the beach with varying degrees of attention to design and impact mitigation. In some cases, emergency shoreline structures have even been placed on top of previously required public access dedications. Unfortunately, once riprap is placed on the beach, it becomes very difficult to either remove it or require alternative shoreline protection design.

The Coastal Commission's regulations define an "emergency" as "a sudden unexpected occurrence demanding immediate action to prevent or mitigate loss or damage to life, health, property or essential public services." Applying this definition is difficult, particularly when an emergency permit request is being made during a major storm event. Emotions are running high, and given the lack of time to address important planning questions, such as the real risks to the structure, or the proper emergency response, pressures bear on the commission and local governments to take a cautious approach and allow the emergency shoreline structure. After the storm has passed, it may become apparent that the risks were not as great as originally perceived, or that the response went well beyond that needed to address the emergency event. The commission is also subject to pressure to give emergency approvals for revetments in cases where, it is argued, there is insufficient time to analyze alternatives or to design and build a more appropriate shoreline protective device before the next big winter storms, or where funding would be lost if approvals are not given. These so-called emergencies are really planning failures on the part of the project proponent.

Implementing rational coastal hazard or protection policy is particularly difficult in areas where local governments themselves have significant stretches of public land along the coastline. For example, the cities of

Figure 8.2   Riprap placed under emergency conditions along the West Cliff Drive area of Santa Cruz during the 1997–98 El Niño. Photo © 2002–2004 Kenneth and Gabrielle Adelman, California Coastal Records Project, www.Californiacoastline.org.

Carmel-by-the-Sea and Santa Cruz have significant public access to the shoreline and also cliffs that are subject to erosion. Although great efforts have been made by both communities to protect these resources, responses have sometimes occurred in an ad hoc fashion, through emergency response. Along West Cliff Drive in Santa Cruz, significant amounts of riprap have been placed on pocket beaches through emergency permitting (Figure 8.2). These revetments protect cliff-top recreation areas, but they have also resulted in the cumulative loss of beach area and impacts to the scenic character of West Cliff. In recognition of this dilemma, the City has an LCP policy that requires a comprehensive shoreline management plan to better anticipate and reconcile the competing coastal resource issues along West Cliff. For example, such a plan might identify those areas where relocation or redesign of bluff-top amenities might be feasible and preferable over the long run to the placement of rock on the beach below. Similarly, the plan could identify those areas likely to need shoreline protection in the near future; and the planning, design, and funding of vertical structures that would minimize beach encroachment and aesthetic impacts might be pursued well in advance of the inevitable winter storm events and associated erosion.

Comprehensive coastal protection plans require considerable financial and political resources, however, and it is not easy for local governments to support such efforts. The lack of comprehensive planning and funding for more expensive, tailor-made erosion responses leads to a situation where the easiest and least expensive response for local governments is to place rock or riprap in those locations where bluff-top amenities are about to succumb to coastal erosion. More generally, the problem of emergency shoreline response highlights a deeper tension in the Coastal Act between the need to avoid shoreline protection structures by limiting them to identified "dangerous" situations, and the need to plan ahead for appropriate coastal erosion response. Although emergency response is not a good way to maximize protection of resources, there may be some truth in the observation that waiting to approve shoreline structures until there is an imminent threat to a structure makes it more likely that agencies will be responding to coastal erosion in an emergency situation.

IMPACT ASSESSMENT AND MITIGATION FOR SHORELINE STRUCTURES

The Coastal Act requires that the impacts of shoreline structures be avoided, minimized, and mitigated. Some of these impacts are fairly easy to address, either through design changes or mitigation measures. For example, to reduce the beach encroachment of a structure, the commission frequently requires that a vertical seawall, not a sloping revetment, be used where feasible. Another common mitigation measure to address the visual impacts of a seawall or bluff stabilization is to require that they be colored and textured to look like a natural bluff. Recent projects in Pebble Beach (Figure 7.20) and Santa Cruz (Figure 7.19) illustrate how far the technologies for replicating natural bluff features in shoreline stabilization projects have come.

Other shoreline structure impacts, however, have proven more challenging to address. The effects of shoreline structures on sand supply and beaches are perhaps the most difficult. Coastal Act Section 30235 specifically requires that impacts to local shoreline sand supply be eliminated or mitigated. The commission has identified three types of such impacts that need to be addressed by projects. First, many coastal protection structures, and riprap revetments in particular, encroach onto sandy beaches, resulting in the physical loss of beach area. Second, by design, shoreline structures stop coastal bluff erosion, thus cutting off a potential source of sand that would otherwise feed local beaches in the littoral cell. Third, studies have shown that fixing the position of the back edge of the beach on a retreating coastline with a seawall causes passive erosion and ultimately the loss of the sandy beach in front of the structure, as the shoreline continues to move landward on either side of the structure. Taken together, these three impacts may cumulatively result in the degradation or even loss of entire stretches of beaches in urban areas. In Monterey Bay, approximately 25 acres of beach have been covered by shoreline structures. The specter of global climate change and associated sea-level rise makes the problem of shrinking beach areas that much worse.

The commission has used a methodology to address sand supply impacts that relies on the quantification of the beach sand covered, retained, and subject to passive erosion. For example, a proposed revetment might cover several thousand square feet of sand, retain hundreds of cubic yards of sand behind it that would have otherwise supplied the beach, and cause the long-term loss of sandy beach in front of the structure through passive erosion. Other than the encroachment, these impacts generally cannot be avoided or minimized through design changes and thus must be mitigated for the commission to be able to approve the project under the Coastal Act. The most effective application of this mitigation methodology has occurred in San Diego, where the commission and the San Diego Association of Governments have established an in-lieu fee program. Property owners that build new seawalls pay a fee in an amount that reflects the estimated cost of putting the identified lost sand back into the natural system. These fees are then to be used to finance a comprehensive beach sand replenishment program for the region. The reliance on a mitigation fee, in lieu of an actual individual sand replenishment requirement, also recognizes that sand replenishment mitigation projects on the scale of the individual home site are difficult to design, and of questionable value in terms of actually accomplishing beach replenishment.

Other areas of California do not have established regional sand supply replenishment programs; to do so requires significant funding, technical study, and political support for regional or subregional beach erosion response. Beach replenishment programs raise significant questions about whether effective shoreline response is even possible in any particular circumstance. Because of this, the commission has not been able to effectively use the sand supply impact methodology and in-lieu fee approach in other regions of the coast. In a few circumstances mitigation fees have been collected, but the implementation of an actual sand supply project has not yet occurred. Again, this is in part because of recognition that an individual, one-time sand supply effort to mitigate the effects of a single seawall would likely prove ineffective in addressing the actual long-term sand supply impacts of a project unless they were based on more comprehensive coastal erosion studies and analysis of sand supply dynamics in the specific area. It also is infeasible for such studies and projects to be completed by individual landowners.

Because of the difficulties in addressing sand supply impacts of individual projects, these impacts have sometimes gone unaddressed in regulatory actions because no feasible mitigation was available. Some argue that this should not be a concern because the impacts of shoreline structures on sand supply are minimal relative to the impacts of sediment sources such as rivers and streams. Although more study is needed, recent research has shown how the importance of bluff sand may vary. In the Santa Barbara littoral cell, for example, cliff erosion contributes less than 1 percent of the total sand supplied to the beaches of the cell, because of the presence of four large sand-producing rivers. In contrast, bluff erosion contributes approximately 12 percent of the beach sand in the Oceanside cell. Without further study, though, it is difficult to dismiss the potential cumulative impact of shoreline structures on the beach environment in some littoral cells. The lack of comprehensive analysis and establishment of regional or subregional sand supply mitigation programs to address impacts to beaches is a major challenge to successful implementation of the Coastal Act's shoreline structure policy.

THE BIG PICTURE: SOCIAL COST/BENEFIT ANALYSIS AND PRESERVATION OF THE NATURAL SHORELINE

The issues discussed thus far are ultimately rooted in the fundamental conflict created by our decisions to put homes, buildings, and roads along coastlines that are eroding. From a social history standpoint, California's urban coastal development is relatively young. Many of the eroding bluff areas that are now at the center of conflicts concerning new seawall development were relatively undeveloped as little as 50 years ago (Figure 8.3). But many coastal urban areas are close to being built out, and few vacant lots remain. Redevelopment of aging structures in hazardous areas is now a common concern for the commission and local development. Maintaining Highway 1 and other scenic coastal roads in the face of coastal erosion is a continuing and growing challenge.

Increasingly we seem to be at a crossroads along the coast in terms of "big picture" decisions on how to respond to coastal hazards. From a social investment standpoint, significant sums of money have been and continue to be directed to development in hazardous coastal areas. Federally subsidized insurance is available for communities that have development in hazardous areas. Federal monies also are available for shoreline protection projects with defined goals of *stopping* coastal erosion. Current California law says that the commission shall permit shoreline protective measures for structures in danger from erosion if there is no other feasible, less environmentally damaging alternative.

Figure 8.3   (A) East Cliff Drive area of Santa Cruz in 1928, showing a lack of cliff-top development. (B) East Cliff Drive area of Santa Cruz in 1975, with all oceanfront property developed with homes. Courtesy of County of Santa Cruz.

But there is a social cost to the economic investment in shoreline development, and this is the cumulative degradation of our coastline through the proliferation of protection structures and all of the impacts that follow. Recent studies have established the immense economic value of California's coastline and beaches in terms of direct tourism dollars coming to local communities, indirect positive impacts on California's economy, and associated contributions to federal tax revenues. Although it is difficult to quantify, there is an impact on the social value of California's coastline, and specifically its beaches, from shoreline structure development. Few would disagree, for example, that the intrinsic value of an unaltered natural shoreline, with a full sandy beach, is greater than that of the urban beach covered with an eclectic array of riprap, seawalls, concrete debris, and other relics of our ongoing battle with coastal erosion (Figure 7.1). Protecting our natural coastlines by avoiding shoreline armoring is thus an important part of the social economics of beaches. More fundamentally, how we choose to respond to coastal erosion relates directly to our quality of life along the shoreline. Are urban areas destined to be fully armored, with little or no natural bluff, beach, or other shoreline features? Or are there ways to maintain natural coastlines where most of us live and work, through erosion responses that do not involve artificial shoreline structures?

The Coastal Commission has taken a strong proactive stance on the restoration of natural shoreline features and processes in situations where it seems feasible. For example, when the U.S. Army proposed the demolition of Stillwell Hall, an aging building located on a eroding bluff on Monterey Bay, the commission asked that as part of the demolition project, the Army also clean up the beach below, where tons of riprap have been dumped over the years in a failed attempt to save Stillwell Hall from the sea (Figure 7.25). In cases involving the placement of emergency revetments to protect Highway 1 (Figure 8.4), the commission has required that the California Department of Transportation pursue long-term highway realignment projects or bridge alternatives, to move the highway out of the active erosion zone and thus enable restoration of the natural shoreline. These cases represent policy decisions to require social investment in shoreline restoration and planned retreat, so that the coastline can continue to evolve and function naturally, without interference from human-made structures.

Urban areas present a significant challenge to policymakers with respect to "big picture" choices and social economics. For some, it is easiest to focus on the need to protect existing private and public investments on the shoreline, and simply accept the eventual armoring of most of our urban areas. Under this view, little "natural" coastline will remain in these areas over the long run, and large-scale beach replenishment or the construction of groins

Figure 8.4   Riprap being placed to protect California State Highway 1 south of
Pescadero Creek in San Mateo County. Photo © 2002–2004 Kenneth and Gabrielle
Adelman, California Coastal Records Project, www.Californiacoastline.org.

may be the only options available to maintain sandy beaches in front of
shoreline protective devices. Another vision, though, would focus on the
possibilities of investing in planned retreat and shoreline restoration at the
community, state, and federal levels. In this long view, economic incentives
and commitments, as well as legal requirements, would need to be changed
significantly to establish an economic and social context more conducive to
a policy of shoreline retreat.

The typical shoreline erosion project analysis focuses on the erosion risks
to existing public or private development, and the feasibility of alternatives
to a shoreline protective device. The analysis presumes a social objective of
protecting the existing development and then limits the potential ways to
achieve this objective by eliminating "infeasible" alternatives. Although feasibility analysis often includes a technical component (i.e., whether an alternative technically can be accomplished), it also often boils down to a
conclusion that certain alternatives are not economically feasible because
they cost too much for the project proponent to implement. For example,
the alternative of removing part of a residence at risk, although technically
feasible, is usually deemed too costly and otherwise unreasonable to ask of
private homeowners.

More fundamentally, presuming that the protection of existing development is the social policy objective and eliminates the alternative of planned

Figure 8.5   Area proposed for bluff stabilization along the Pleasure Point area of East Cliff Drive in Santa Cruz. Photo © 2002–2004 Kenneth and Gabrielle Adelman, California Coastal Records Project, www.Californiacoastline.org.

retreat from the start, then a comparison of the social costs and benefits of planned retreat versus shoreline armoring is never really developed in full. Although the regulatory process of the commission and local governments presumably mitigates the individual project impacts or "social costs" of a seawall, piecemeal implementation of shoreline protection projects does not really allow for more comprehensive analysis of the cumulative social costs and benefits of long-term coastal erosion response alternatives.

For example, in 2003 the U.S. Army Corps of Engineers and the Redevelopment Agency of Santa Cruz County proposed a shoreline protection project for a portion of East Cliff Drive along northern Monterey Bay. The project proposed the armoring of approximately 1,100 linear feet of bluff to limit long-term erosion risks to the existing East Cliff Drive and public utilities (Figure 8.5). Although the proposed bluff stabilization project would entail significant environmental impacts, it would also protect the public space between the bluff edge and the first row of private homes, in addition to the roadway and the sewer and water lines. If the structure were not built, it was argued, the eventual result would be not only the loss of these public amenities, but also the construction of seawalls, ultimately, to protect the inland private development when it should become endangered—the same end result of an armored shoreline without the public benefits.

The Army Corps completed a requisite cost-benefit analysis of project alternatives in order to meet the project objective. The analysis assumed that protection of East Cliff Drive was the project objective, and planned retreat was not included as an alternative for more detailed analysis. The cost-benefit analysis was limited to comparing the costs of project construction and maintenance on one hand, and certain quantifiable benefits on the other (the avoidance of costs to relocate public utilities in the roadbed, and the costs of delay and additional travel distances for motorists from necessary detours once East Cliff Drive was closed).

Within the limited universe of alternatives considered to meet the project objective, the Army Corps was able to identify a full armoring alternative that maximized the net economic benefits. This is not the same exercise, however, as analyzing and comparing the social costs and benefits of planned retreat, or a beach replenishment project, with the armoring alternatives and the no-project alternative. Such an analysis would need to identify not only the relatively quantifiable costs and benefits of each alternative, but also the costs and benefits that are difficult to quantify, such as the social benefit of maintaining an unarmored shoreline or of maintaining bluff-top access amenities. In the case of a planned retreat, the costs of eventually acquiring and demolishing existing private development inland of East Cliff Drive to allow for long-term erosion, and the relocation of the public bluff-top space would need to be considered.

It is not clear that more comprehensive cost-benefit analyses in cases such as the East Cliff Drive project, if they could be completed at all, would lead to different outcomes in responding to identified coastal erosion problems. But the lack of broader policy analysis certainly means that planned retreat or other potential erosion responses that do not involve armoring will not be fully considered, and thus not taken seriously enough in community deliberations about how to respond to coastal erosion. The first step in pursuing a planned retreat policy is to identify more completely the range of feasible long-range alternatives, and to analyze more fully their social costs and benefits, even if this analysis relies on broad assumptions. Such analysis may in fact help communities better identify project objectives that maximize environmental benefits to the community while minimizing the exposure of private development to coastal hazards.

CONCLUSION: DIRECTIONS FOR POLICY REFORM

The major challenges faced by the Coastal Commission and local governments in implementing the California Coastal Act suggest potential reforms that might improve the policy response to coastal erosion in California. The goal of these suggested policy changes would be to increase the preservation and restoration of natural shorelines while reducing the exposure of development to coastal hazards.

- *Strengthen the "line in the sand" against new shoreline structures.* Clarification of existing law is needed to ensure that new shoreline protective devices are limited to developments that exist currently. This could be accomplished through amendments to LCPs declaring that new development approved after the date of enactment of the amendments shall not be eligible for permanent shoreline protection. LCPs also should require that new development abide by a "no future seawall" condition for as long as it is in existence.

- *Clarify geotechnical analysis requirements.* LCPs should require that geotechnical reports submitted in support of applications for new shoreline development and shoreline structures address the risks associated with development proposals, and specify whether and to what degree structures are in danger. Training seminars for the professional geotechnical consulting and local planning community concerning the needs of regulatory decisionmakers and the requirements of the Coastal Act and LCPs would be useful in this regard.

- *Strengthen restrictions on the redevelopment of structures.* Financial incentives to maintain private development in hazardous areas must be minimized. Further restricting redevelopment options for nonconforming structures and other development along the shoreline through planning and zoning will facilitate planned retreat by limiting the increase in value of coastal properties located in hazardous areas. Minimal redevelopment of shoreline structures should be allowed absent full conformance with setback and strict engineering requirements. Design alternatives for new development that would facilitate planned retreat, such as relocatable or movable structures, should be evaluated. Developments approved with an assumed economic life should be required to be retired, through explicit legal agreement, at the end of the identified time period.

- *Conduct comprehensive subregional planning.* The California Department of Boating and Waterways has identified regional planning as a critical step in developing meaningful beach replenishment projects tailored to the unique conditions of California's various shoreline areas. The California Resources Agency has also recently recommended that a California Coastal Sediment Master Plan be developed, which would include regional identification and assessment of erosion risks and mechanisms to protect shorelines and beaches. This master plan approach is under way, jointly sponsored by the Resources Agency and the Army Corps of Engineers, with the participation of the Coastal Commission and other agencies. Regional knowledge concerning sand supply, littoral drift rates, and long-term changes in beach width, beach dynamics, and cumulative impacts is necessary to establish effective mitigation programs and measures for individual shoreline protection projects. This should include identification of public access and recreation improvement projects that could be supported through in-lieu fee programs. Better knowledge of regional and site-specific erosion trends would support more specific planning for necessary shoreline response to minimize the need for emergency actions. Subregional planning would also provide a context for better evaluating "big picture" options for urban areas with eroding coastlines, as well as a context for improved governmental coordination and decisionmaking. LCPs provide a useful framework for such planning.

- *Increase public investment in restoration of natural coastlines and processes, and planned retreat.* Financial assistance is needed to support identified shoreline restoration and planned retreat opportunities, such as relocating Highway 1 in areas where it is under attack from erosion and where it is feasible to do so. Money is needed to support local government shoreline planning, design, and construction of state-of-the-art projects that minimize impacts to other coastal resources. Restoration funding should be pursued to support relocation or elimination of existing development in hazardous areas.

- *Eliminate financial incentives for hazardous development and armoring.* A corollary to increased restoration funding is to eliminate financial incentives that promote continued location of, or investment in, development in hazard zones. Changes to the National Flood Insurance Program and opportunities for increased pre-hazard mitigation (planned retreat) should be pursued. Federal and state beach restoration programs should be required to fully evaluate and, where appropriate, provide funding for nonstructural beach erosion response alternatives.

- *Investigate beach replenishment strategies to avoid shoreline armoring.* Sediment management for the purpose of beach replenishment is potentially a critical component of a comprehensive response plan to coastal erosion, and more research and planning is needed to determine whether it is a feasible alternative to armoring of the shoreline. Large volumes of sand may be necessary, there may be significant impacts associated with exploiting the sand sources themselves (such as removal of upstream dams), and the feasibility, life span, and thus long-term effectiveness of beach replenishment are uncertain. Nonetheless, further investigation is necessary if the larger social choice issues surrounding coastal erosion policy are to be addressed. Where it is possible to do so, the beneficial reuse of sand materials available for replenishment (such as clean dredging materials) should be considered, consistent with other coastal resource protection objectives.

These recommendations represent only some of the major areas where policy reform is needed toward the improvement of coastal hazards policy response and implementation in California. The next phase of California's coastal hazards policy will need to better address the big picture of social investment and alternative responses to coastal erosion. There are indications that California is moving in this direction. Ultimately, though, the protection of public beaches and natural shorelines will require significant changes in the legal requirements, financial incentives, and social perceptions that currently shape our desire to live and build on the coast. The greater public interest in managing our coastlines will need to be better articulated and evaluated by those responsible for implementing coastal erosion policy so that natural coastlines are not lost through incremental, parochial, or short-range decisions to protect existing developments. At the same time, the socioeconomic and political context that supports continued private investment in hazardous coastal areas will need to change if we are to seriously pursue planned retreat and other policies that promote the public's interest in natural shorelines over the longer term. Thus, the tension between private and public interests will continue to lie at the heart of the coastal hazards policy debate as we search for new ways to live along our eroding coastline.

CHAPTER NINE

THE NORTHERN CALIFORNIA COAST

The Oregon Border to Shelter Cove

**LAURET SAVOY, DOROTHY MERRITTS,
GARY GRIGGS, AND DEREK RUST**

Heavy winter rains, steep coastal cliffs, and forests of redwood and Douglas fir distinguish California's northwestern edge, the traditional homeland of the Tolowa, Yurok, Chilula, Wiyot, and Mattole tribal peoples.

The earliest European explorers sailed the north coast waters in search of new land and navigable harbors during the sixteenth and seventeenth centuries. Spain showed the greatest interest in exploring the northwest coast of the New World, and the Cabrillo-Ferrelo voyage of 1542–43 is generally considered the first attempt to reach the coastline of Alta California (Upper California). Many historians believe that members of this voyage sighted and named Cape Mendocino (after Antonio de Mendoza, Cabrillo's patron and Mexico's viceroy) and perhaps ventured as far north as southern Oregon. Although few records remain of the voyages, some researchers believe that Sir Francis Drake in 1579 and Sebastian Rodriguez Cermeno in 1595 landed near what is now Trinidad Head in Humboldt County. The last of the early explorers, Sebastian Vizcaino, probably reached Cape Mendocino in 1603.

Nearly 200 years passed before renewed interest in northern California exploration brought European and Russian vessels to this area. The Spaniards Juan Francisco de la Bodega y Cuadra and Bruno Heceta "discovered" the protected anchorage of Trinidad Bay on Trinity Sunday in 1775, and named it Puerto de la Trinidad. Russian fur traders began to exploit the sea otters in the Bering Sea in 1740 and then moved progressively down the west coast of North America. Trade ships for the Russian American Fur Company "discovered" the entrance to Humboldt Bay in 1806. Thinking the bay unnavigable, they did not report their finding until 1852, after it had been encountered by a land-based expedition. In the early 1800s, exploitation of sea otters in northern California resulted in competition among Russian, English, and American vessels, all of which tried to avoid the Spanish. The Spanish retaliated with the seizure of ships and men and with harsher regulations, but controlling the lucrative sea otter fur trade was difficult in those days. In the 50 years between about 1770 and 1820, an estimated 350,000 to 1 million sea otters were killed along what is now coastal California.

Fur trapping brought the first land-based white explorers, led by Jedediah Smith, to the north coast in 1828. It was easier to reach the area by ship in those years than by traveling overland until 1848, when the discovery of gold in the nearby Trinity River brought countless prospectors and fortune seekers. Josiah Gregg led an expedition westward from the Trinity Mountains to the coast to reach Humboldt Bay in 1849, and a ship entered the port from sea the following year. Shiploads of miners and settlers descended on the coast after gold was discovered in the region, establishing port settlements such as Crescent City, Trinidad, Eureka, and Klamath City. Crescent City was laid out in 1853, and within a year it became a major port for shipping gold. Crescent City was even briefly promoted as a site for the state capital. The prosperous redwood logging industry also began to flourish, and with the increasing demand for lumber created by California's booming population, logging eventually surpassed mining as a major industry.

With overland travel nearly impossible, particularly because of the difficulty of crossing the mouths of deep gulches and canyons along the coast, ships exported most of the timber and lumber. Sheltered harbors are few along the Mendocino coastline, so nineteenth-century mill operators constructed landings and chutes to transfer lumber and cargo to small schooners that tied off to rocks offshore. At least 60 of these landings, referred to as "dog holes" (nineteenth-century slang for an unattractive location), existed along the coast between Bodega Head and Humboldt Bay, and many can still be identified by metal hooks and other relict structures on coastal cliffs.

Throughout the twentieth century, logging and milling decreased in economic importance in the region, with many mills closed by the 1980s. Although logging, along with commercial fishing, is still a principal industry, economic figures for Mendocino County in 2002 indicated that wine grapes brought in $81 million in revenue, while lumber dropped to the number 2 official crop, with only $54 million in revenue.

PHYSICAL SETTING AND DEVELOPMENT

Much of the rugged northwestern coast of California is still sparsely populated, with the total population of Del Norte and Humboldt counties in

2003 being only 156,000. Humboldt County has the longest coastline of any California county, and most of it is protected in the form of national and state parks and beaches. Coastal development and population in this region are concentrated in Crescent City and the Smith River mouth in Del Norte County, and in Eureka in Humboldt County. The southern portion of Del Norte County south of Crescent City and much of northern Humboldt County recently became protected within the 50-mile-long Redwood National Park (approximately 108,000 acres).

Crescent City, a town of about 28,000, lies on the southwest corner of a broad coastal lowland nestled between wooded ridges approximately 20 miles south of the Oregon border. The Smith River drains the lowland at its northern end, flowing across a wide floodplain. To the north of the river mouth, a low, flat coastal terrace is cut at the base of the mountains. Rugged ridges rising abruptly from the sea at Enderts Beach in Redwood National Park mark the southern terminus of the lowland. Coastal development prior to the 1960s was limited to scattered homes near the cliffs along Crescent City's western shoreline. By the 1960s houses and a fronting roadway (Pebble Beach Drive) lined most of this oceanfront.

Between Crescent City and Eureka are several wildlife preserves and state and national parks that protect coastal forest habitats and beaches. These include the following, from north to south: Redwood National Park, Del Norte Coast Redwoods State Park, Redwood National Park (continued from the north), Prairie Creek Redwoods State Park, Humboldt Lagoons State Park, Patrick's Point State Park, Trinidad State Beach, Little River State Beach, Humboldt Bay National Wildlife Refuge, and the King Range National Conservation Area, which extends southward to the Mendocino County border.

Eureka, a city of about 26,000, lies on the southern corner of a broad coastal lowland that extends 19 miles south-southwest from Trinidad Head to the mouth of Humboldt Bay, the second largest bay in California. This lowland is backed by low marine terraces and bayside bottomlands. Humboldt Bay itself has a narrow, shallow mouth located approximately midway between two long coast-parallel sand spits. Today, the bay's mouth is protected by a jetty on each side, referred to as the North and South jetties. Protected behind the dune-covered spits are the larger Humboldt Bay to the north and a smaller part of Humboldt Bay to the south that is referred to as the South Bay. Eureka was settled to the east of the northern part of Humboldt Bay, with its first settlers arriving only months after Josiah Gregg's party encountered the bay in 1849. By 1853, nine mills operated around Humboldt Bay and shipped 20 million board feet of lumber to the booming city of San Francisco to the south. Although the spits protecting the bay were historically used in part as landings for lumber ready to be shipped, they are in large part undeveloped today. The Eel River, the third largest of California's rivers, reaches the coast about 2.5 miles south of Humboldt South Bay. The river's broad, sandy mouth is very dynamic, having shifted its position northward nearly 2 miles between 1988 and 1995.

Several miles south of the Eel River mouth, the terrain changes dramatically. The low sand spits and barrier islands characteristic of the coast to the north are replaced first by narrow beaches with some dunes, and then eventually by rocky cliffs beveled occasionally by flights of marine terraces. Even more common along this stretch of the coast are immense landslides, one of which is the Centerville Beach slide. Very little development has occurred along this sparsely settled part of the coast, and most of the inhabitants to this day are descendants of a relatively small number of nineteenth-century ranch families. Because the coast is pristine and isolated, it is frequented by large marine mammals, including a sizable colony of sea lions that breed just north of Cape Mendocino, the westernmost point in California.

GEOLOGICAL SETTING

Most of the coastal area is underlain by rock known as the Franciscan mélange (French for mixture), which is exposed in the seacliffs at the Smith River mouth, along Point St. George and part of the Crescent City shoreline, in innumerable offshore rocks and small islands, and along the rugged coastal stretch that begins several miles south of the Eel River mouth and extends southward to Point Delgada at the Humboldt-Mendocino county border. The melange, by definition, is a disrupted mixture of boulders "floating" in a muddy matrix. These hard boulders of mostly sandstone are relatively resistant blocks that form many prominent points and offshore sea stacks, as waves easily erode away the surrounding weaker matrix. Along the mile-and-a-half length of Pebble Beach to the south of Point St. George, layered mudstones and siltstones occur at the shoreline. Loosely consolidated and erodible sands and clays overlie the mélange, the siltstone, and the sandstone along much of this coast.

The rugged, sparsely settled region between Cape Mendocino and Point Delgada straddles a tectonic feature known as the Mendocino triple junction, the place where three major tectonic plates meet. These plates are the North American Plate to the east, the Pacific Plate to the southwest, and the Juan de Fuca Plate to the northwest. The Juan de Fuca Plate is subducting beneath the North American Plate along the Cascadia subduction zone, which extends from Punta Gorda northward into Canada. The Pacific Plate slides against the North American Plate along the San Andreas Fault, a

Figure 9.1 Low wave-cut bluff along Highway 101 between the Oregon border and the mouth of the Smith River with scattered development, including this modular home park, which is partially protected by bedrock exposed on the beach. Photo © 2002–2004 Kenneth and Gabrielle Adelman, California Coastal Records Project, www.Californiacoastline.org.

major plate boundary that extends from Punta Gorda southward into Mexico. The triple junction region is one of the most seismically active in North America, and it has one of the highest rates of coastal uplift in the world, rising at an average rate of 0.08 to 0.2 in./yr. This uplift has produced the King Range, an uplifted block of deep-sea sediments (now sedimentary rocks) that is the youngest member of the Franciscan mélange.

THE OREGON BORDER TO SMITH RIVER

Between the Oregon border and the Smith River mouth, the coast is characterized by a sand and cobble beach backed by bluffs up to 40 feet in height. Along this entire reach the more resistant bedrock occurs as offshore sea stacks and outcrops along the shoreline, where it provides some protection for the low, sandy bluffs (Figure 9.1). With the exception of a state beach and local park, the oceanfront property is subdivided. Many of these bluff-top parcels are now developed with single-family homes as well as one modular home development.

Along most of this 4-mile segment, the nonresistant sands and clays that cover the mélange form the bluffs and are exposed to wave attack. The occurrence of large driftwood logs on the back beach indicates that waves

A. Resistant rocky coastline with offshore seastacks and rocky beaches.

B. Mouth of the Smith River has migrated north against Pyramid Point and has created a large estuary.

C. Wide sandy beach backed by a broad, vegetated sand dune system that extends for 12 miles from the Smith River mouth to Point St. George.

D. Back dune property has been subdivided and some roads exist but very little development to date.

E. Beach and dunes backed by two large fresh/brackish lakes, Lake Earl and Lake Talawa.

F. Wide sandy beach backed by broad and vegetated dune field.

G. Point St. George consists of rocky seacliffs and offshore seastacks. Coves backed by eroding seacliffs occur between more resistant points.

H. Low eroding sandstone cliffs fronted by narrow beach. Roadway runs along an uplifted marine terrace at the top of the cliff. Approximately 1000 feet of riprap has been placed to protect the clifftop coastal road from erosion.

I. Road follows top of low cliff along edge of marine terrace. Homes occur immediately landward of road. Riprap occurs in scattered locations along base of cliff where erosion has begun to threaten roadway.

J. Rocky shoreline with offshore seastacks and small islands. Rocky beach fronts low bluff. Houses line cliff edge and road and some houses potentially threatened by erosion.

K. Crescent City breakwaters protect a harbor.

L. Motel development on low bluff armored by riprap.

M. Sandy beach backed by low bedrock bluff. Beginning of Redwood National Park.

N. Highway 1 follows the shoreline just above beach level at False Klamath Cove and has been completely protected with riprap.

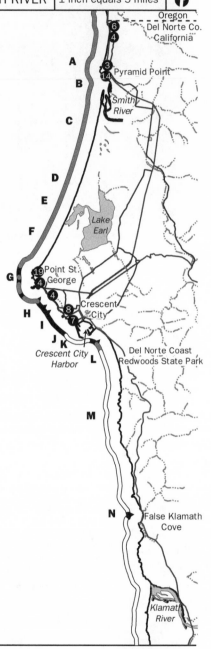

Hazard Level
☐ Stable: low risk
▨ Caution: moderate risk
■ Hazard: high risk

Erosion Rates
① Inches per year

Armoring
▲▲ Riprap
▨▨ Seawall
◙◙ Other

do reach the base of the bluff and have the potential to erode the bluffs during major storms.

Just north of the mouth of the Smith River the coast assumes an irregular outline where the Franciscan mélange is exposed at the beach level. Massive boulders or "knockers" within the mélange form prominent points and sea stacks, including Pyramid Point and Prince Island. The mouth of the Smith River has migrated northward and is now constrained by this rocky headland. With limited development, this area has so far experienced few erosion problems, although large storm waves coincident with high tides can erode these weak bluffs. Any bluff-top development should give careful consideration to the erodibility of these cliffs and long-term recession rates, which range from a few inches to a foot per year along this stretch of coast.

SMITH RIVER TO CRESCENT CITY

Unlike the beaches to the north of the Smith River, which are backed by developed, low, wave-cut bluffs, the 12-mile-long beach between the Smith River and Point Saint George is wide and backed by extensive sand dunes, wetlands, and two large fresh- to brackish-water lagoons (Lake Earle and Lake Talawa). Vegetated and stable, the dunes extend several thousand feet inland and were formed from sand discharged by the Smith River. Much of this dune and back dune area was subdivided in the 1960s, but as of 2003 no development has occurred near the shoreline (Figure 9.2). The shoreline resumes an irregular, cliffed configuration at Point Saint George. This prominent headland and the cliffs between Pebble Beach and Battery Point are exposures of the resistant mélange bedrock. East of Point Saint George the coastline consists of a low cliff eroded into the outer edge of an uplifted marine terrace. Pebble Beach Drive closely parallels the bluff edge along this stretch. Houses are confined to the inland side of the cliff-top road, although continued erosion of the low cliffs has led to the placement of several extensive areas of riprap (Figure 9.3). Residential development becomes more extensive farther south along Pebble Beach Drive near Crescent City.

During the El Niño winters of 1978 and 1983, storm waves and high tides combined to accelerate cliff retreat along the 2 miles of Pebble Beach Drive north of Crescent City. One home situated at the bluff edge was threatened in 1978, and an emergency permit was required for construction of additional foundation support, with an additional permit granted for riprap construction. Erosion has threatened the roadway at a number of locations during the storms of the past several years. Limited timber cribbing and riprap were emplaced in attempts to arrest erosion along Pebble Beach Drive and are now being repaired and rebuilt. In 1973–74, the Harbor

Figure 9.2 Area of extensive sand dunes between the mouth of the Smith River and Crescent City. Although this area was subdivided in the 1960s, development to date has been minimal. Photo © 2002–2004 Kenneth and Gabrielle Adelman, California Coastal Records Project, www.Californiacoastline.org.

District dredged 600,000 cubic yards of sediment from Crescent City Harbor and dumped it at the base of the bluff in an attempt to slow the recession of the beach and bluffs fronting the Seaside Hospital north of Battery Point. Much of the dredged material consisted of mud and silt, which, because of the small grain size, was quickly removed by wave action. Within two years approximately 80 percent of the sediment was removed, once again exposing the bluffs to wave attack.

Between Crescent City and the boundary of Redwood National Park, the shoreline is defined by an undeveloped beach and marsh area known as South Beach. Highway 1 lies immediately above the beach on a low bluff. Riprap has been emplaced to protect a motel in this area. Farther south the road turns inland, the coast becomes inaccessible, and the low terrace changes to the steep rocky cliffs of Redwood National Park (Figure 9.4). These cliffs extend almost continuously to the mouth of the Klamath River. Highway 1 intersects the coast at False Klamath Cove, which lies just above beach level and is completely armored with riprap. The coastline southward to the mouth of the Klamath River consists of steep, rocky cliffs with a few scattered rocky beaches.

The coastline from the Klamath River to Redwood Creek is protected in a combination of national and state parkland. The northern and southern

Figure 9.3   Low cliff along Pebble Beach Drive just north of Crescent City where riprap has been emplaced to protect the road. Photo © 2002–2004 Kenneth and Gabrielle Adelman, California Coastal Records Project, www.Californiacoastline.org.

Figure 9.4   The rugged eroding cliffs of Redwood National Park. Note the remains of an earlier roadway along the cliff face. Photo © 2002–2004 Kenneth and Gabrielle Adelman, California Coastal Records Project, www.Californiacoastline.org.

sections of this 17-mile stretch of coastline consist of steep, rugged coastal cliffs of the Franciscan Formation. Large slumps and slides are common, as are rocky outcrops on the beach and offshore sea stacks. Between these two rugged segments is an approximately 5-mile-long stretch of coast known as Gold Bluffs, which consists of a narrow coastal plain and a wide sandy beach

A. Coastline of Redwood National Park: steep rocky cliffs with scattered narrow rocky beaches.
B. Mouth of Klamath River with long sand spit extending north to rocky headland.
C. Park land with steep rocky cliffs and narrow rocky beaches.
D. Beach begins to widen as heavily forested coastal terrain becomes less steep. Large rock outcrops scattered along the beach and occasional seastacks offshore.
E. Gold Bluffs: Wide sandy beach with low active dunes backed by vertical cliffs. Road traverses coastline at base of cliffs.
F. Rocky vegetated cliffs fronted by beach of variable width. Rocky headlands and points serve as natural groins leading to sand accumulation and wider beach.
G. Redwood Creek: sand spit extends from the south such that the discharge point is typically against the rocky headland on the north side of the stream valley.
H. Wide sandy beach backed by low dunes and Freshwater Lagoon. State Highway 1 traverses sand spit separating Freshwater Lagoon from the ocean.
I. Freshwater Rocks and Sharp Point are resistant headlands that form natural groins trapping littoral sand forming wide beaches.
J. The headland between Dry Lagoon and Big Lagoon is a massive slump in the Franciscan Formation. Downslope movement has exceeded 9 feet per year during wet winters. In 1973 part of Highway 1 was abandoned and relocated inland at a cost of over $10 million (in 2003 dollars).
K. Homes at Big Lagoon have been built on a bench above a rapidly eroding bluff. Roads have been lost, homes have been relocated, and existing oceanfront development is at risk from future bluff erosion.

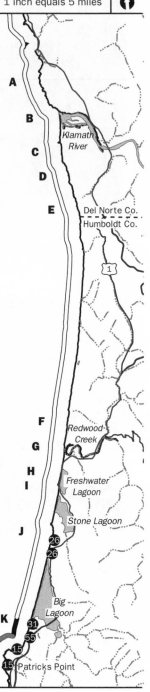

Hazard Level

☐ Stable: low risk
▨ Caution: moderate risk
■ Hazard: high risk

Erosion Rates

① Inches per year

Armoring

▲▲ Riprap
▨▨ Seawall
⊡⊡ Other

Figure 9.5   South end of Freshwater Lagoon, where Highway 101 turns inland at a resistant headland. Photo © 2002–2004 Kenneth and Gabrielle Adelman, California Coastal Records Project, www.Californiacoastline.org.

backed by vegetated sand dunes. The dunes end in a near-vertical cliff 100 to over 200 feet high consisting of weakly consolidated sedimentary rock.

South of Redwood Creek the coastline changes markedly as three large lagoons (Freshwater, Stone, and Big lagoons) form the dominant coastal landforms for the next 9 miles. A fourth lagoon, Dry Lagoon, was drained in the early 1900s to serve as farmland. These lagoons are fed by small streams and separated from the ocean by wide sand spits with low dunes (Figure 9.5). The sandbars impounding these lagoons are periodically breached, either by very large waves during extreme high tides or when the water level in the lagoons rises to 10 to 13 feet above mean sea level in the ocean. All of this coastal land is parkland to the southern end of Big Lagoon.

Highway 1 crosses the sand spit between Freshwater Lagoon and the ocean and then heads east to follow the inland edge of the two southern lagoons. The headland between Dry Lagoon and Big Lagoon is made up of a massive, slow-moving earth flow that is most active during periods of prolonged heavy rainfall. Movement has been recorded at rates as high as 8 feet per year during winters with high rainfall, such as in 1964, when over 30 inches of rain fell in this area over a six-day period. In 1973 a portion of Highway 101 in this area was abandoned and rerouted farther inland at a cost of over $10.4 million (in 2005 dollars).

Figure 9.6 Homes at Big Lagoon on an eroding sandy bluff where a number of structures have been removed. Note the loose material at the base of the bluff and the distance between the bluff edge and the homes. Photo © 2002–2004 Kenneth and Gabrielle Adelman, California Coastal Records Project, www.Californiacoastline.org.

The coastline again changes at the south end of Big Lagoon as a low bluff consisting of loosely consolidated, very erodible sands replaces the sand spit. The bluffs increase in height from 20 to 25 feet near the development at Big Lagoon (Figure 9.6) to several hundred feet at Agate Beach. The beach fronting these bluffs varies in width. Erosion of these loose bluffs can occur rapidly during particularly stormy winters when large waves and high tides combine to remove the beach sand.

The problems at Big Lagoon were readily identifiable and demonstrate the importance of conducting a geological analysis prior to initiating coastal construction. The protective beach at Big Lagoon has presented a deceptive impression of security, and in 1929 about 60 beach cottages were built, 15 of them within 30 to 40 feet of the bluff edge. During the winter of 1935–36 and again in the winter of 1939–40, unfavorable combinations of storm waves and high tides removed the entire beach. All of the most vulnerable cottages were gone by November 1941, probably relocated farther inland. Up to 60 feet of bluff were lost between 1931 and 1941. During the 1941–42 winter, the lack of a protective beach encouraged further bluff retreat of up to 30 feet by February 1942.

These disastrous winters were followed by a long period of generally benign conditions, from the mid-1940s to the late 1970s, now understood as a La Niña period, with less intense storm wave activity along the California coast. In 1962 lots for the Big Lagoon/Ocean View subdivision were laid out

A. Homes along Oceanview Drive are threatened by continuing retreat of the loose, sandy bluffs during severe wave attack. Very high bluff retreat rates.

B. Active slumping of cliffs from south end of Agate Beach to Mussel Rocks.

C. Rocky point with offshore rocks and sea stacks. Cliff failure common within the coves.

D. Coastline south of Patricks Point is rugged and rocky with numerous resistant headlands, offshore rocks and coves. Landslides are common and erosion rates vary widely.

E. Rugged rocky coastline with offshore rocks and sea stacks. A few scattered homes set well inland.

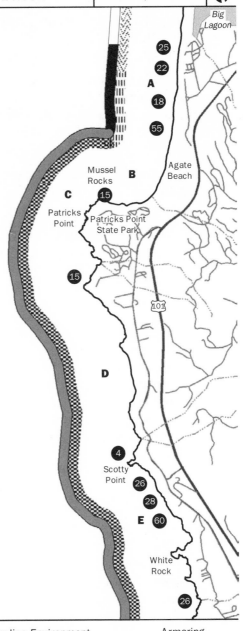

Hazard Level
- Stable: low risk
- Caution: moderate risk
- Hazard: high risk

Erosion Rates
- Inches per year

Shoreline Environment
- Sandy beach
- Beach backed by marsh
- Beach backed by dunes
- Cliff or bluff fronted by beach
- Rocky coast with no beach

Armoring
- ▲ ▲ ▲ Riprap
- Seawall
- Other

Figure 9.7   Wave run-up eroding the bluff at Big Lagoon during the El Niño winter of 1997–98. Photo by Tom Stephens.

along the top of the bluffs to the south of the original community (Figure 9.6). In subsequent years additional homes were built to the south along the bluff top on Oceanview Drive. Long-term average erosion rates of the bluffs south of the Big Lagoon development are very high, ranging from 1.5 to 4.5 feet per year.

With the beginning of a period of more intense and frequent El Niño events in 1978, erosion of the bluffs again increased. The 1982–83 winter waves cut back the cliffs up to 33 feet in another catastrophic erosion event. Two more cottages had to be moved and sand was stripped away below the cliffs bordering the new subdivision, threatening many new homes on the cliff top. Erosion continued during 1983–84, with up to 30 feet of additional erosion at the north end of the community. One cottage that sat over 88 feet from the bluff edge in 1931 was within 10 feet of the edge by 1985, an average retreat rate of 18 inches per year for over 50 years.

During the 1997–98 El Niño, 11 additional homes were threatened by accelerated bluff failure as a 50-foot stretch of one yard dropped into the ocean (Figure 9.7). Humboldt County tagged several structures as unsafe to occupy.

Although the beach fronting the bluffs will accrete and erode depending on wave and tidal conditions and sediment supply, the long-term trend is clear from the bluff retreat rates and the accumulation of talus at the base of the slope (Figure 9.6). The bluffs will continue to retreat and houses will

A. Trinidad Head is a tombolo and traps littoral drift from the north, forming an upcoast beach.

B. Between Trinidad Head and Moonstone Beach, the coastline is rocky with offshore rocks and sea stacks and small sand and gravel pocket beaches. The road that follows the coast has been repeatedly damaged by slope failures.

C. At the mouth of the Little River the coastline changes dramatically to the Humboldt Bay lowlands, a 33-mile-long area characterized by sand spits, dunes, marshes and estuaries. The northerly 2 miles of the sand spit is protected in parkland.

D. Very wide beach backed by broad dune field with Highway 101 behind the dunes at the base of the McKinleyville Bluffs.

E. Riprap protects road and base of bluffs landward of dunes. Sandy bluffs landward behind road subject to failure.

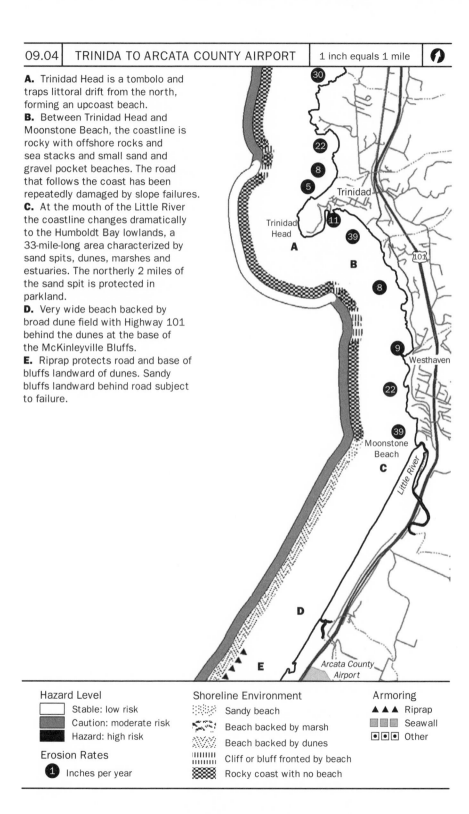

Hazard Level
- ☐ Stable: low risk
- ▨ Caution: moderate risk
- ■ Hazard: high risk

Erosion Rates
- ① Inches per year

Shoreline Environment
- Sandy beach
- Beach backed by marsh
- Beach backed by dunes
- Cliff or bluff fronted by beach
- Rocky coast with no beach

Armoring
- ▲▲▲ Riprap
- ▨▨▨ Seawall
- ⊡⊡⊡ Other

eventually have to be relocated or demolished. No additional building on oceanfront lots in this area should be permitted.

THE ROCKY CLIFFS FROM AGATE BEACH
TO TRINIDAD HEAD

The character of the 10-mile-long coastline from Agate Beach to Trinidad Head is defined by the chaotic Franciscan mélange and by sands and gravels. Mélange forms the rugged seacliff exposures between Patrick's Point and Trinidad Head, with numerous resistant blocks, ranging in size from boulders to Trinidad Head itself, floating in a muddy matrix. Attacking waves easily erode the relatively weak matrix, forming numerous embayments, such as College Cove, while leaving the isolated knockers and the more resistant rocks as headlands and sea stacks (Figure 9.8).

The coastal cliffs, with few exceptions, retreat by landsliding and slumping during the wet winter months. Infiltration of rainwater and runoff into the ground substantially weakens the slopes. Segments of the cliff edge break away in slumps and then move downslope, breaking up into a slowly moving mass. Winter waves quickly remove the material at the base of the cliff, keeping the slopes unstable and encouraging continued failure. Large earthquakes can also induce slope failure, and about 30 earthquakes of magnitude 6 or greater have affected north coastal California since 1871.

Wave heights in a typical winter storm season are likely to exceed 20 feet and can be much higher. As one extreme example, seawater from waves breaking against Trinidad Head in December 1914 reportedly rose to the lamp housing on the lighthouse, at an elevation of 196 feet above sea level.

Because of the variability in the resistance of the rocks exposed along the coastline, erosion rates change dramatically from point to point. A good example is the range in average annual erosion rates between Scotty Point and White Rock, which reflects the following factors:

1. Shortness of the time period used to determine the average annual erosion rates, in comparison with the frequency of major coastal slope failures. For example, the cliff top just north of White Rock retreated at an average rate of about 3.5 feet per year between 1870 and 1942, and was then stable from 1942 to 1974 during a relatively calm climatic period. Many cliff localities that appear to be stable now are simply waiting for heavy rainfall, seismic shaking, or wave and tidal conditions extreme enough to cause failure.

2. The effect of localized protection of slopes from waves by large offshore rocks or sea stacks.

3. The formation of deeply indented bays that are relatively protected from the dominant northwesterly waves, such as College Cove.

Figure 9.8   The small community of Trinidad is built on a low terrace with Trinidad Head, a resistant block of Franciscan rock, forming a headland on the right. Photo © 2002–2004 Kenneth and Gabrielle Adelman, California Coastal Records Project, www.Californiacoastline.org.

4. The difference in retreat rate between the top and bottom of coastal cliffs.

5. Temporary seaward displacement of the shoreline caused by slumped material.

CASE STUDY OF THE COASTAL ROAD BETWEEN TRINIDAD AND MOONSTONE

From Trinidad Head south to the mouth of the Little River and the edge of the Humboldt Bay lowlands, the coastline is low and rocky with only a few small, coarse-grained pocket beaches. A road crosses this coastal area with scattered single-family home developments. This road exemplifies the problems encountered when development ventures too close to an unstable coastline. The road was constructed in 1885 along the coastal slopes as a county wagon road. By 1908, however, the road was abandoned due to landslides, some of which were probably aggravated by the road construction, and a new road was opened farther inland. This road followed a longer route through Westhaven.

However, road engineers turned their attention back to the coastal route in 1922–23, and the California Highway Department built the present coastal road. This construction oversteepened the coastal slope in several places, resulting in renewed landsliding. In 1939, the slopes just south of Honda Cove were reconstructed to allow rebuilding of a section of the road that had been lost.

In 1956–58 the state constructed the present route of Highway 101, parallel to the coast but farther inland, and turned the troublesome coastal road over to the county. Since then the coastal road has required constant maintenance, and parts of it are only one lane. Just north of Tepona Point large gates were set up to allow a section of the coastal road notorious for failing to be closed off easily for repairs. Several houses bordering the road have also been damaged. During the wet 1982–83 winter, landslide movement increased and the road was again disrupted in several places. These sites were temporarily repaired, but at that time no attempt was made to restore the pavement because of excessive costs. Instead the county proposed that Caltrans (California Department of Transportation) resume responsibility for the road as a scenic alternative to Highway 101.

MOONSTONE TO CENTERVILLE BEACH:
THE HUMBOLDT BAY LOWLANDS

At Moonstone the rocky coast changes abruptly to long, linear beaches backed by dunes and low, sandy cliffs. These beaches and dunes extend 35 miles to Centerville Beach and are fed by the great volumes of sand brought to the coast by rivers in the area such as the Little, the Mad, the Van Duzen, and the Eel. The combination of high rainfall, unstable bedrock and erodible soils, steep slopes, and the effects of over a century of logging on vegetation cover, leads to very high sediment production from these coastal watersheds. The Eel River, for example, transports more sediment for each square mile of its drainage basin than any other river in the coterminous United States. Each year on average, the Eel River discharges about 4 million cubic yards of sand (about 400,000 dumptruck loads) to the coastline, much of which replenishes the beaches and dunes of this area.

CASE STUDY OF THE MCKINLEYVILLE CLIFFS

Between the Little and Mad rivers beach and sand dunes are backed by a line of low cliffs cut in relatively weak, loose sands. These cliffs give the appearance of stability because of the width of the beach and dunes. Many houses sit along the cliff top (Figure 9.9).

Wave action and the meandering of the Mad River mouth up and down the coast erode the cliffs, however. The mouths of many of the north coast rivers move around in response to river discharge, wave attack, littoral drift, and tidal conditions. Observations over the last several decades suggest that the normal northerly migration of the Mad River reverses and migrates

A. Homes constructed on bluff top. Extensive failure of sandy bluffs. Lagoon occupies back beach at base of bluff along former course of Mad River.

B. Mouth of Mad River with lagoon extending north from mouth along base of bluff. A broad sand spit separates river and ocean.

C. Broad sandy beach backed by wide expanse of vegetated and active sand dunes. Lowlying marshy area (Mad River Slough) landward of dunes.

D. Sand dunes widen and are higher proceeding southward. Lower vegetated dunes immediately landward of the beach. Higher active dunes further inland.

E. Highway 255 crosses the Mad River Slough and dunes and provides access to development on the spit at Samoa, Manila, and Fairhaven. Active dunes increase in elevation proceeding southward.

F. Housing developed in back dune area. Large pulp mill on bay side of spit.

G. High wave energy at this location has made construction and maintenance of the jetties and entrance to Humboldt Bay a challenge since 1888, when the first stabilization project was initiated. On average, 465,000 cubic yards are dredged each year to maintain the channel.

H. Wide beach backed by sand dunes and then Humboldt Bay.

I. Sandy beach backed by narrow dunes and low bluff fronting an uplifted marine terrace.

J. Sandy beach backed by vegetated dune field and slough/marshland. The Eel River commonly floods this area during the winter months.

K. Narrow sand spit with some dune development backed by tidal flat and slough of Eel River mouth. Shoreline position migrates in response to seasonal changes in wave conditions and sand input.

L. Eel River mouth marked by shifting sand bars and migrating channel.

M. Sandy beach backed by low vegetated dunes and marshy lowland.

N. Beach narrows, eroded dune scarp. Backed by extensive marsh area and some grazing land.

O. Beach backed by near vertical bluff undergoing massive recent slump failure. Head of slump could soon threaten Naval Oceanographic Station.

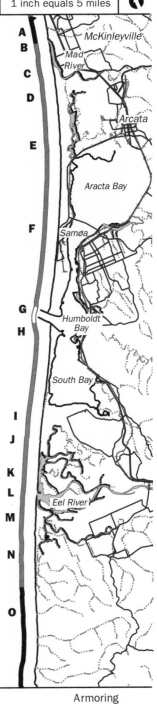

Hazard Level
- ☐ Stable: low risk
- ☐ Caution: moderate risk
- ■ Hazard: high risk

Erosion Rates
- ① Inches per year

Armoring
- ▲▲ Riprap
- ■■ Seawall
- ⊡⊡ Other

Figure 9.9   Homes built on bluffs at McKinleyville that have been threatened from time to time by the northward migration of the Mad River mouth. Photo © 2002–2004 Kenneth and Gabrielle Adelman, California Coastal Records Project, www.Californiacoastline.org.

south after El Niño years. Beginning about 1970 and continuing until migration was halted in 1991 by the emplacement of riprap, the mouth of the Mad River moved northward over 3 miles to Vista Point, west of the Arcata County Airport. A large area of high dunes fronted the entire length of the McKinleyville Bluffs, and the migration of the river progressively removed these 3 miles of dunes, allowing large waves to attack the base of the cliffs. Although approximately 4.4 million cubic yards of dune sand were added to area beaches by dune erosion, the protection offered to the bluffs by the dunes was eliminated. Two homes built on the cliff top in 1979 overlooked a protective beach and sand dunes almost 1,200 feet wide, yet by 1982 waves could reach the cliffs below the houses. Near Vista Point the dunes were cut back up to 575 feet during the 1981–82 and 1982–83 winters. Since 1983, the river mouth has moved about 3,000 feet to the south, and the beach and dunes are slowly being reestablished. However, these bluffs should be regarded with caution, including the area north of Vista Point, where frequent landsliding and gully erosion in the weak sands has occurred.

HUMBOLDT BAY COAST

Humboldt Bay lies at the southern end of a 33-mile-long lowland protected from the ocean by a wide, sandy beach and a large expanse of sand dunes

Figure 9.10 Humboldt Bay, showing jetties and extensive sand accumulation and dune development more pronounced on the north spit (1972). Humbodlt Bay Army Corps of Engineers.

(Figure 9.10). It is the only major harbor for more than 200 miles along the Pacific Coast. Before jetty construction began in 1888, a sandy shoal known as Breaker Flats extended across the seaward side of the 3,000–3,500-foot-wide natural inlet, forming a bar. Early surveys show that the Humboldt Bay entrance channel migrated from north to south over about a five-year cycle, and when it reached the southern extremity, the inlet would break through at the north end. Vessels were often trapped inside the harbor due to channel shifts. The high wave energy made jetty construction for entrance channel stabilization very difficult. Multiple and repeated efforts have been made to construct and maintain the jetties, which are 4,500 to 5,100 feet long, and the entrance channel is dredged regularly to maintain a shipping channel 40 feet deep. Waves as high as 30 feet battered the jetties during the winter of 1978, moving a 20-ton boulder 25 feet across the top of one jetty. Total damages to the jetties exceeded $1.5 million (in 2005 dollars). In the winter of 1983, significant shoreline erosion within Humboldt Bay took place from Fairhaven to the North Jetty, and a Coast Guard pier and moorage were also destroyed on the North Spit.

Sand brought to the shore by rivers in the area, the Eel being the most significant, feeds the beaches and sand dunes of the North Spit and South Spit that enclose Humboldt Bay. Despite the wide beach and massive quantities of sand stored on the beaches and in the dunes, extreme seasonal or cyclical changes in shoreline position still occur in response to winter storm waves and sand discharge from the major rivers. Sand discharge from the Eel River, for example, has ranged from a low of about 4,200 cubic yards in 1977 to 35 million cubic yards in 1965, with a long-term annual average of about 4 million cubic yards. Any proposed development in the dunes here or anywhere else in California must be informed by the long-term migration of the shoreline and then conditioned with the use of conservative buffer zones that would allow for temporary coastal retreat during periods of intense storm wave activity. South of the Humboldt Bay entrance, large winter waves may wash over portions of the spit north of the Eel River mouth.

Within Humboldt Bay, the coastline between the Elk River and King Salmon has eroded dramatically in historic time, with almost 1,400 feet lost at Buhne Point from 1854 to 1955. This erosion, which was largely caused by the effects of jetty construction and dredging at the entrance to Humboldt Bay, has now been stabilized by riprap. On average, 465,000 cubic yards of sand are dredged each year from the entrance channel. In the winter of 1978, the ground floors of about 20 homes in King Salmon were inundated with 12 to 20 inches of water during very high tides.

THE MOUNTAINOUS CAPE MENDOCINO COASTLINE FROM CENTERVILLE BEACH TO CAPE MENDOCINO

The sand bars, dunes, and wetlands that characterize the Humboldt Bay lowlands change markedly at Centerville Beach, several miles south of the Eel River mouth. For the next 10 miles the coastline is very rugged, nearly inaccessible, and characterized by steep, rocky cliffs with very little development. A few roads, a naval observatory, and isolated farm buildings are the extent of the development. Beginning directly below the bluff-top Naval Oceanography Station on Centerville Road, large slumps have disrupted the coastal slopes (Figure 9.11) and continue all the way to Cape Mendocino. Most of these slumps are very large, extending to the ridge tops and involving the entire seaward slope. Some of these massive slumps are very fresh, whereas others have been deeply dissected. It is a spectacular coastline but virtually impossible to access from land. A flight of low marine terraces extends a short dis-

Figure 9.11　Large, active slump at Centerville. A Naval Oceanographic Observatory is right above the edge of the slump behind the trees on the upper left of the photograph. Photo © 2002–2004 Kenneth and Gabrielle Adelman, California Coastal Records Project, www.Californiacoastline.org.

tance north and south of the mouth of Bear River, but the rest of the coastline is very steep. What beaches do exist are narrow and rocky.

CAPE MENDOCINO TO SHELTER COVE:
THE LOST COAST

Cape Mendocino is the westernmost point in California and consists of resistant rock of the Franciscan Formation. At this point the coastline changes orientation from north–south to northwest–southeast and a flight of low (less than 60 feet elevation) uplifted marine terraces forms the coastline for the next 7 miles. All of the low terraces are very young, having formed during the modern sea-level highstand of the past 10,000 years. The lowest of these terraces was uplifted above the water about 3 feet in 1992 during a magnitude 7.1 earthquake along the southern tip of the Cascadia subduction zone, just offshore. Narrow beaches front the low bluff eroded into the outer edge of the terrace complex, and farther to the south, sand dunes occur landward of the beach. The 30 miles of coastline from Cape Mendocino to Shelter Cove are completely undeveloped, and the southern half is preserved in the King Range National Conservation Area.

A. Narrow sandy beach backed by low bluff cut into outer edge of uplifted marine terrace. Mattole Road traverses the terrace.

B. Terrace narrows such that road is on edge of the low eroding bluff. Riprap and seawalls have been constructed continuously along an extensive stretch of bluff to protect road from wave attack.

C. Terrace widens, wide beach with sand dunes backed by moderately sloping grassy hills.

D. Steep rocky cliffs with large slumps and landslides fronted by a narrow rocky beach.

E. Steep rocky cliffs with large landslides and slumps.

F. Mattole River mouth with sand bar extending north toward Mattole Point.

G. Narrow rocky beach backed by steep cliffs subject to landslides and slumps.

H. South of Punta Gorda the beach widens and is backed by moderately sloping mountains.

I. The Lost Coast: very steep, rocky, failure-prone cliffs with no beach developed.

J. Narrow beach backed by low uplifted coastal terrace that widens to the south. Streams have deposited alluvial fans where they discharge from steep mountains.

K. Steep, rocky, slide-prone cliffs and very narrow rocky beach.

L. Beach is backed by low, wave-cut bluff at base of steep rocky cliff.

M. Near, vertical cliffs with offshore sea stacks and rocks, coarse pocket beaches, and some home development on cliff top.

N. High vertical cliffs with houses and an inn perched close to cliff edge.

O. Upper portion of cliff (loose terrace deposits) near end of runway has been armored to protect from wave overtopping.

P. Scattered offshore rocks, low rocky cliffs with homes along cliff edge.

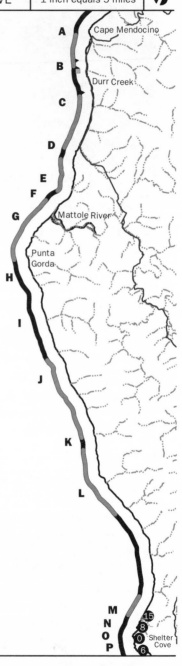

Hazard Level
- ☐ Stable: low risk
- ☐ Caution: moderate risk
- ■ Hazard: high risk

Erosion Rates
- ① Inches per year

Armoring
- ▲ ▲ Riprap
- ☐ ☐ Seawall
- ⊡ ⊡ Other

The Mattole Road, which cuts through the mountains just south of Cape Mendocino and traverses the low coastal terraces for about 7 miles before turning inland north of the Mattole River, serves as an excellent example of how some protection efforts are no match for attacking waves along this rugged, high-energy coast. Mattole Road was built as a wagon road in the 1860s, and eventually connected the town of Petrolia—which holds the distinction of being the site of the first oil strike in California—with towns in the Humboldt Bay area. By the early 1930s the road had eroded so severely that the Humboldt County Civilian Conservation Corps built log cribs, filled them, and replaced the road on top. Since 1935, storm-induced erosion has caused maintenance problems during several winters by exposing the log cribbing to direct wave attack. Once the timber cribs were damaged, the unconsolidated fill underlying the road was easily eroded. For many years the crib wall structures were repaired and replaced, often with riprap dumped at the base of the road. The destructive El Niño storms in 1978 undermined, damaged, or destroyed 5,700 feet of the road, completely closing it for 10 days.

In 1980, the preexisting wood cribbing was replaced by concrete cribbing at a cost of $10.4 million (in 2005 dollars). In only 12 days, however, waves damaged the structure to the extent that another $3.3 million was spent on repairs and additional riprap. Concrete cribbing and riprap now armor nearly the entire roadway where the marine terrace is narrow and the road is close to the shoreline (Figure 9.12). Further repair work has been required several times since then, and this area remains a maintenance headache for the Humboldt County Road Department.

At the narrow, southern end of Singley Flat, a broad complex of low marine terraces just south of Cape Mendocino, an active alluvial fan surface requires more attention than most other parts of the Mattole Road. The source of this fan is a steep slope rising about 1,000 feet above Singley Flat that is prone to nearly continuous landsliding (Figure 9.13). Debris flows from the slide complex move down along a small, ephemeral stream channel that often becomes clogged with coarse, muddy debris. Frequent shifting of this channel has built up a steeply sloping alluvial fan across Singley Flat. The Humboldt County Road Department has attempted to control this actively growing fan where it crosses the Mattole Road. Their strategy has been to create a larger channel in which to contain the debris flows, and to construct a bridge that spans the increasing width of the fan's active channel area. The bridge has required numerous repairs since the 1970s in response to particularly heavy storms, including those of the 1978, 1982–83, and 1997–98 El Niño years.

South of Mussel Rock the Mattole Road turns inland and the coast becomes relatively inaccessible for about 50 miles. Much of this area south

Figure 9.12 Concrete cribbing and riprap have been used to protect the Mattole Road. Photo © 2002–2004 Kenneth and Gabrielle Adelman, California Coastal Records Project, www.Californiacoastline.org.

Figure 9.13 A small, ephemeral stream draining a massive landslide on the steep slope at the rear of Singley Flat is producing an actively growing alluvial fan. Mud, rock, and water in the stream exceed channel capacity and spread onto the fan surface debris. Photo © 2002–2004 Kenneth and Gabrielle Adelman, California Coastal Records Project, www.Californiacoastline.org.

of Punta Gorda is appropriately known as the Lost Coast, for no one has ever attempted to maintain a permanent road here near the coast. South of the Mattole River the coastline for the next 50 miles or so is protected within the King Range National Conservation Area (60,000 acres) in the north in Humboldt County, and the Sinkyone Wilderness State Park (7,400 acres) in the south across the border in Mendocino County. Together, this protected land is the largest span of pristine shoreline on the Pacific Coast within the coterminous United States.

This rapidly uplifting coastline is very rugged and consists primarily of very steep, rocky cliffs that descend to the shoreline fronted by narrow, rocky beaches. A dozen of the King Range peaks exceed 2,000 feet in elevation, and the highest, Kings Peak, exceeds 4,000 feet. Scattered along the coast are numerous low, relatively narrow, uplifted marine terraces flanked by narrow beaches and low wave-cut bluffs. The terraces have local names, such as Spanish Flat and Big Flat. Steep coastal streams deposit bouldery alluvial fans on the surfaces of the marine terraces where they emerge from the steep mountain fronts. In 1998 the Bureau of Land Management, which manages the King Range National Conservation Area, closed the Black Sand Beach just north of Shelter Cove to off-road vehicle traffic. Although this closing will protect habitat along the beach, it is unlikely to affect rates of erosion, as the beach is remarkably dynamic as a result of long-term geological factors, including frequent earthquakes, steep slopes, and rapid uplift.

SHELTER COVE

Point Delgada is an anomalous flat terrace along the otherwise steep, rugged Lost Coast (Figure 9.14). At its southern end, the coast curves inland to produce a small, relatively sheltered cover, appropriately named Shelter Cove. The small town developed there has the same name. As one of the few places that offer protection between Bodega Bay and Humboldt Bay, Shelter Cove is an important harbor for small boats. The San Andreas Fault trends northwest along the Lost Coast but comes onshore and passes between Point Delgada and the rest of the Coast Ranges. During the 1906 "San Francisco" earthquake, the San Andreas Fault ruptured from the Santa Cruz Mountains south of San Francisco to at least as far north as Point Delgada. Geological and local reports from that time indicate that intense ground shaking during the main event and numerous aftershocks caused thousands of landslides in the Shelter Cove region. Many of the slides buried former beaches and the low marine terraces. Large blocks of rock still present along the beaches might well date to this event.

Figure 9.14 The community of Shelter Cove sits on an elevated marine terrace alongside the San Andreas Fault. (The fault runs along the base of the hills on the right side of the photograph.) Photo © 2002–2004 Kenneth and Gabrielle Adelman, California Coastal Records Project, www.CaliforniaCoastline.org.

Figure 9.15 Homes and a hotel have been built on the edge of the marine terrace at Shelter Cove adjacent to the seacliff. Photo © 2002–2004 Kenneth and Gabrielle Adelman, California Coastal Records Project, www.Californiacoastline.org.

Subdivision and development began at Point Delgada in 1965, and many of the oceanfront and cliff-top lots are now developed with homes as well as visitor facilities (Figure 9.15). The slopes of the seacliffs in the area are precipitous, and retreat is driven by both landsliding and winter wave attack. Several eroding cliff areas are protected by riprap and gunite. One area at the north end of the airport runway used gunite protection of the overlying terrace deposits, presumably weakened from winter waves overtopping the cliffs. In 1967 heavy rainfall caused about $5.2 million (in 2005 dollars) in damage to roads in the subdivision. Much of the damage was associated with the relatively slow movement of deep-seated rotational slumps that have formed along the San Andreas Fault zone. These slumps continue to move during relatively wet years, causing ubiquitous ground cracking that affects roads, pipelines, and other structures along the Point Delgada coast. For these reasons, at least 1,000 of the original 4,500 lots in the 1965 subdivision either are deemed undevelopable or no longer exist.

The low headland area on the west side of the San Andreas Fault that includes Point Delgada is made up of resistant rocks of the Franciscan Formation. These rocks also form a protective fringe of emergent reefs in the surf zone that limit coastal erosion on the western side of Point Delgada. Even in normal conditions, however, the waves around these rocky areas are hazardous; between 1972 and 1978, nine people were swept away by waves and drowned. The 1983 storms overtopped and damaged the breakwater at Shelter Cove.

CHAPTER TEN

POINT DELGADA TO POINT ARENA

DOROTHY MERRITTS, LAURET SAVOY, GARY GRIGGS, AND ROBERT WALKER

POINT DELGADA TO CAPE VIZCAINO

In contrast to the 5 miles of Black Sand Beach along the Lost Coast to the north of Point Delgada and Shelter Cove, the next 27 miles of coast to the south is one of the most inaccessible and rugged stretches of coastline in the state (Figure 10.1). Cliffs that form the coast are very steep and unstable, and existing beaches are narrow and usually rocky. Several parts of the coast are inaccessible at high tide, including the notorious Dead Man's Pass, just south of Shelter Cove, and Point No Pass, a mile farther to the south. The northern segment of this stretch of coast is protected within the King Range National Conservation Area, and the southern segment is part of the Sinkyone Wilderness State Park.

Established in 1977, the Sinkyone Wilderness State Park was expanded by 3,000 acres to the south after a contentious battle between conservationists, environmentalists, and the Georgia-Pacific Corporation logging company in the 1990s. Today the Sinkyone Wilderness State Park covers 7,400 acres and, together with the King Range National Conservation Area (60,000 acres) to the north, provides 40 miles of pristine beach and shoreline. In addition, another 4,000 acres of coastal land purchased from Georgia-Pacific Corporation became the nation's first intertribal park in 1996. Substantial efforts at ecological and watershed restoration are ongoing at the Sinkyone Intertribal Park. Prior to selling the land, however, Georgia-Pacific heavily logged the region in the 1980s and early 1990s, as it still held vast groves of virgin redwood, fir, and spruce. Soil erosion in the region since the time of

Figure 10.1   Rugged, inaccessible coastline north of Cape Vizcaino. Photo ©
2002–2004 Kenneth and Gabrielle Adelman, California Coastal Records Project,
www.Californiacoastline.org.

that logging has been extreme, with many of the waterways becoming
clogged with sediment and organic debris (including whole trees).

The mountains adjacent to the Shelter Cove–Sinkyone coast rise steeply
to elevations of over 2,500 feet. A small country road leads to Shelter Cove
from Highway 1, and a few, mostly unpaved, roads perilously skirt the high
coastal ridges between Shelter Cove and Hardy Creek, where Highway 1
emerges from the Coast Ranges to follow the coastline south. The entire
stretch is relatively uninhabited, with the exceptions of the small community at Shelter Cove and some caretakers at the Sinkyone Wilderness State
Park. Many of the steep, narrow roads following the coast north of Cape
Vizcaino were originally built over 100 years ago to haul redwood and fir
timber to the scattered mills and landings, and some have changed little
since then.

Uplifted marine terraces from 40 to 300 feet in elevation become the
dominant coastal landform about 4 miles south of Cape Vizcaino.
Highway 1, for example, traverses the coastline along a nearly flat terrace,
although a major road cut-and-fill project and riprap were required to
stabilize a portion of the highway at Hardy Creek (Figure 10.2). The first
sizable coastal communities south of the Humboldt Bay area, including
the older towns of Westport, Fort Bragg, and Mendocino, and newer

A. Very steep, high eroding cliffs with active slides and very narrow beach at base of cliff. Coastal areas inaccessible and part of King Range National Conservation area.

B. Very steep, high eroding cliffs with very narrow beach or no beach at base of cliff. Coastal area is part of Sinkyone Wilderness State Park.

C. Coastal terrace fronted with eroding bluff and rocky beach; dirt road traverses back edge of terrace.

D. Inaccessible, steep cliffed coast with active sliding. Narrow or non-existent beach.

E. Mouth of Usal Creek with sandy beach. Cliffs immediately downcoast fronted by beach.

F. Steep, actively eroding and failing cliffs fronted by narrow rocky beach. Highway 1 approaches within a mile of the coastline.

G. Low wave swept terrace.

H. Mouth of Cottaneva Creek with sandy beach.

I. Highway emerges at the coast and crosses a cut and fill slope above the beach protected by riprap. Steep slopes are actively eroding.

J. Very steep, high eroding cliff with Highway 1 traversing narrow bench. Very narrow rocky beach.

K. Marine terrace widens and gets lower to the south. Campgrounds along terrace.

L. Marine terrace traversed by Highway 1 and fronted by rocky cliff with narrow beaches.

M. Beaches at mouths of Sages and DeHaven Creeks. Cliff fronted by beach with offshore rocks and seastacks.

N. Community of Westport built on marine terrace. Most of homes are set back from cliffs; evidence of cliff failure.

O. Highway crosses steep cliff with eroding slopes.

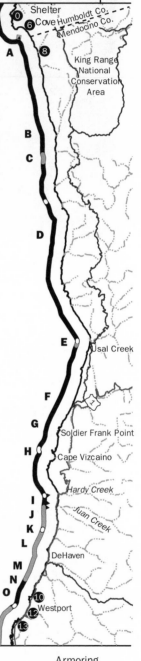

Hazard Level
- Stable: low risk
- Caution: moderate risk
- Hazard: high risk

Erosion Rates
- ❶ Inches per year

Armoring
- ▲ ▲ Riprap
- Seawall
- ▣ ▣ Other

Figure 10.2   State Highway 1 first emerges on the coast at Hardy Creek, south of Cape Vizcaino, where extensive grading and riprap was used to stabilize the roadway. Photo © 2002–2004 Kenneth and Gabrielle Adelman, California Coastal Records Project, www.Californiacoastline.org.

scattered subdivisions (e.g., Ocean Meadows) and developments have been built on this terrace.

Sandstone and shale of the Franciscan complex form most of the coastline between Shelter Cove and Point Arena. Unlike the Franciscan terrain north of Cape Mendocino, this coastline does not exhibit prominent knobby blocks or knockers forming innumerable sea stacks, although the coastline is still irregular and embayed. There are still offshore rocks and sea stacks that result from differing erosional resistance of the rock types exposed along the shoreline. The shale-rich rocks in this region are very prone to mass movement, particularly where shearing occurs along numerous northwest-trending faults along the coast that are part of the San Andreas Fault system (e.g., the Whale Gulch fault zone). The Anderson Cliffs, for example, are a mile-long span of nearly continuous landslides.

Many of the cliffs along this stretch of coast fail regularly through large-scale sliding and slumping, as well as through gullying from rainfall, runoff, and wave erosion at the base of the cliffs. Beaches are very narrow and rocky and provide little protection to the cliffs from wave attack, with the exception of the places backed by low marine terrace benches. The many northwest-trending points and coves reflect the variations in the resistance of the cliff

A. Highway 1 south of Chadbourne crosses steep cliff on cut with slope stability and stabilization problems.

B. Broad marine terrace with only scattered development, including some newer homes set back from edge of cliff. Cliff with irregular embayed shoreline and offshore rocks.

C. Group of homes built on Ten Mile River Terrace. Most are set back from cliff edge but several are adjacent to steep unvegetated cliff with only narrow cobble beach at base.

D. Wide sandy beach backed by extensive dune field.

E. Low terrace with dunes and scattered homes fronted by low bluff and sandy beach with offshore rocks and seastacks.

F. Low marine terrace fronted by rocky cliff and no beach. Coastal strip is protected within MacKericher State Park.

G. Fort Bragg is built on a broad marine terrace fronted by low rocky bluffs with intermittent pocket beaches. No development near bluff edge.

H. Mostly undeveloped subdivision on headland with new homes being constructed but with large setback from cliff edge. Some moderate erosion in 1982-83 winter.

I. Large new homes being built on top of rocky, deeply embayed cliffs with no protective beach.

J. New subdivisions and developments with large homes and other structures being built close to edges of deeply embayed rocky cliffed coast. Erosion rates are unknown.

K. Rugged rocky points and deep coves characterize the coastline. New homes being built on outer edge of terrace within small community of Caspar. Cliffs appear stable and resistant. Many new homes have been built in subdivision on south side of Caspar Anchorage on terrace. Narrow rocky beach. Erosion rate of cliffs unknown but homes to date have not encroached close to cliff edge.

L. Low resistant rocky cliff capped by weaker marine terrace deposits that have been eroded back by wave overtopping. Small development with houses built along edge of rocky cliffed embayment.

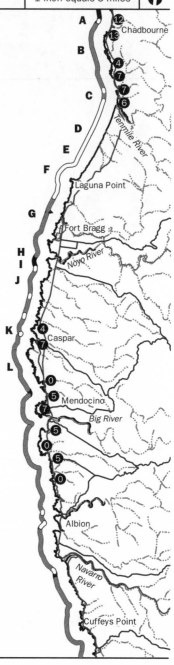

Hazard Level

- ☐ Stable: low risk
- ▨ Caution: moderate risk
- ■ Hazard: high risk

Erosion Rates

① Inches per year

Armoring

- ▲▲ Riprap
- ■■ Seawall
- ⊡⊡ Other

rock. Little erosion can be measured on many of the resistant headlands. With few beaches to act as buffers, waves attacking weaker zones in the cliffs have formed innumerable coves, sea caves, and natural bridges. Rockfalls commonly occur where wave action has undercut the base of the cliffs.

CAPE VIZCAINO TO TEN MILE RIVER

Between Cape Vizcaino and the Ten Mile River, the coastal terrace widens and contains one sizable community, Westport, and a few rural hamlets that were once productive mill towns and shipping points in the late 1800s. At the north end of Westport some homes have been built close to the edge of the coastal cliff. Limited home development has taken place on the outer terrace recently at Kibesillah, although well back from the cliff edge. A group of homes has also been built close to the cliff edge on the rocky headland known as Ten Mile River Bluff, just north of the mouth of the Ten Mile River. Many of these homes are built on narrow rocky points.

South of the Ten Mile River mouth, the Ocean Meadows subdivision has recently been developed on the coastal terrace seaward of Highway 1. Homes have been built on the outer edge of the terrace fronted by a vegetated bluff, a sandy beach, and offshore rocks and sea stacks (Figure 10.3). Highway 1 follows the shoreline along much of this reach, extending inland no more than a quarter mile, and along certain sections precariously skirts the cliff edge. Storm-related slumping of portions of the roadway has been common, particularly just south and north of Westport, where the road crosses steep cliffs. After a February storm in 2002, slumping 6 miles north of Westport required $550,000 of highway repairs and slide stabilization.

Along this coastline, beaches are few and exist only in small pockets between headlands, with the exception of a wide beach north of the Ten Mile River. The highest erosion rates occur in the weak rock in cove interiors and where small streams and gullies channel through the seacliffs. Prominent points such as Bell Point, and vegetated bluffs fronted by a large beach near the river mouth, have exhibited little change over the past 100 years. For a distance of nearly 4 miles south of the Ten Mile River a large field of sand dunes, up to nearly a mile wide in places, has formed along the terrace, most likely derived from the large volume of sand discharged by the Ten Mile River. Nearly all of the dune fields within MacKerricher State Park are protected.

FORT BRAGG AND VICINITY

Until recently, Fort Bragg was the largest of the few remaining mill towns that dot the north coast. The Georgia-Pacific lumber mill and a few

Figure 10.3 The Ocean Meadows subdivision immediately south of the mouth of Ten Mile River. The cliffs here appear to be relatively stable and are partially protected by offshore rocks and a seasonal beach. Photo © 2002–2004 Kenneth and Gabrielle Adelman, California Coastal Records Project, www.Californiacoastline.org.

municipal facilities still occupy most of the city's shoreline, but in 2003 the last remaining active mill shut down. Much of the shoreline north of Fort Bragg is contained in MacKerricher State Park. New coastal residences and strip malls are being built around Todd's Point at the southern edge of the city. This development extends about 2 miles farther south along the outer marine terrace, where the coastline is characterized by eroding coves and resistant headlands oriented northwesterly, resulting from rock weaknesses (Figure 10.4). The seacliffs are 40 to 50 feet high and drop vertically to a rocky shoreline.

Winter damage associated with storms and wave erosion is common in the Fort Bragg area. The only hard coastal protection structure on the Mendocino County coast is the breakwater at the mouth of Noyo Harbor, at the southern end of Fort Bragg. In 1978 winter waves removed a portion of a seawall below the lumber mill site and damaged part of the Noyo Harbor breakwater. In January 1983 waves caused minor damage to the harbor and slight bluff erosion at Todd's Point and the mouth of Pudding Creek. The U.S. Army Corps of Engineers has begun studying the feasibility of adding two more breakwaters to the cove entrance to better protect it from high-energy waves.

Figure 10.4 Development of new homes on the rocky and embayed coastline south of Fort Bragg. Some homes are very close to the cliff edge. Photo © 2002–2004 Kenneth and Gabrielle Adelman, California Coastal Records Project, www.Californiacoastline.org.

CASPAR TO POINT ARENA

The coastline is rugged along most of the reach from Caspar to Point Arena, with rocky cliffs 40 to 60 feet high dropping abruptly to the ocean at the terrace edge. Along this cliffed coastline, only a few small pocket beaches occur in the large coves. South of Fort Bragg streams emerge from the Coast Ranges through steep-walled canyons. Most of the coastal communities built along this reach were associated with the lumber business in the mid-1800s; the town of Mendocino is by far the largest, and many are now little more than hamlets. Homes and businesses in the town of Mendocino have been built well back from the cliff edge although newer homes north of the town in the Mendocino Headlands/Point of View subdivision have encroached on the cliff edge, where slope instability is evident (Figure 10.5). Much of the land is in agricultural use or falls within various state parks, beaches, and reserves (Mendocino Headlands, Russian Gulch, and Van Damme state parks; Caspar, Elk, and Manchester state beaches; and Jug Handle State Reserve).

Although the area is sparsely populated, many new subdivisions, such as Caspar Headlands, Mendocino Headlands, Sea Fair, Coast Highlands, and

A. Homes on outer edge of marine terrace, fronted by rocky cliffs with offshore rocks and seastacks.

B. Mendocino Headlands State Park and Point of View subdivision. Homes built on edge of steep cliffs with clear evidence of recent slumping. Historic town of Mendocino with older homes and commercial buildings set well back from cliff edge.

C. On east side of Mendocino Bay, homes and other structures built on outer edge of terrace close to cliff edge. Erosion rates are unknown but setbacks are small.

D. Very broad uplifted marine terrace fronted with low rocky beach and eroding terrace deposits.

E. Large homes built close to cliff edge on high rocky embayed coastline north and south of Van Damme Beach State Park.

F. Houses scattered along edge of near vertical rocky cliffs.

G. Houses and a restaurant built on outer edge of terrace. Steep eroding bluffs fall to narrow rocky beach.

H. Development of homes on Salmon Point on outer edge of marine terrace. Cliffs are steep and show evidence of active slumping.

I. Steep unstable and undeveloped cliffs at outer edge of marine terrace.

J. Homes on outer edge of narrow terrace fronted by steep, eroding cliff.

K. Highway follows narrow marine terrace with steep rocky unstable cliffs.

L. Small community of Elk developed on outer edge of marine terrace. Houses built along edge of steep rocky cliffs with scattered narrow rocky beaches.

M. Scattered homes on edge of steep, rocky cliffs.

N. Small pocket beaches fronting a steep, rocky seacliff with small landslides. Highway 1 runs along the flat marine terrace and skirts the cliff edge near Mallo Pass Creek.

O. Wide beach backed by active dunes and wetlands at Manchester Beach State Park.

P. Point Arena was considered as a site for a nuclear power plant in the 1970s but was eliminated due to its proximity to the San Andreas fault.

Hazard Level
- ☐ Stable: low risk
- ▨ Caution: moderate risk
- ■ Hazard: high risk

Erosion Rates
- ① Inches per year

Armoring
- ▲▲ Riprap
- ▨▨ Seawall
- ⊡⊡ Other

Figure 10.5 The Point of View subdivision at Mendocino Headlands, where newer homes have been built at the edge of an unstable cliff. Note the landslide scarps at the cliff edge on the right side of the photograph. Photo © 2002–2004 Kenneth and Gabrielle Adelman, California Coastal Records Project, www.Californiacoastline.org.

Irish Beach, are being developed at the terrace edge. Much of the coastline consists of nearly vertical rocky cliffs and offshore rocks and sea stacks. Seacliff erosion poses a long-term threat to structures situated too close to the cliff edge, particularly on weak rocks that occur in the cove interiors (Figure 10.5). Much of the bedrock itself appears to be somewhat resistant to erosion, but the overlying terrace deposits are much weaker and are susceptible to erosion from surface runoff and, where the terrace is lower, to wave overtopping.

South of Mendocino the coastline continues to be very irregular, with rocky headlands and sea stacks alternating with embayments or coves and rocky pocket beaches. Homes are scattered and clustered along the outer edge of the nearly continuous marine terrace. At some locations— north and south of Van Damme Beach, for example, and south of the Navarro River mouth—many homes are precariously perched above the waves on rocky headlands (Figure 10.6) or on the edges of steep, unstable seacliffs.

The shoreline between Alder Creek and Point Arena contains over 5 miles of low, sandy beach backed by extensive sand dunes and wetlands.

Figure 10.6 Home construction on a rocky point directly above the waves south of Van Damme State Park. Photo © 2002–2004 Kenneth and Gabrielle Adelman, California Coastal Records Project, www.Californiacoastline.org.

Much of this area is protected within Manchester State Beach. Similar to the lowland and dune fields at Ten Mile River, this dune field is associated with downwarping of the lowest marine terrace. The San Andreas Fault, which cuts across the coastline and heads seaward at the mouth of Alder Creek, is associated with the deformation of the terraces.

CHAPTER ELEVEN

POINT ARENA TO SAN FRANCISCO

LAURET SAVOY, DOROTHY MERRITTS, KAREN GROVE, AND ROBERT WALKER

The coast between Point Arena and San Francisco is the traditional home of many Native peoples, such as the Pomo and Miwok. In 1579, almost 40 years after the Cabrillo-Ferrelo voyage along the then-uncharted coast, the English pirate Francis Drake sailed along coastal northern California, his ship filled with booty seized in raids on Spanish ports and vessels along the west coast of the southern Americas. Drake stopped to make repairs in a convenient bay, which many historians believe is the bay named for him that lies in the lee of Point Reyes. In 1595 Sebastian Cermeno reached the coast of northern California in his return from a trade voyage to the Orient. His Manila galleon, laden with cargo for Spain, supposedly anchored in Drakes Bay for needed repairs. A violent winter storm drove the ship ashore, wrecking her completely. With the entire cargo and several crew members lost, the survivors sailed south in an open long boat and, after several months, reached Acapulco. Seven years later, Sebastian Vizcaino sailed to northern or Alta California in search of a sheltered harbor in which Spanish trade vessels could safely anchor. For a century and a half after this voyage, Spain's interest in the exploration of Alta California waned, until other countries, such as Russia and England, began extending their "holdings" to the western portion of the continent. In the mid-1760s, in the hope of maintaining its stake in California, Spain began establishing permanent settlements in much of the region.

In October 1775 Don Juan Francisco de la Bodega y Cuadra apparently found a sheltered anchorage in the lee of Bodega Head. In 1805 Russian traders moved south from Alaska in search of otter and fur seals. They established a village near the present town of Bodega Bay, and in 1812 opened a trading post/fort, which they named Ross (for Russia), 13 miles north of the Russian River. At that time the Russians referred to the river as the Slavianka, meaning Slav woman. The Russian American Fur Company, as it was called, prospered for nearly 30 years until the near depletion of the sea otter population. After the Russians departed, the Mexican government monopolized the coastal lands with land grants up to the Gualala River, until California obtained statehood in 1849.

Development of the region, particularly north of the Russian River, was shaped in the mid- to late 1800s primarily by the numerous lumber operations scattered along the coast. Most of the small coastal communities began as mill and/or schooner shipping points in coves or "dog holes," including Fisk Mill, Stewarts Point, Timber Cove, Iverson Landing, and Anchor Bay.

POINT ARENA TO FORT ROSS

PHYSICAL SETTING AND LAND USE

The 57-mile-long stretch of coastline from Point Arena to Fort Ross is quite rugged and scenic, with innumerable points, coves, and offshore rocks. Nearly vertical seacliffs, 20 to 100 feet high, are backed by a relatively flat plateau or marine terrace. Beaches are scattered and rocky. Much of this coastal area remains undeveloped and is used chiefly for grazing and agricultural purposes. Many communities, such as Stewarts Point and the town of Point Arena, were once productive lumber mill sites and shipping points.

Highway 1 follows the coastline along the entire reach, and housing developments range from individual lots flanking the highway to large-scale rural residential developments such as Sea Ranch. There are three state holdings along the coast: Kruse Rhododendron State Reserve, Salt Point State Park, and Fort Ross State Historic Park, all of which are in Sonoma County.

GEOLOGY

From Point Arena to Fort Ross, the coastline is part of the Gualala block of rocks that lies to the west of the San Andreas Fault. The San Andreas comes ashore from the south at Fort Ross and continues inland subparallel to the coast along the Gualala and Garcia rivers before going offshore again to the north at the mouth of Alder Creek (just north of Manchester Beach). The Gualala block consists predominantly of sedimentary rocks (conglomerate, sandstone, and shale) that formed in a marine shelf environment contemporaneous with deposition of the deeper-water sediments that now form the Franciscan complex of mélange rocks. At some locations, including

A. Steep, wave-eroded cliffs along edge of wide coastal terrace. Occasional small pocket beaches. Bluff erosion over the past several years has been noted, including that along the cliffs to the south of the lighthouse. Storm waves seriously damaged the Point Arena Harbor area in January, 1983.

B. Highway 1 is close to shore along this part of the coast, and has required extensive maintenance. At this location, the road was reduced to a single lane across a metal bridge as of 2002-2003.

C. Uplifted marine terrace with a few scattered homes on top of rocky cliffs with no fronting beach.

D. A few houses are built on the marine terrace, but they are set well back from the cliff edge. Offshore rocks and seastacks are prevalent in this area.

E. Campground threatened at beach level.

F. Scattered houses built on the edge of a steep cliff.

G. A small group of homes built on cliff top terrace. Town of Gualala is built on a rocky uplifted terrace.

H. North end of Sea Ranch development on coastal terrace.

I. Sea Ranch is an expensive development of homes along an uplifted, wide marine terrace.

J. Low, narrow, mostly undeveloped marine terrace fronted by sea cliff and scattered rock beaches.

K. Landslides and slumps along sea cliff.

L. Massive rock falls along sea cliff.

M. Steep rocky cliffs mark the edge of the terraced shoreline. Development is limited. Cliff erosion is prevalent where weak rocks form the cliffs, such as within coves. Houses and roads built too near the cliff edge may be endangered by erosion.

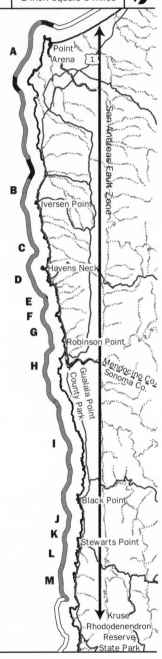

Hazard Level	Erosion Rates	Armoring
☐ Stable: low risk	**①** Inches per year	▲▲ Riprap
☐ Caution: moderate risk		■■ Seawall
■ Hazard: high risk		⊡⊡ Other

much of the coastal stretch from Point Arena to the town of Gualala, the Gualala block rocks consist of white to buff-colored shales and fine-grained sandstones that are highly deformed, producing dramatic cliffs of contorted rocks that resemble chalk from a distance.

EROSION

Sandstones, mudstones, and limited occurrences of volcanic rocks form the seacliffs along this coast. These cliffs rise vertically from the ocean, and few beaches exist, other than within coves, to protect the cliffs from attacking waves. These rocks have been folded and faulted, but usually are not as extensively deformed as the rocks of the Franciscan Formation to the north of Point Arena and to the south of Fort Ross. These cliffs tend to erode actively where the rocks have been weakened by cracks or faults. The many coves and embayments occur where such structurally weakened rocks form the seacliffs, whereas the more resistant rocks form the prominent points. An example of a cove with numerous rockfalls and landslides is Timber Cove, with a small commercial development on the low marine terrace.

Although information is limited, apparently few erosion problems exist along this rugged, undeveloped coast, and many of the rocky points have shown insignificant changes over the past century. Cliff retreat, primarily in the form of landslides and rockfalls, has been observed along the cliff line around the lighthouse at Point Arena, but the actual amount of land (around 21 acres) has been reduced little since the station was established on the broad marine terrace at this site in 1870. The present lighthouse keeper, working with local geologists to assess erosion, reports that the dominant erosional process is associated with large blowholes (similar to sinkholes in appearance) and small caves formed by wave action below cliff overhangs (Figure 11.1). Air is forced up as waves pound the steep, rocky seacliffs, resulting in blowholes at the terrace surface.

Coastal retreat also occurs where sections of Highway 1 and a few residential structures are situated at the bluff edge in coastal embayments. The winter storms of 1983 did not cause significant erosion, but waves were large enough to destroy one-third of the Arena Cove pier and flood several buildings there, temporarily trapping patrons in a wharf cafe. South of Point Arena, near Schooner Gulch, cliff retreat during the 1990s cut into the Highway 1 roadbed. A metal bridge has been added as a temporary measure until the road can be realigned and stabilized (Figure 11.2). Although waves destroyed all of the beach access stairways at Sea Ranch and locally eroded the bluffs, no homes there were threatened.

Figure 11.1 The Point Arena lighthouse, showing a large collapse area to the left, where a blowhole has collapsed. Photo © 2002–2004 Kenneth and Gabrielle Adelman, California Coastal Records Project, www.Californiacoastline.org.

Figure 11.2 Coastal cliff failure south of Schooner Gulch, where a metal bridge structure has been emplaced to convey State Highway 1. Photo © 2002–2004 Kenneth and Gabrielle Adelman, California Coastal Records Project, www.Californiacoastline.org.

FORT ROSS TO BODEGA HEAD

PHYSICAL SETTING AND LAND USE

The coastline between Fort Ross and the Russian River is more rugged than the adjacent stretch to the north, with precipitous, slide-prone cliffs over 1,000 feet in elevation plunging to the sea. Most of this area is undeveloped and privately owned, and the steepness of the slopes prevents access to the shoreline. Southward from the Russian River to Bodega Head, a distance of 11 miles, the coastline is less mountainous and is characterized by low marine terraces, but still is rugged and scenic.

Land use along the Sonoma Coast is primarily recreational, as most of this shoreline is state beach. The Sonoma Coast state beaches, including several small beaches such as Shell Beach, Wright Beach, and Portuguese Beach, line the shore, and the more extensive Salmon Creek Beach lies to the north of Bodega Head. Two small bluff-edge subdivisions are located at Ocean View near Duncan's Point and along Gleason Beach. One large ranch holding still exists just north of Duncan's Point and extends nearly all the way to Goat Rock. Small pockets of ranch land even farther to the north, to Fort Ross, are still used for grazing, and cattle sometimes are seen along Highway 1.

GEOLOGY

The entire coastal strip lies in close proximity to the San Andreas Fault. Just south of Fort Ross, near Timber Gulch, the San Andreas Fault goes offshore, but it clips the coastline at Bodega Head. Bodega Head lies to the west of the fault, on the Pacific Plate. During the 1906 San Francisco earthquake, Bodega Head moved 15 feet to the north with respect to the mainland. Bodega Harbor and Bodega Bay are lowlands that lie along the fault trace. Bodega Head was the proposed site for a large nuclear power plant in the mid-1960s. The project was terminated when foundation excavation revealed a branch of the San Andreas Fault passing through the construction site. The fault again clips the coastline at Point Reyes and forms Tomales Bay, now a submerged lowland along the fault trace. Any development in these areas should take the possibility of damage from a major earthquake into careful consideration.

Franciscan rocks, including some that are highly metamorphosed, form the coast from south of Fort Ross to near Bodega Head. Where the fault crosses the coast near Fort Ross, the rock types and topography change abruptly. Sandstones and mudstones are found to the north and west of the fault. To the south and east, grassy yet steep slopes disrupted by numerous landslides characteristic of the inherently unstable Franciscan Formation occur. Any plans for development in this region should also take into consideration the possibility of strong seismic shaking and surface rupture during a large earthquake.

At Bodega Head, uplifted rocks west of the San Andreas Fault juxtapose granitic rocks to the west with Franciscan rocks to the east. The Bodega Head granite is the northernmost exposure of the Sierran granite that lies west of the fault. Most of the bedrock on Bodega Head is overlain by unconsolidated sand and mud deposited in sand dunes, streams, bays, and beaches.

EROSION AND PROTECTION

Landsliding within the Franciscan zone is quite hazardous to coastal development, with slides varying in both magnitude and speed; some creep slowly, and others fail rapidly. Many slumps occurred during storms in 1983, 1992, 1997–98, and 2002. Where the San Andreas Fault goes offshore just south of Fort Ross, deep-seated rotational slumps in the highly sheared rocks extend from the drainage divide to the coastline. Highway 1 crosses these slumps at several localities.

One of the most challenging parts of Highway 1 to maintain is the stretch between Fort Ross and the mouth of the Russian River, at the town of Jenner. Called the Jenner Grade, the highway climbs to the drainage divide in order to avoid the extensive, deep-seated slide mass. In January 2002 Caltrans initiated a roadway project about 2.8 miles south of Fort Ross in response to the threat of roadway loss after significant storm damage and slope failure during the 1990s. Initially, $6.8 million was allocated to grade the slope and construct a soil-nail retaining wall. During construction, the scope of the failure was discovered to be much larger than first recognized, and the nature of the soil material required more work than anticipated. An additional $5.5 million was devoted to the project in October 2002. Finally, in February 2003, an additional $1.95 million was allocated to complete the project and to provide additional rock slope protection along the coast. The total cost to maintain just a short stretch of this vital two-lane link between the Mendocino and Sonoma county coasts was more than $14 million. Farther south, 5 miles south of Fort Ross, about $1.2 million was spent to repair Highway 1 following storm damage from the 1997–98 El Niño winter.

Seacliff retreat threatens several homes south of Duncan's Point at the Gleason Beach development. Although most of the cliff-top homes at the Gleason Beach subdivision are protected by concrete seawalls of various designs, erosion of the bluff face is common (Figure 5.3). At this particular location, due to the cliff-top location and lack of flat land on each parcel, septic tanks have been placed at the base of the bluff behind the seawalls. Several houses either have been destroyed or have suffered serious founda-

Figure 11.3   Continuing bluff failure at the Gleason Beach development has destroyed and damaged many homes. Septic tanks in this development were built in perched fill behind seawalls on the back of the beach. Photo by Gary Griggs and Kiki Patsch.

tion problems as unstable slopes have failed from rain and groundwater seepage (Figure 11.3). The neighboring Pacific View development did not receive much damage because the bluff is fronted by a wide protective beach. After cliff erosion from storm damage in the 1990s, FEMA and the California State Office of Emergency Services allocated $3.4 million to "hard-hit" Sonoma County residents for recovery from landslide and flood damage.

BODEGA HEAD TO POINT REYES

PHYSICAL SETTING AND LAND USE

The coast between Bodega Head and Point Reyes is less mountainous than the stretch farther to the north, but is still beautiful, with its rugged cliffs and numerous offshore rocks. Sonoma Coast state beaches, including Shell Beach, Wright Beach, and Salmon Creek Beach, line the shore. Land use is primarily recreational, and development is restricted to a small residential community at Dillon Beach in Marin County, just north of Tomales Bay. Homes situated on the cliff edge at Dillon Beach are threatened by erosion. The rugged coastline southward to the Bolinas Peninsula is part of the Point Reyes National Seashore.

The Point Reyes Peninsula extends south from Tomales Bay to Bolinas Lagoon. The peninsula has steep, rugged cliffs along its northern and southern

A. Rugged shoreline with rocky cliffs, scattered pocket beaches, and innumerable sea stacks. Development is limited. Structures and road situated near the cliff may be endangered by cliff erosion, particularly where weak, fractured rocks form the cliffs. These weak areas may erode by creep, landslides, or rockfalls.

B. In January 1983, a house at Timber Cove was damaged by high waves. Houses built along cliff edge around Ocean Cove.

C. The active trace of the San Andreas fault comes onshore just north of Timber Gulch and passes beneath Highway 1. The fault continues offshore for 75 miles southward to Salmon Creek, near Bodega Bay. At this locality, ~12 feet of offset occurred across the fault in 1906.

D. From the mouth of Russian Gulch northward to Timber Gulch, the coast is characterized by a steep, deeply dissected slope formed on highly sheared rocks along the eastern side of the San Andreas fault. Substantial road work is done every year along the stretch of Highway 1 between Russian and Timber gulches.

E. Wide beach at the mouth of the Russian River. Sand bar connecting bedrock island to mainland is protected with riprap.

F. Gleason Beach homes built on unstable cliff seaward of Highway 1. Several homes have collapsed and others are threatened by continuing erosion. A variety of seawalls were constructed in an attempt to protect these homes.

G. Mouth of Salmon Creek. The San Andreas fault rift zone comes onshore near the mouth of Salmon Creek.

H. Bodega Head was the proposed site for a nuclear power plant in 1964 that was abandoned because of proximity to the San Andreas fault.

I. Two jetties protect the inlet to Bodega Bay, an important harbor north of San Francisco.

J. Rugged seacliffs with numerous coves, pocket beaches, and offshore rocks. Homes along the cliff edge at Dillon Beach are threatened by erosion.

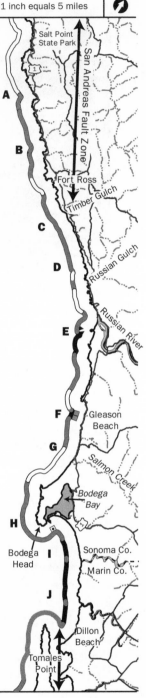

Hazard Level
◻ Stable: low risk
▨ Caution: moderate risk
■ Hazard: high risk

Erosion Rates
1 Inches per year

Armoring
▲▲ Riprap
▨▨ Seawall
⊡⊡ Other

parts, where Inverness Ridge lies next to the sea, and along the western edge at Point Reyes. Between Inverness Ridge and Point Reyes is a low-lying area with extensive sand dunes on the northern coast and shallow coastal lagoons (Drakes and Limantour esteros) on the southern coast. Most of the peninsula is included in the Point Reyes National Seashore and is protected from development. Historic dairy farms are part of the cultural landscape and are still active within the park. The southernmost end of the peninsula remains in private ownership and is developed with houses and small businesses that are part of the Bolinas village.

GEOLOGY

The Point Reyes Peninsula lies entirely west of the San Andreas Fault, an active transform fault for the last 20 million years that, therefore, separates distinctly different rocks. The peninsula consists of granitic and metamorphic rocks, which are offset parts of the southern Sierra Nevada, overlain by a thick sequence of marine sedimentary rocks that are offset from a similar sequence in the Santa Cruz Mountains on the east side of the fault. The topography of the peninsula, with the steep slopes of Inverness Ridge and Point Reyes along its eastern and western edges, and a broad gentle lowland in between, reflects the underlying structure of a land that has been compressed and crumpled into a large U shape. Flat surfaces, which are terraces formed at sea level and subsequently uplifted, tell the story of the peninsula's recent emergence from the sea.

The San Andreas Fault zone, along the eastern boundary of the peninsula, is over a mile wide and extends from Tomales Bay through Olema Valley to Bolinas Lagoon. The largest amount of horizontal offset (over 16 feet) during the great 1906 San Francisco earthquake was measured in the Olema Valley. The rocks in the fault zone are slices of Franciscan complex types overlain by a variety of unconsolidated stream and bay deposits. East of the fault zone are diverse rock types of the Franciscan complex.

POINT REYES TO SAN FRANCISCO

PHYSICAL SETTING

South of Bolinas Lagoon the coastal cliffs are extremely steep and rugged. Landslides are common along this coast. In 1989 shaking during the Loma Prieta earthquake caused slumping of the Lone Tree Landslide and closure of State Highway 1. The highway was reopened two years later after extensive grading and relocation to a position farther from the cliff edge. The steep shoreline is interrupted by several small valleys that create points of access at Muir Beach and Rodeo Lagoon/Cronkite Beach. Most of this coast

A. Tomales Bay is a long, narrow embayment that follows the San Andreas fault.

B. Steep cliffs with large scale landsliding.

C. Steep cliffs fronted by a narrow rocky reef or platform. Several landslides have occurred along cliff face and road near cliff edge has been partially removed by erosion.

D. Steep cliffs fronted by narrow beach. Several landslides have occured along cliff face. Portions of Ocean Parkway have been destroyed, and many structures have been either damaged or threatened by erosion and high wave conditions.

E. Sandy beach and low sand dunes. Homes at Stinson Beach are built on a sand spit separating Bolinas Lagoon from the ocean. Many homes have been either threatened, damaged, or destroyed during large storms such as the winter of 1983.

F. Mountainous coastline with highway cut into slope. Scattered houses perched on ridge are threatened during large storms. Occasional landslides on the steep slopes.

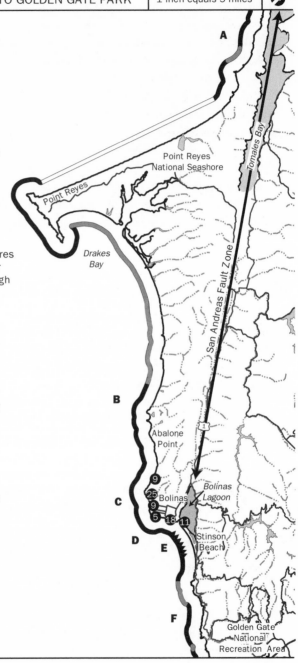

Hazard Level
- ☐ Stable: low risk
- ▨ Caution: moderate risk
- ■ Hazard: high risk

Erosion Rates
- ❶ Inches per year

Armoring
- ▲▲ Riprap
- ▨▨ Seawall
- ⊡⊡ Other

is part of the Golden Gate National Recreation Area and is protected from development.

GEOLOGY

Various rock types of the Franciscan complex, including shale and serpentine, are exposed along the coastline and are susceptible to frequent landsliding. The main strand of the San Andreas Fault goes through Bolinas Lagoon and last moved in the great San Francisco earthquake of 1906. The western boundary fault (western edge of the San Andreas Fault zone) goes through Bolinas and separates mudstone on the west from Merced Formation sand and mud in the fault zone. The Merced Formation consists of slices that have been separated from parent materials located along the coast south of San Francisco and east of the fault. A 5.0 magnitude earthquake on the western-boundary fault in 1999 was the first known historical movement on that fault.

BOLINAS—CASE STUDY OF AN UNPLANNED DEVELOPMENT

Located approximately 25 miles north of San Francisco, the Bolinas Peninsula forms the southernmost extension of the larger Point Reyes Peninsula. Duxbury Point represents the southwestern tip of the Bolinas Peninsula. The Duxbury reef extends south from the point and disperses waves entering Bolinas Bay from the northerly through westerly directions. The bay, like other crescent-shaped bays along the coast, such as Half Moon Bay, is exposed to severe southerly winter storms. Most of Bolinas Peninsula is a broad, flat plateau or terrace that is bordered by steep seacliffs 140 to 200 feet high.

Bolinas Beach, which fronts the southeast-facing bluffs of the peninsula, has been a major recreational area in Marin County for over 100 years. Originally, Bolinas and neighboring Stinson Beach were part of the Baulenes Rancho land grant, which was first settled and used as a cattle ranch in the 1830s. In 1880 the first summer home subdivision was developed in what is now the town of Bolinas. Since that time seacliff erosion has been a never-ending hazard.

EROSION AND PROTECTION

Development is limited along the area north of Duxbury Point to residences along the bluffs near the point. Farther to the north the mesa is primarily open space, with the exception of the site of an old RCA transmitting station currently leased by the Commonweal Association. The entire seacliff appears to be a continuous zone of landslides. Resistant, wave-truncated outcrops of shale at beach level create a narrow, shingle-covered tidal platform. Duxbury Reef is a partially submerged offshore extension of this rock platform. As is common throughout the area, most retreat occurs during the wet winter months. Groundwater seeps can be seen along the base of the seacliff during winter and spring. Most of this water originates from rainfall percolating through the fractured bedrock. Heavy rains also tend to cause gullying on the cliff face.

An RCA engineer recorded the relative positions of the terrace edge and cliff bottom on a base map every few years from 1913 until 1950. A survey of the present position of the seacliff indicates that along the few sections where the mudstone is less disrupted and supports more vegetation, the edge has receded an average of 0.4 to 7 inches per year since 1913. This average rate increases southward to over 36 inches per year where the mudstone is more fractured and less competent. Between Bolinas and Duxbury points, the narrow, wave-cut platform and beach are often inundated by waves. Along this west-facing shore, attacking waves cause landslides by undercutting the base of the cliffs. Failure is facilitated by increased subsurface flow of water and saturation due to septic effluent from cliff-top homes as well as winter rainfall. Along this reach, the average rate of recession along the cliff base ranges from 6 to 24 inches per year.

Numerous homes line the cliff edge directly to the northeast of Duxbury Point. Since the area was initially subdivided in 1927, many of these oceanfront lots and the fronting Ocean Parkway have been either partially removed or damaged by cliff erosion (Figure 11.4). Between Duxbury Point and Terrace Avenue, the mudstone is deeply weathered and exhibits extensive fracturing and shearing, particularly near the area where the San Andreas Fault slices through the bluff. The instability of the cliff is evident in the numerous landslides. Seeps commonly occur along the seacliff base during the winter and spring. This water probably originates from rainfall and septic system leakage, as residences on Ocean Parkway and Terrace Avenue rely on individual septic tanks.

The combination of inherent bedrock weakness, groundwater seepage, and wave attack leads to rapid cliff recession. Seacliff failure occurs in a variety of forms, from flows of loose earth to massive landsliding. East of Duxbury Point, landslides have been extensive and destructive enough to damage portions of Ocean Parkway and remove several homes from their foundations.

The cliffs just west of the entrance to Bolinas Lagoon support the most densely developed residential area in Bolinas, with 20 houses located on the brow of the cliff, some of which are precariously perched on the face (Figure 1.2). Fronting these bluffs is the most heavily used stretch of beach. With

Figure 11.4 Extensive landsliding along the coastal bluffs in the Bolinas area has destroyed portions of Ocean Parkway. Photo by Gary Griggs and Kiki Patsch.

each winter the position of the homes, road, and utilities located at the terrace edge becomes increasingly dangerous. This is not a new problem; the hazards resulting from wave attack and erosion have threatened Bolinas for over 100 years.

The materials forming the seacliffs are very erodible and susceptible to failure, particularly during winter months. Cliff retreat commonly has occurred in a variety of forms, from debris flows to large rock falls. The first bulkheads and groins were emplaced in the 1880s beneath the cliffs backing the beach, in an attempt to halt cliff erosion and stabilize the beach. These early wooden structures were regularly maintained by the community for many years, and provided a beach of sufficient dimensions to shield the cliff base from wave attack. Permanent and seasonal dwellings were constructed on the beach during this time. During the harsh winter of 1912–13, storm waves destroyed many of the protective structures in addition to the homes. The groins and seawalls were subsequently rebuilt and well maintained until the 1930s. By 1942, however, severe storms had damaged the structures to the point of uselessness. Within three years most of the beach sand had been removed, exposing bedrock. In 1947 the community constructed a reinforced

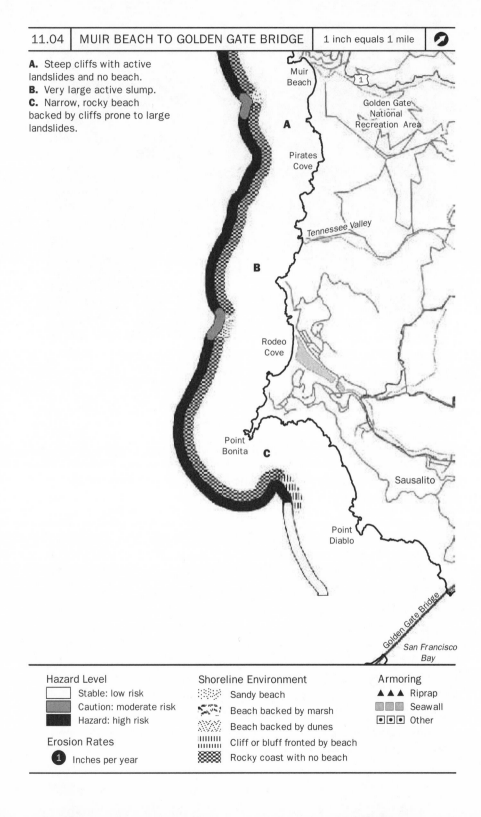

A. Steep cliffs with active landslides and no beach.
B. Very large active slump.
C. Narrow, rocky beach backed by cliffs prone to large landslides.

Muir Beach

Golden Gate National Recreation Area

A

Pirates Cove

Tennessee Valley

B

Rodeo Cove

Point Bonita   C

Sausalito

Point Diablo

Golden Gate Bridge

San Francisco Bay

Hazard Level
- Stable: low risk
- Caution: moderate risk
- Hazard: high risk

Erosion Rates
1 Inches per year

Shoreline Environment
- Sandy beach
- Beach backed by marsh
- Beach backed by dunes
- Cliff or bluff fronted by beach
- Rocky coast with no beach

Armoring
- ▲▲▲ Riprap
- Seawall
- Other

concrete groin near the mouth of Bolinas Lagoon, and within a year the beach sand was restored. The groin functioned well until the storm waves of the past several winters severely damaged it and substantially decreased its effectiveness. The few remaining bulkheads are damaged and scattered. Although a few individual property owners have placed riprap and other protective structures below their cliff property, much of the bluff remains unprotected. Groin failure and the lack of protection at the cliff base have resulted in increasing damage to structures on the terrace edge every winter.

During the prolonged and intense rains of the January 1982 storms, significant cliff failure occurred. Four homes along Terrace Avenue above Bolinas Beach were irreparably damaged and subsequently condemned. With the winter wave attack of 1983, one cliff-top home was completely destroyed and several others suffered damage (Figure 1.2). The effects of cliff erosion have been nothing short of disastrous in this community. Had the geological conditions been evaluated prior to development, much of the erosion "problem" could have been forestalled through better planning and setbacks.

Erosion has proceeded at an average rate of 6 to 20 inches per year since 1939. An 1882 plot of the Terrace Avenue area shows the road at one location as situated 88 to 127 feet inland of the cliff edge; the same road is now at the edge of the cliff.

STINSON BEACH

Stinson Beach lies on a narrow sand beach southeast of Bolinas Lagoon. The beach is exposed to the full brunt of winter southerlies, and it is no surprise that waves commonly inundate the beach and threaten the homes built on the low spit.

Originally subdivided in 1906, the spit at Stinson Beach contains two separate developments: Stinson Beach Village, at its eastern terminus, and the newer, exclusive Seadrift subdivision, closer to the western tip. Houses in both developments are situated either directly on the beach or on the sensitive dunes behind the beach.

The result of either poor planning or a lack of appreciation of the dynamics of a beach/sand spit can be seen in the damaging and costly effects of storms over the last several decades. Storm waves and high tides during the 1977–78 winter endangered nine beachfront homes at Seadrift as 10 to 90 feet of protective sand dunes were eroded. After first attempting to protect the beach from further erosion with sandbags, residents installed 6,000 feet of Longard tubing (polyethylene tubes filled with sand slurry) along the beach. In time this tubing was undermined and torn by debris. Subsequently, riprap

Figure 11.5   Wave inundation of homes along the sand spit northwest of the community of Stinson Beach in 1983 led to the emplacement of a continuous riprap revetment. Photo © 2002–2004 Kenneth and Gabrielle Adelman, California Coastal Records Project, www.Californiacoastline.org.

Figure 11.6   Low-lying homes built on the sand at Stinson Beach were damaged by wave impact and inundation during the 1983 El Niño event. Photo by Gary Griggs and Kiki Patsch.

was emplaced at significant cost (Figure 11.5). Total protection and repair costs exceeded $525,000 (in 2005 dollars), much of this at public expense. The county assessor estimated that each of the oceanfront lots lost approximately $147,000 (in 2005 dollars) in market value at that time.

The storms of 1982–83 were particularly disastrous to Stinson Beach Village. One home was completely washed out to sea, approximately 15 homes were destroyed, and over 50 more houses were damaged (Figure 11.6). Estimated damage to these homes approached $1.9 million (in 2005 dollars). Stinson Beach homes again suffered damage in the El Niño year of 1997–98. The lesson learned here is painfully clear. Before any construction on vulnerable beach and dune areas is allowed, the hazards of erosion and wave overwash, as well as the long-term history of the site, should be carefully considered.

CHAPTER TWELVE

THE SAN FRANCISCO COASTLINE

GARY GRIGGS, KIM FULTON-BENNETT, AND LAURET SAVOY

San Francisco Bay and Fisherman's Wharf, rather than the city's ocean coastline, typically come to mind when one thinks of the San Francisco waterfront. The exposed Pacific coastline extends for 8 miles from Golden Gate Bridge to the San Mateo County line. Steep, frequently crumbling cliffs fronted by narrow beaches characterize the shore between the Golden Gate and Seal Rocks. Most of this area is federal or state property (military and park facilities), with the only private development occurring along a short stretch of coast where expensive homes line the edge of the cliffs above Bakers and Phelan state beaches. The long, sandy Ocean and Fort Funston beaches, which are part of Golden Gate National Recreation Area, lie to the south of Seal Rocks. The Esplanade, which is protected by an extensive seawall, and the Great Highway run along the shoreline behind the beach.

HISTORY AND DEVELOPMENT

For over 200 years after European explorers first sailed along the northern California coast, San Francisco Bay and the peninsula remained unknown to them, but not to the Native peoples who continued to live around the bay. In 1769, an overland expedition led by Don Gaspar de Portola accidentally reached the bay and peninsula while searching for Monterey Bay. During the next several years the bay area was further explored, and Captain Juan Bautista de Anza chose sites for the mission and presidio on the peninsula in 1776. The original mission was named San Francisco de Asis for the patron of the Franciscan order, and the presidio served as the northernmost such post in Alta California until it fell into disrepair. These remained the only major settlements on the peninsula until the 1830s.

By the early nineteenth century, with the growth of maritime trade in the Pacific, Yerba Buena Cove on the bayside of the peninsula had become a major port of call for ships from various countries, including Russia and the United States. The town of Yerba Buena grew around this bayside port, and in 1847, after the Bear Flag Revolt, the town was officially renamed San Francisco. Not a trace of Yerba Buena Cove can be seen today. The cove was later filled in with the rapid expansion of San Francisco, and if you were to follow Market Street to the Ferry Building, you would stand on what was once the center of the cove.

Through the rest of the nineteenth century and into the twentieth century, the city of San Francisco grew around Yerba Buena Cove and Telegraph Hill. During this time the western side of the peninsula, with the exception of the military posts and the popular Golden Gate Park and Ocean Beach, remained what many residents considered dreary, barren sand hills.

GOLDEN GATE TO POINT LOBOS (CLIFF HOUSE)

The spectacular rocky cliffs between the Golden Gate and the Cliff House are made up largely of serpentinite, which gives the cliffs a greenish color. Some parts of these cliffs contain relatively solid rocks with few joints or weaknesses, making them quite resistant to wave attack. These resistant rocks occasionally erode when large blocks break loose and roll or fall into the surf, but in general, they have a relatively low long-term average rate of retreat. Other sections of cliff contain broken down and weathered serpentinite rocks. These areas are prime locations for landslides, especially where water runs over or through the cliffs. Some large slides are on public land in the area, but they can occur wherever the cliffs are made up of fractured and weathered serpentinite (Figure 12.1).

Houses have been built on the sandy bluffs and rocky cliffs above Bakers Beach. This beach is somewhat protected from large surf, but most of the oceanfront properties directly above the beach are now protected by low concrete seawalls (Figure 12.2). Despite these seawalls, and numerous retaining walls higher up on the slope, some house foundations are threatened in this area, primarily by slow slumping of loose cliff materials. Such slumps may be triggered as much by surface runoff and subsurface water as by wave erosion.

Between Bakers Beach and Seal Rocks the steep rugged cliffs are undeveloped and form the seaward portion of Lincoln Park. The remains of the historic Sutro Salt Water Baths can still be seen near the southern end of this

Figure 12.1   Steep, unstable cliffs eroded into the Franciscan Formation just south of the Golden Gate Bridge. Much of this section of seacliff consists of large, active slumps. Photo © 2002–2004 Kenneth and Gabrielle Adelman, California Coastal Records Project, www.Californiacoastline.org.

Figure 12.2   Development on the sandy bluffs and cliffs above Bakers Beach. Note the low seawalls along the base of the bluff. Photo © 2002–2004 Kenneth and Gabrielle Adelman, California Coastal Records Project, www.Californiacoastline.org.

Figure 12.3 The historic Cliff House sits on resistant granite at the north end of Ocean Beach, which is on the right side of the photograph. The remains of the Sutro Saltwater Baths are to the left. Photo by Gary Griggs and Kiki Patsch.

area, just below the Cliff House, which is perched directly above the waves on a resistant rock outcrop (Figure 12.3).

THE CLIFF HOUSE TO FORT FUNSTON

The continuous stretch of sandy beach and dunes known as Ocean Beach, stretching from the Cliff House to Fort Funston, is entirely in public ownership. A massive concrete seawall has protected the northern portion since 1929. This immense structure, almost half a mile long and extending 20 feet above and below sea level, cost about $600,000 when it was first built. It was recently extended southward at a cost of over $9,350 per linear foot (in 2005 dollars); thus, constructing the original half-mile of concrete seawall today would cost almost $25 million. Although it has experienced some tremendous storms in its 75-year life, this wall was very carefully engineered and remains in relatively good condition.

South of the seawall is a region of actively migrating sand dunes, which have occasionally threatened to bury the Great Highway. This wide highway runs along the dunes, seaward of any permanent buildings. The south end of the Great Highway is periodically endangered by erosion of the dunes on which it is built. At the southern end of this section of roadway, seaward of

A. High cliffs eroded into serpentine of the Franciscan Formation, prone to sliding.

B. Houses built on sandy bluffs and rocky cliffs protected by concrete seawalls.

C. Ocean Beach— seasonally wide sandy beach backed by massive concrete seawall, the O'Shaughnessy seawall, built in 1929.

D. Sandy beach varying in width seasonally, backed by narrow dunes and the Great Highway.

E. Riprap added recently to protect highway.

F. Narrow beach backed by high, unstable sandy bluffs.

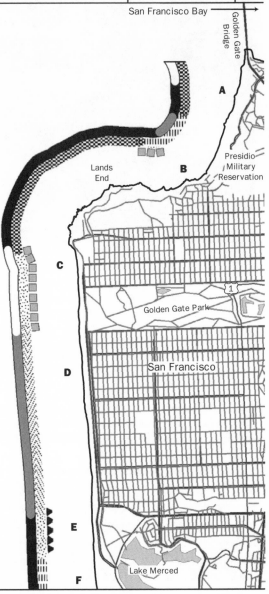

San Francisco Bay

Golden Gate Bridge

A

Lands End

B

Presidio Military Reservation

Golden Gate Park

San Francisco

Lake Merced

Hazard Level

 Stable: low risk

Caution: moderate risk

Hazard: high risk

Erosion Rates

 Inches per year

Shoreline Environment

Sandy beach

Beach backed by marsh

Beach backed by dunes

Cliff or bluff fronted by beach

Rocky coast with no beach

Armoring

▲ ▲ ▲ Riprap

Seawall

Other

the Fleishacker Zoo, riprap was placed following the 1983 winter to protect the highway from wave attack.

Even farther south, toward the San Francisco County line, the high sand bluffs of Fort Funston and the Olympic Golf Club provide beautiful views for park visitors but are relatively unstable. The original Ocean Shore Railroad ran along these bluffs from 1907 until 1920. In the 1930s a major road was built along the same route. By 1957 this highway was closed because it cost too much to maintain. The bluffs are highly unstable, and large slumps and landslides underlie much of this coastal area.

CHAPTER THIRTEEN

SAN FRANCISCO TO AÑO NUEVO

GARY GRIGGS, JERRY WEBER, KENNETH R. LAJOIE, AND SCOTT MATHIESON

INTRODUCTION AND REGIONAL SETTING

The Pacific coastline of San Mateo County extends about 56 miles from the city and county of San Francisco on the north to Santa Cruz County on the south. In general, the northern part is a continuation of the urbanized San Francisco coastline, whereas the southern section is rural and agricultural. This roughly north–south-trending rocky coastline consists of nine distinct segments that reflect local variations in coastal processes and geological conditions, primarily topography and rock type. The distinctive character of each coastal segment generally reflects the local balance between erosional processes and the resistance of cliffs or bluffs to wave attack or other processes. Overall, this coastline is sand deficient, and only locally do broad beaches protect the seacliffs from wave erosion.

In most of the relatively straight or linear coastal segments (Daly City, Pacifica, and Half Moon Bay, for example), weak sedimentary rocks with little erosion resistance make up the bluffs. Some protection from low or moderate wave attack is offered by the beaches fronting the bluffs, but the potential for rapid beach and seacliff erosion in these areas is very high if large waves arrive during very high tides or if the natural wave patterns are altered suddenly, which occurred when the breakwater was constructed at Half Moon Bay.

In most of the irregular cliffed coastal segments (Devil's Slide, Montara, and Pescadero to Franklin Point), rocks highly resistant to erosion are exposed along the shoreline. Seacliff erosion rates in these areas are generally low, typically less than 2 inches per year.

In the remaining coastal segments (Point San Pedro, Moss Beach to Pillar Point, Miramontes Point to Pescadero, and Point Año Nuevo), rocks with moderate erosion resistance are exposed to wave attack along much of the coastline. Rocky headlands separated by small pocket beaches dominate the coastal landforms, and sea caves and arches characterize the steep cliffs. Rates of seacliff erosion vary widely, from less than an inch per year to over 4 feet per year.

Slope failure, though not uniquely a coastal process, contributes significantly to the formation of many coastal landforms and is a major contributor to geological hazards along the San Mateo coast. Large coastal landslides, which form where weak or highly fractured rocks are exposed in high coastal bluffs, occur at Thornton Beach, Mussel Rock, Devil's Slide, Seal Cove, and Tunitas Creek. Minor landslides and rockfalls occur along steep seacliffs and bluffs that are retreating at low to intermediate rates under persistent wave attack. These conditions exist at Pacifica, Moss Beach to Pillar Point, Miramontes Point to Pescadero State Beach, and the south shore of Point Año Nuevo. The only parts of the San Mateo County coastline with no slope failures are the high, stable granite cliffs just south of Devil's Slide and the low, stable bluffs between Pescadero State Beach and Franklin Point.

Historical maps and photographs, and old structures such as early railroad and highway grades, provide useful information for measuring long-term (75–150 years) rates of cliff erosion in this area. Average rates of seacliff erosion are reported in inches per year, but, as stated in Chapter 5, we must take care in interpreting or using this data. Cliff erosion is very irregular in both space and time, so we should not expect the yearly increment of erosion along any particular segment of coastline to equal the value reported here. An average long-term bluff erosion rate of 12 inches per year doesn't mean that the bluff erodes 12 inches each year, but is more likely the result of a combination of erosion events of different magnitudes over time. Over the long term, these episodic increments average out to the rates reported here. In some locations, where historical data is sufficiently detailed, we can document this episodic nature of cliff erosion (Seal Cove, Princeton, and El Granada, for example). The storms of 1981–82 and especially 1982–83 caused severe cliff erosion in many parts of coastal San Mateo County that had undergone little or no observed erosion over the previous 100 years.

CULTURAL HISTORY AND DATA SOURCES

The early European explorers of the California coast (1543–1776) left no detailed charts or maps of the San Mateo coastline. However, the written records of some early expeditions contain useful information. For example, on January 3, 1603, the Spanish explorer Vizcaino sailed past the prominent marine headland at the southern tip of San Mateo County and named it Punta del Año Nuevo (New Year's Point). The written account of this voyage makes no mention of the island that lies off the point today, which suggests that it was connected to the mainland in 1603. Thus, Año Nuevo Island was probably formed by rapid seacliff erosion sometime after 1603.

In 1769 the Gaspar de Portola expedition, traveling overland and using only a mariner's navigation handbook of the Alta California coastline, set out to map the route from Baja California to Monterey Bay. Expecting to find a "fine harbor, sheltered from winds," they concluded that the rough seas they met at the mouth of the Salinas River could not possibly be their goal. They continued north and ended up at what is now Pacifica. Although running short of food and supplies, with many of the party sick from scurvy and having to be carried by litters, they persisted and reached the top of Montara Ridge on October 31, 1769. From there they could see Point Reyes and the Farallon Islands, and they realized they had come too far north. Before turning back, however, they camped and rested in a peaceful valley (now Linda Mar) to rest, explore, and scout for much-needed food.

Two years after gold was discovered at Sutter's Mill (1848), California became the thirty-first state. The population of central California expanded rapidly in the 1850s, and detailed navigational charts were needed to guide the growing maritime traffic that supplied the gold fields through the port of San Francisco. In response to this navigational need, the U.S. government commissioned a series of hydrographic charts and topographic maps of the California coast at a scale of 1:10,000. The seven detailed topographic maps (1853–66) in this series that cover the San Mateo Coast provide an excellent means of assessing coastal changes and for making quantitative measurements of long-term rates of seacliff erosion.

During the late 1800s coastal San Mateo County was sparsely populated, mainly because it was physically isolated by the Santa Cruz Mountains. Most of coastal San Mateo County south of Half Moon Bay is still sparsely populated for the same reason. The mountainous barriers to overland transportation were so great that lumber and farm produce from this area were shipped to San Francisco on small coastal schooners that were loaded by wire cables from long piers or from staging areas on high coastal bluffs. The scant remains of a few of these cargo-loading facilities provide some information on seacliff erosion.

In 1879 the first coastal road in the area, a dirt trail, was built to the coast over San Pedro Mountain. In 1905 the Ocean Shore Railroad Company was incorporated in San Francisco to build a railroad along the coastline of San Mateo and Santa Cruz counties that would connect San Francisco and Santa Cruz. This railroad was never completed, but the railbed along the entire proposed route was excavated and rail was laid between San Francisco and Tunitas Creek, about halfway down the San Mateo coast. During the 1906

Figure 13.1 A very large landslide typical of the cliffs between Thornton Beach and Mussel Rock. The remains of the old Ocean Shore Railroad railbed is still visible about one-third of the way up the bluff. Photo © 2002–2004 Kenneth and Gabrielle Adelman, California Coastal Records Project, www.Californiacoastline.org.

San Francisco earthquake, railway cars, tracks, and other equipment were tossed into the ocean by the strong shaking. The section of the railroad from San Francisco to Tunitas was used for freight and passenger service from 1908 to 1921. The Ocean Shore Railroad Company went bankrupt in 1921, due in large measure to the high costs of building and maintaining the railbed across the steep, unstable coastal bluffs south of Thornton Beach and near Point San Pedro in the northern part of San Mateo County. The abandoned railbed across the bluffs between Thornton Beach and Mussel Rock was converted to highway use and carried automotive traffic from the late 1930s to 1957, when landslides caused by a local earthquake closed the grade (Figure 13.1). One short section of railbed across the unstable bluffs south of Point San Pedro (Devil's Slide) was converted to highway use in 1939 and is still occupied by State Highway 1 today, although requiring constant maintenance. A new plan would replace this hazardous stretch of roadway with a controversial tunnel through the mountain. Parts of this old railroad grade, especially those along the steep, unstable coastal bluffs in the northern part of the county, are still visible today and provide excellent data for measuring coastal erosion and assessing slope stability.

In the aftermath of the San Francisco earthquake and fire of 1906, there was widespread belief that many residents would leave the city out of fear that a similar catastrophe would happen again. This belief was incorrect. The access afforded by the new Ocean Shore Railroad led to speculative land development along the San Mateo coast. Between 1906 and 1908 several new coastal communities along the rail line's route were incorporated and subdivided. Most of these communities never developed, but in some of them, streets were excavated and concrete sidewalks were installed. These streets and sidewalks and the detailed plot maps of some of these old subdivisions provide valuable reference marks for quantitatively measuring local rates of seacliff retreat over the past 95 years.

Old photographs taken from 1895 to 1950 from early geologic reports, property surveys, and historical archives provide detailed information on seacliff retreat at several localities along the San Mateo coast. The first aerial surveys of this area were made in the late 1920s, and comparisons between these and several sets of later aerial photographs provide time-lapse analyses of coastal changes over the past 75 years.

As mentioned previously, most of southern coastal San Mateo County is sparsely populated, and the northern part has been built up extensively since the 1950s. This recent development took place with little regard for potential geological hazards. Consequently, much of our data on short-term erosion rates and landslide movement is derived from recently built structures that are now being damaged or threatened. Also, some of the most rapid seacliff erosion along this coast was artificially induced by construction of the breakwater at Half Moon Bay in 1959. Ironically, these recent sources of erosion information best illustrate the coastal hazards that can be avoided with prudent land use policies.

COASTAL HAZARDS AND RISKS: COUNTY LINE TO MUSSEL ROCK

The straight, 3-mile-long coastline between the San Mateo County line on the north and Mussel Rock on the south, essentially the corporate limits of Daly City, consists of high (125–730 feet), steeply sloping bluffs eroded into weak sands and gravels fronted by narrow beach. Very large active landslides characterize much of this segment of coastline. A large block landslide forms a prominent step in the bluff face at Thornton Beach State Park in the northernmost part of this coastal segment, and a second very large landslide occupies a broad, bowl-shaped topographic depression in the southernmost part, where the San Andreas Fault intersects the coastline just north of Mussel Rock. A number of additional active slides occur on the steep bluff face between these two very large failures (Figure 13.1). A very large section of bluff failed in this area during the 1989 Loma Prieta earthquake, threatening homes at the top of the bluff (Figure 5.9), even though the epicenter for this event was nearly 60 miles away.

The narrow beach in this area does not protect the base of the bluff from seasonal wave attack. Winter wave erosion typically oversteepens the lower part of the bluff and triggers numerous landslides and debris slides that propagate upslope and are slowly destroying the old railroad grade. During heavy storms in January 1983, wave erosion and groundwater saturation caused the large slump block of Thornton Beach State Park to move. At the southern end of the landslide, a drop of 20 feet destroyed the access road to a parking lot situated on the landslide itself. Headward expansion of the slide's northern end threatens Skyline Boulevard near the San Mateo County line. During the 1983 storms, wave erosion destroyed two pedestrian access routes from the parking lot to the state beach at the toe of this landslide.

Wave erosion contributes to the movement and enlargement of the large landslide at Thornton Beach and others to the south, and the resulting landward expansion of each landslide has damaged streets and homes since the area was developed in the late 1950s. Homes in a residential subdivision built between 1956 and 1966 along the top of the landslide above Mussel Rock continue to be removed, while adjacent homes are damaged from continued upslope failure.

Between 1866 and 1956, the head of the large gully bisecting the bluffs in the central part of this coastal segment eroded inland at an average rate of 20 inches per year. Severe erosion in this gully caused by heavy rains during the winter of 1981–82 threatened several homes in this area. An extensive landfill and riprap were emplaced at the base of the slope in 1983, however, which helped to stabilize the lower portion of this slide. About 1,600 feet of shoreline at the seaward edge of the landfill up coast from Mussel Rock are now completely armored with riprap.

In summary, the primary coastal hazard in this area is movement and headward or landward expansion of the large landslides that threaten numerous homes along the crest of these high, unstable coastal bluffs. Wave erosion, heavy and sustained rainfall, and seismic shaking can all contribute to the movement of these slides.

MUSSEL ROCK TO POINT SAN PEDRO

The 5.6-mile-long coastline between Mussel Rock and Point San Pedro, which corresponds to the corporate boundaries of the City of Pacifica (which, when incorporated in 1957, consolidated the communities of Edgemar, Pacific Manor, Westview, Sharp Park, Fairway Park, Vallemar, Rockway Beach, and Linda Mar), consists of a low, narrow coastal plain in

A. Very large active landslide with movement caused by heavy rainfall and wave erosion of the base.

B. Very large debris slide took place during the 1989 Loma Prieta earthquake.

C. Old Ocean Shore Railroad grade about 150 feet above beach level is progressively being destroyed by large landslides along this entire section of bluffs.

D. San Andreas Fault passes through the bluff top subdivision and heads out to sea at this location. Large-scale landsliding of the unstable bluff has led to progressive removal of homes along the head scarp or top edge of the slide. Sanitary landfill along base of slide now armored with riprap.

E. High bluffs actively eroding and threatening apartment complex. Emergency riprap added in 2003.

F. Ten homes along Esplande threatened by bluff erosion and demolished in April 1998.

G. Seawall and riprap protecting mobile home park damaged in 1983 and 1998 El Niño winter. Ocean front trailers removed during 1983 winter.

H. Low unstable bluffs in Sharp Park completely armored with concrete seawall and riprap, covering the beach.

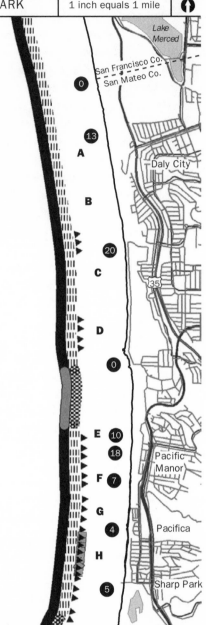

Hazard Level

☐	Stable: low risk
▦	Caution: moderate risk
■	Hazard: high risk

Erosion Rates

① Inches per year

Shoreline Environment

Sandy beach

Beach backed by marsh

Beach backed by dunes

Cliff or bluff fronted by beach

Rocky coast with no beach

Armoring

▲ ▲ ▲ Riprap

■ ■ ■ Seawall

◉ ◉ ◉ Other

its northern half and two small coastal valleys flanked by narrow ridges in its southern half. This coastal terrace slopes gently southward from an elevation of about 160 feet near Mussel Rock to sea level at Laguna Salada, a natural lagoon north of Mori Point. The straight shoreline along the seaward margin of this narrow plain consists of a sandy beach backed by vertical bluffs eroded into loosely consolidated sands and gravel (Figure 13.2).

Pacifica has become more densely populated over the past 30 years and now has numerous homes, apartments, businesses, and a mobile home park built close to the bluff edge on a marine terrace that is 25 to 100 feet above sea level (Figure 13.2). Photographs from the 1970s show eroding bluffs, but the only armor at that time was a retaining wall and riprap protecting the Pacific Skies Mobile Home Park and some scattered concrete slabs that had been dumped at the base of the bluff at several locations. Houses along the Esplanade were close to the bluff edge at that time, and evidence existed for recent bluff failure. Some rock had recently been dumped on the beach below a house by its resident.

During the repeated high tides and storm wave attack during the El Niño winter of 1983, waves seriously eroded these weak bluffs, threatening a number of structures. A row of homes along the Esplanade were endangered by bluff erosion, and within a year following the 1983 winter, an engineered revetment had been constructed at the base of the bluff (Figure 7.6). A short distance south, a seawall, an access road, and a number of foundations at a mobile home development in Sharp Park sustained heavy damage (Figure 3.6). The history of erosion and protection at this site provides useful perspective for the entire Pacifica coastline. Between 1941 and 1970 about 30 feet or more of bluff retreat occurred, with trailer park construction taking place in 1966–67. By 1969, riprap and concrete rubble that had been placed at the base of the low bluff had been washed away by wave action. Around 1970 a wood piling bulkhead covered with gunite and fronted by riprap was constructed, and around 1973 additional rock had to be placed at the north end due to outflanking. In February 1983 the timber-gunite wall was breached, extreme beach scour occurred, and the riprap eroded, followed by bulkhead failure. Wave attack, bulkhead destruction, and bluff erosion continued, and by March 1983 23 mobile homes were moved landward to prevent further losses. The bluff, which consists of poorly consolidated sands and gravel, was cut back 40 feet at the southern end of the mobile home park and 80 feet at the north end. Just to the north, a large three-story building, which was almost undercut by the 30 feet of bluff erosion, was also moved to prevent damage.

Along the low cliffs of Beach State Park in southern Sharp Park, beach and bluff erosion threatened a public road during the January 1983 storms. Riprap was quickly emplaced to protect the road and homes along its

Figure 13.2 (A) Apartments built in the early 1970s close to bluff edge in Pacifica. 1972 photo courtesy of the California Department of Boating and Waterways. (B) By 1979 there had been minor bluff retreat but the entire area has been more intensively developed. Photo courtesy of the California Department of Boating and Waterways.

landward side. To the south, heavy surf severely eroded the broad beach and inundated the windbreak along the spit at Laguna Salada. Waves toppled several trees and washed sand over a large part of the golf course. At the southern end of the spit, just north of Mori Point, beach erosion exposed a low cliff to wave attack in 1983. About 60 feet of bluff retreat

Figure 13.2 *(Continued)* (C) Wave erosion in the winter of 2002. Photo ©
2002–2004 Kenneth and Gabrielle Adelman, California Coastal Records
Project, www.Californiacoastline.org.

occurred, threatening a house, followed by 20 feet of additional erosion
the next winter. Historical records suggest that for about 100 years prior to
1983 there had been virtually no cliff erosion in this area.

Following the 1983 El Niño storm damage, most of the developed sites
along the Pacifica coastline where significant erosion took place were
armored. An engineered revetment was constructed below the houses along
the Esplanade (Figure 7.6), and a new 650-foot-long structure consisting of
concrete piles spaced 48 inches apart backed by riprap was built in front of
the mobile home park (Figure 13.3).

The Esplanade area has a well-documented history of coastal change that
provides a useful long-term perspective. According to older maps, this fragile coast had been relatively stable from 1853 to 1946. When 12 houses were
built on the seaward side of Esplanade Drive in 1949, the street still lay 150
feet east of the 65-foot-high bluff. Although by 1973 this distance had been
reduced by half, it changed little thereafter, even in the 1982–83 El Niño
storms (Figure 13.4). During the subsequent years, the revetment installed
to protect homes along the Esplanade settled, despite the use of filter cloth
and careful layering of the rocks. By the time the high tides and storm waves
of the 1997–98 El Niño arrived in early 1998, there was little of the revetment
left to protect the weak bluffs. High tides and large waves combined to
remove about 40 feet of bluff, despite the efforts of the homeowners to

Figure 13.3   A mobile home park on the bluffs at Pacifica showing the combined concrete pier and riprap structure installed to armor the bluffs from wave attack. Photo © 2002–2004 Kenneth and Gabrielle Adelman, California Coastal Records Project, www.Californiacoastline.org.

Figure 13.4   Comparison of historical positions of the coastal bluff edge along the Esplanade, showing changes from the late 1800s to 1998. Photo by Ken Lajoie, U.S. Geological Survey.

install additional rock. Heavy equipment working under emergency conditions on the beach was damaged when caught by high tide. Continued bluff erosion through March and April led to undercutting of house foundations (Figures 3.8, 7.9), condemnation of all but 3 homes, and the ultimate demolition of 10 homes in April 1998 (Figure 1.4).

Losses at this location were among the most dramatic during the 1997–98 winter and provided evidence of the potential limitations of riprap or revetments placed on deep sand foundations, as well as evidence of the ability of

Figure 13.5   The Esplanade in September 2002, showing the three remaining bluff-top homes and the new riprap. Photo © 2002–2004 Kenneth and Gabrielle Adelman, California Coastal Records Project, www.Californiacoastline.org

heavy surf to strip sand from narrow beaches and directly attack soft bluff materials already weakened by groundwater saturation. Breaking waves undercut the bluff face, inducing block falls and slumps in the oversteep-ened bluff. Following demolition of the homes, additional rock, funded by FEMA, was placed at the base of the bluff along the entire Esplanade even though most of the houses were now gone (Figure 13.5). The concrete piling structure protecting the mobile home park was damaged as some of the piles rotated seaward. Along Shoreline Avenue, bluff retreat threatened many homes; one was red-tagged as unsafe to occupy, and four emergency permits were issued for repair and expansion of revetments.

Winter waves during a weak El Niño in the winter of 2002–3 eroded 20 to 30 feet of the high bluffs at the north end of Pacifica and began to threaten a large apartment complex on the bluff top. Emergency rock was brought in to protect the toe of the bluff (Figure 7.26) below the apart-ments, and a request was made in early 2003 for a permit to place 2,500 tons of 6–8-ton stones along the base of the eroding bluffs at the Land's End townhouses, directly to the north (Figure 13.2).

By the summer of 2002, almost the entire bluff frontage of Pacifica, over 1.5 miles of coastline from just south of Land's End to Laguna Salada, had been armored with either riprap, seawalls, or a combination of the two. These high, weak bluffs are very unstable under wave attack, yet the bluffs have been nearly completely urbanized. Erosion problems can be expected here in the future, and more property and structures will be threatened and probably lost.

Mori Point, a rocky headland, marks the down-coast end of the developed portion of Pacifica. A large development was proposed some years ago for a former rock quarry on the south side of this headland. Calera Valley and San Pedro Valley, south of Mori Point, end at the shoreline in broad pocket beaches flanked by rocky headlands. Dense commercial development in the community of Rockaway Beach extends down to the surf zone at the mouth of Calera Valley. Over the past 15 years wave erosion has periodically threatened and damaged a parking lot on the seaward side of a restaurant at Rockaway Beach. A concrete seawall and riprap now protect this and more recent commercial development on the berm of this small pocket beach. However, waves have cut deeply into the low cliffs south of this seawall in recent years, and the 1997–98 El Niño undermined a parking lot, street end, and walkway at Rockaway Beach. Requests for emergency permits for placement of additional rock followed that event.

The abandoned rock-cut grade of the Ocean Shore Railroad forms a prominent bench around Rockaway Point between Calera and San Pedro valleys and along the steep north-facing bluffs of Point San Pedro. Over the past 80 years erosion and landslides have destroyed much of the abandoned railroad grade around Rockaway Point, but the rocky shoreline below the grade has not receded measurably during the same period.

In San Pedro Valley, dense residential and commercial development is restricted to the east side of State Highway 1, which runs along the back edge of the beach, well behind the surf zone. However, a few homes and businesses lie on the beach berm west of the highway. At the south end of this valley, several homes and fishing-related commercial businesses have been built along the back edge of the beach against a steep low bluff.

During the heavy storms in January 1983 beach erosion and wave inundation threatened the homes at the back edge of the beaches at San Pedro Valley and the small community at Shelter Cove. During these storms, cliff erosion caused a major slope failure that destroyed a section of the access road (old railroad grade) to Shelter Cove and threatened homes at the top of the steep cliff above the road. The homes at Shelter Cove, situated a few feet above the back beach, have been completely isolated, and several houses there have been abandoned (Figure 13.6).

In summary, the main hazard in the coastal segment between Mussel Rock and Point San Pedro is wave erosion of the weak but intensively developed bluff tops and periodic damage to structures in Pacifica and Sharp Park, as well as wave inundation and damage to oceanfront struc-

Figure 13.6   The small community of homes at Shelter Cove has been damaged by large waves in recent years. Vehicle access was cut off permanently by landslides during the 1983 El Niño winter, leading to the abandonment of several homes. 1989 photo by Gary Griggs and Kiki Patsch.

tures at Rockaway Beach, the southern part of San Pedro Valley, and Shelter Cove.

SAN PEDRO MOUNTAIN

The 2.8-mile-long coastline between Point San Pedro on the north and Montara Beach on the south consists of steep coastal bluffs up to 900 feet high that plunge precipitously into the ocean along the western flank of San Pedro Mountain (Figure 13.7). The bluff face in the northern half of this area consists of deformed and pervasively fractured sandstones and shales and is traversed by the remnants of the Ocean Shore Railroad grade 200 to 300 feet above sea level. The rock-cut grade of State Highway 1, built in 1939, traverses the southern part of this bluff face above the railroad grade, but joins it in a deep notch behind the small headland called Devil's Slide.

Numerous debris chutes scar this broad, unstable slope, and a large, complex landslide occurs at its southern end, just north of Devil's Slide. Wave erosion removes landslide debris from the base of these bluffs and oversteepens the lower part of the slopes by cutting into the fractured bedrock. The cumulative effect of many small slope failures is slowly destroying the old railroad grade, and periodic movement on the large

A. Concrete seawall and riprap protect oceanfront commercial development.

B. Point south of Rockaway still has remnants of Ocean Shore Railway bed, although erosion and landslides have removed

C. Homes built on narrow beach subject to wave damage.

D. Access to small community of Shelter Cove permanently cut off by damage to access road. Houses damaged by waves, including boulders washed into homes in 1983 winter.

E. Devil's Slide is an area of very high, steep, and unstable cliffs. Some segments of the Ocean Shore Railway bed can still be seen, and Highway 1 has a constant maintenance problem as it crosses the upper portion of the slide area.

F. 33 feet of bluff erosion during 1983 storms threatened new restaurant at top of bluff. Riprap subsequently installed along with concrete grout.

G. Rocky granitic coastline with houses built along outer edge of terrace. Granitic bedrock exposed at beach level is resistant but overlying sands and gravels are more easily eroded. House undercut by wave and gully erosion. Restaurant at back edge of beach destroyed by storm waves in 1931. Road at top of bluff lost in 1957. Riprap and gunite installed to protect residences.

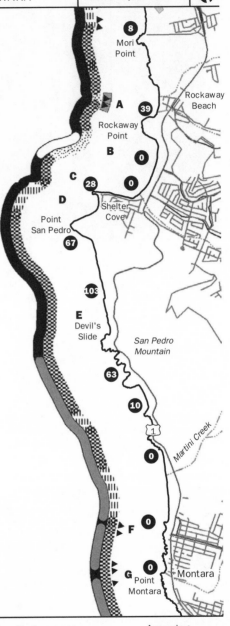

Hazard Level

- ☐ Stable: low risk
- ▒ Caution: moderate risk
- ■ Hazard: high risk

Erosion Rates

- ① Inches per year

Shoreline Environment

- Sandy beach
- Beach backed by marsh
- Beach backed by dunes
- Cliff or bluff fronted by beach
- Rocky coast with no beach

Armoring

- ▲▲▲ Riprap
- ■■■ Seawall
- ▣▣▣ Other

Figure 13.7 The steep cliffs between Point San Pedro and Devil's Slide are traversed by the remains of the Ocean Shore Railroad grade on the lower left and State Highway 1 in the upper right. This area is subject to constant slope failure. Note the large slide mass in the center of the photograph that has destroyed the railroad grade and bulged out the coastline. Photo © 2002–2004 Kenneth and Gabrielle Adelman, California Coastal Records Project, www.Californiacoastline.org.

landslide damages the existing highway. Groundwater infiltration related to prolonged rainfall has caused most of these slope failures.

Historically, San Pedro Mountain and its steep coastal bluffs restricted access from the north to the central and southern parts of coastal San Mateo County. As noted earlier, the Ocean Shore Railroad penetrated this barrier for 13 years before its bankruptcy in 1921. The present-day highway is closed periodically because of slope failures, especially where the roadbed crosses the active landslide just north of Devil's Slide. The California Department of Transportation has installed horizontal drainpipes into the slide to lower groundwater levels, and has also drilled along the roadway and installed long steel rods in an attempt to increase the slope stability and reduce road failure. Caltrans now has an approved plan to permanently tunnel through San Pedro Mountain, bypassing the Devil's Slide area completely to ensure reliable access to the residential communities to the south as well as to the many other users of Highway 1.

The steep, irregular bluffs between Devil's Slide and Montara State Beach consist of deeply weathered but stable granitic rocks. Here Highway 1 mainly follows the rock-cut grade of the Ocean Shore Railroad as it winds through the steep road cuts. The highway also lies on artificial fill across

Montara Valley. No development exists in this area other than one parking area and a steep coastal trail to Gray Whale Cove State Beach. We can document no historical changes along these precipitous but stable bluffs. However, rapid runoff from heavy rains in the winter of 1981–82 washed out the fill and highway grade across Montara Valley.

In summary, the main hazards in the coastal segment between Point San Pedro and Martini Creek are the slope failures that periodically damage and close Highway 1 in the Devil's Slide area.

MONTARA VALLEY TO PILLAR POINT

The 4.4-mile-long convex coastline between Montara Valley on the north and Pillar Point on the south consists of low, rolling hills, with a flat terrace developed near the coastline that is truncated by low to moderately high seacliffs. This area is moderately developed, with most of the population concentrated in the residential communities of Montara and Moss Beach. These two communities, and the smaller community of Seal Cove in the southern part of the area, were subdivided in 1908 but have acquired most of their population only in the past 40 years. Many homes and a few businesses have been built close to, or directly on, the eroding seacliffs.

The geologic diversity of this area is expressed by the variable nature of the coastline, with three distinct segments along this stretch of coast. The straight 0.7-mile-long northernmost segment is a broad, sandy beach (Montara Beach) backed by low bluffs cut into loose sediments that cover the floor of Montara Valley. These sediments offer little resistance to wave erosion, but the long-term rate of cliff retreat is low because the broad beach generally protects the cliffs from all but the largest storm waves. However, in January 1983, El Niño storm waves combined with the highest tides in decades, cut into these cliffs, and threatened a restaurant built too close to the cliff top near the south end of the beach. Riprap was subsequently installed from the beach to the top of the bluff and along several hundred feet of coastline (Figure 13.8). Runoff from heavy rains during the winter of 1982–83 washed out the artificial fills beneath Highway 1 where it crosses two deep gullies near the north end of the beach.

The 1.3-mile-long central coastal segment consists of a rocky shoreline backed by irregular cliffs cut into highly jointed but stable granitic rocks that are overlain by weaker sands and gravels. Although the granitic rocks are highly resistant to wave erosion, the overlying deposits yield locally to wave attack during heavy storms. A few homes in the southern part of this area were threatened, and one was badly damaged, by erosion of these weak materials during heavy winter storms in 1982 and 1983.

Figure 13.8   Bluff erosion during the winter of 1983 threatened this bluff-top restaurant. Extensive riprap was emplaced to protect the restaurant from future bluff failure. Photo by Gary Griggs and Kiki Patsch.

The 2.4-mile-long southern coastal segment consists of narrow, rocky beaches backed by steep seacliffs 30 to 130 feet high cut mainly into mudstone, which is covered by weak sand and gravel. The southern end of the linear ridge into which these cliffs are cut is the prominent headland Pillar Point. Two active landslides indent the seaward side of this narrow ridge, and the community of Seal Cove lies partly within the northern landslide depression.

At the northwestern end of this ridge at Moss Beach, cliff retreat due to wave erosion has been very rapid. At one point the cliff has retreated about 150 feet in 120 years, which yields a long-term average erosion rate of 15 inches per year. In 1971 the planning department rejected permit applications to develop three residential lots on the cliff top because of the potential erosion hazard. A restaurant built at the back of this beach in 1905 was destroyed by storm waves in 1957. Subsequently, wave erosion destroyed the public road at the top of the cliff behind the restaurant, and now threatens several homes close to the edge of the cliff. Severe cliff erosion, particularly during the heavy storms in January 1983, has led to the installation of riprap, covering of the bluff with gunite (Figure 13.9), and other measures to reduce bluff erosion.

Because beaches do not protect the unstable seacliffs south of Moss Beach, long-term rates of seacliff retreat are fairly high, up to 8 inches per year over a 95-year period. Several residential lots and three homes near the

Figure 13.9 Gunite was sprayed over the loose bluff materials at Moss Beach in an attempt to halt erosion. Continued retreat on the right side of the photograph had led to undermining of the gunite. Photo by Gary Griggs and Kiki Patsch.

top of the cliffs in Seal Cove have been damaged within the past 20 years by slope failures; one home was moved in 1983 to avoid complete destruction (Figure 6.13). However, even though cliff erosion is a serious problem in Seal Cove, the primary hazard is seaward movement and headward expansion of the large landslide on which part of the community is built. This slope failure has badly damaged a few homes and threatens several more. The large landslide south of Seal Cove is also active and is expanding northward; here blocks up to 200 feet wide are sliding seaward (Figure 5.10). Wave erosion at the base of the steep cliffs and groundwater infiltration from above contribute to the instability in this area. Directly offshore from Pillar Point is the now well-known big-wave surfing area of Mavericks.

In summary, the greatest hazards in the coastal segment between Montara Valley and Pillar Point are in the southern part of the area and are related to rapid seacliff erosion and large-scale landsliding.

HALF MOON BAY

Half Moon Bay is an embayment located between Pillar Point to the north and Miramontes Point to the south. The 6.5-mile-long, smoothly arcuate, or hook-shaped, shoreline of this open embayment is a broad, sandy beach backed by a low coastal terrace. The northern part of Half Moon Bay has

A. Severe cliff erosion from wave attack and slope failure. Large active landslide underlies portion of Seal Cove community and continues to displace road and crack homes. Two houses have been removed from cliff edge.

B. High unstable eroding bluffs. World famous big wave surfing area, Mavericks, is just offshore.

C. Half Moon Bay Breakwater, constructed in 1959 provides protected anchorage for fishing and pleasure boats but affected wave refraction patterns leading to increased erosion at east end of breakwater.

D. Erosion rates of low weak bluffs increased from a few inches per year to as much as 80 inches per year following breakwater construction. Road and sewer line lost and riprap repeatedly installed.

E. Houses and restaurants built along low weak bluff at front edge of marine terrace, now completely armored with riprap.

F. Apartment house armored with riprap has become a peninsula.

G. Stable shoreline with sandy beach backed by dunes or low bluff.

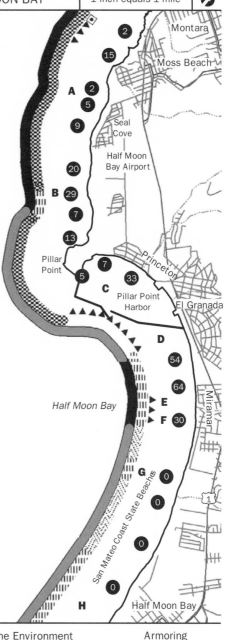

Hazard Level
☐	Stable: low risk
▨	Caution: moderate risk
■	Hazard: high risk

Erosion Rates
- ① Inches per year

Shoreline Environment
- ▨ Sandy beach
- ▨ Beach backed by marsh
- ▨ Beach backed by dunes
- ▥ Cliff or bluff fronted by beach
- ▨ Rocky coast with no beach

Armoring
- ▲▲▲ Riprap
- ■■■ Seawall
- ⦿⦿⦿ Other

become more heavily developed in the last several decades, with most of the population concentrated in the old but rapidly growing communities of El Granada and Half Moon Bay. At Princeton an expanding marina and harbor facility provides berths for many recreational boats and commercial fishing vessels. The area immediately south of the town of Half Moon Bay and continuing down the coast had been sparsely populated until the last several decades. Agriculture has historically been the primary land use, although mobile home parks, golf courses, subdivisions, and hotels are becoming more common.

Between Pillar and Miramontes points, the low coastal bluffs fronting the marine terrace expose loose, nonresistant sands and gravels that are exposed to wave erosion. The prevailing northwest waves bend or refract around the Pillar Point headland and erode the weak sediments between the confining headlands into a simple, gradually tightening curve (log-spiral curve) that evenly distributes wave energy along the shoreline. In this configuration the crests of the breaking waves are nearly parallel to the beach within the bay, and through the process of refraction around the point the wave energy has been spread out over a larger area and therefore reduced. Sand periodically eroded from the weak bluffs accumulates to form a broad permanent beach that protects the low bluffs from all but the largest storm waves from the northwest, and from waves approaching from the west or southwest, which lose little energy through refraction. Under these natural conditions the rate of seacliff retreat was very low, several inches per year, even though the bluffs themselves offer little or no resistance to wave attack.

A long breakwater was built across the northern tip of Half Moon Bay in 1959 to provide a protected harbor for fishing and pleasure boats (Figure 4.23). The breakwater, however, disrupted the equilibrium wave pattern. Wave energy that was formerly spread out or dissipated along the entire shoreline of the bay was now focused on the low, weak bluffs at the eastern end of the breakwater. A county road and the underlying sewer lines were gradually undermined and destroyed, and State Highway 1 was threatened (Figures 5.4 and 13.10). Rates of bluff retreat jumped from 3 inches per year to as high as 80 inches per year on average. Periodic dumping of broken concrete and rock between 1959 and 1981 failed to check this erosion (Figure 13.11). More substantial riprap emplaced after the heavy storms in 1983 provided longer-term protection for Highway 1 where it passes within 30 feet of the present shoreline near the eastern end of the breakwater.

Surprisingly, construction of the breakwater in 1959 did not eliminate erosion of the low, unprotected bluffs inside the enclosed harbor. Roads, utilities, and private property in the harbor community of Princeton have

Figure 13.10 Mirada Road, south of the Half Moon Bay breakwater, and the underlying sewer line were destroyed by increased wave erosion following breakwater completion. 1972 photo by Gary Griggs and Kiki Patsch.

Figure 13.11 The low, weak bluffs south of the end of the Half Moon Bay breakwater continued to erode as waves overtopped the rock and concrete slabs that were hastily dumped on the beach. 1973 photo by Gary Griggs and Kiki Patsch.

Figure 13.12   Riprap has been used to protect this apartment house south of the Half Moon Bay breakwater, where flank erosion is creating a peninsula. Photo by Gary Griggs and Kiki Patsch.

been damaged and are still threatened by bluff retreat that averages 33 inches per year in places. Extensive harbor improvements now protect all but the most quickly eroding part of the low bluffs (at the foot of Broadway Avenue) in this area.

Accelerated cliff erosion extends 1 mile south of the breakwater to the community of Miramar, where another part of the county road, several homes and businesses, and an apartment building are threatened (Figure 13.12). Approximately 1,600 feet of riprap emplaced between 1978 and 1983 offers only temporary protection for these exposed structures. Development of homes continues on this low marine terrace. South of the Miramar the low, eroding bluff is replaced by sand dunes as the effect of the breakwater has been diminished. A few older homes exist on the northernmost of these dunes. Proceeding south to the Dunes Beach, a low bluff composed of weak sands and gravels fronted by a sandy beach reappears to form the coastline. The bluff height increases to the south, and at Miramontes Point a very large hotel and golf course development comes right to the edge of the unstable bluffs. Erosion has exposed the foundation of an older clubhouse, and riprap was recently installed to protect part of the golf course (Figure 13.13).

Figure 13.13   The Ritz Carlton Hotel and golf course south of Half Moon Bay, where riprap has been placed on the beach to protect part of the golf course. Note the foundation exposed by bluff retreat on the right side of the photo. Photo © 2002–2004 Kenneth and Gabrielle Adelman, California Coastal Records Project, www.Californiacoastline.org.

MIRAMONTES POINT TO TUNITAS CREEK

The 6.2-mile-long coastline between Miramontes Point on the north and Tunitas Creek on the south consists of steep-to-overhanging seacliffs (70 to 160 feet high) that back small pocket beaches separated locally by sharp, rocky headlands. Thin, narrow beaches front the high cliffs in many places, but in most areas the beaches do little to protect the cliffs from erosion by winter storm waves. Precipitous cliffs form the southwestern margin of a narrow marine terrace that is backed to the east by rolling hills with broad, open valleys. The streams in these valleys cross the flat coastal terrace in deeply incised ravines and locally plunge over the steep seacliffs as small seasonal waterfalls.

This sparsely populated coastal area is traversed from north to south by State Highway 1, which runs along the landward margin of the coastal terrace, and by the abandoned grade of the Ocean Shore Railroad, which runs midway between the highway and the seacliffs. Artificial fills supporting the railroad grade cross swales and gullies, providing impoundment for stock and irrigation ponds. A rock-cut portion of the railroad grade crosses the sheer cliff face north of Tunitas Creek. Agriculture is the primary land use in this area, with residential and recreational development restricted to scattered homes on the high cliffs near the mouth of Tunitas Creek and at

A. Bluff consisting of weak sediments retreating through both wave undercutting and collapse and also gullying from surface runoff.

B. Weak bluffs eroding immediately north of Miramontes Point. Ritz Carlton Hotel and golf course has been built in recent years. Riprap added to protect portion of golf course. Bluff erosion at Point has exposed an old foundation of a building that was originally set well back from bluff edge.

C. Severe beach erosion during winter storms of 1982-1983.

D. Wave erosion in the late 1990s and the winters of 2001-2002 and 2002-2003 damaged road, resulting in relocation of Highway 1 inland and construction of a riprap revetment.

E. Highway 1 undercut in 2002-2003, necessitating construction of riprap revetment.

F. Large sea cave collapsed in 1971. Rapid enlargement of sea caves in the Monterey Formation in 2000-2003.

G. Approximately 50 to 60 feet of seacliff retreat along this section of coastline between 1976 and 2003. This accelerated erosion follows a 125 year long period of no appreciable erosion.

H. Greyhound Rock beach has thinned between 1975 and 2003. Winter waves are beginning to erode into the toes of talus piles at the base of the seacliffs.

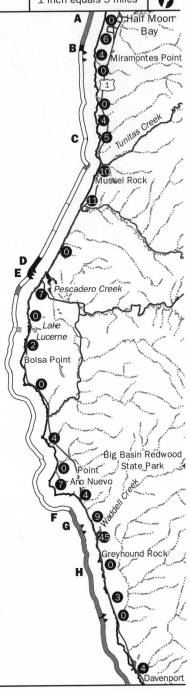

Hazard Level	Erosion Rates	Armoring
□ Stable: low risk	⬤ Inches per year	▲▲ Riprap
▨ Caution: moderate risk		▨▨ Seawall
■ Hazard: high risk		⊡⊡ Other

Martins Beach. Access to most of the isolated pocket beaches along this coastline is both limited and dangerous. Deaths occasionally occur along this section of coastline as a result of falls from cliffs or when beach visitors are swept into the surf by large storm waves.

The sandstones and mudstones that form the seacliffs in this area are cut locally by numerous vertical fractures (joints and small faults) that weaken the rock. Waves selectively erode deep sea caves or small coves along these zones where they intersect the shoreline. Resistant zones and beds in the rocks form a wide variety of arches, sharp headlands, and offshore stacks along this entire coastline. A veneer of sand and gravel up to 65 feet thick covers the older sandstone and mudstone in this area. Groundwater from precipitation and irrigation percolates downward through these permeable sands and gravels, perching on the relatively impermeable bedrock below the sands. Here it flows laterally along the contact between the sands and the bedrock, emerging locally as springs in the vertical seacliffs. These springs can contribute to cliff erosion by weathering (and thereby weakening) the fractured bedrock exposed in the cliff face. However, the seepage along the cliff face also contributes to small slumps and landslides within the soft terrace sands at the top of the cliff that overlie the hard bedrock exposed in the surf zone.

The rates of seacliff erosion in this area are quite variable. Overall, changes in the seacliffs have been minor over the past 80 to 120 years, but locally some changes have been dramatic. For example, about 0.6 miles north of Martins Beach, a large U-shaped cove has formed since 1861, when the first detailed map of the area was made. Erosion may have been initiated by the collapse of a sea cave similar to the ones just south of this site. A long, narrow headland at the north end of Martins Beach has been reduced to an isolated stack over the past 120 years. On the other extreme, the abandoned rock-cut grade of the Ocean Shore Railroad across the vertical cliff face north of Tunitas Creek has not been destroyed by cliff retreat over the past 95 years.

On the shorter time scale, beach erosion and wave inundation have periodically threatened homes at Martins Beach over the past 15 years. At the Miramontes Point golf course a large section of cliff face collapsed during construction of the clubhouse at the top of the seacliff, resulting in the proposed clubhouse being moved to a new site (Figure 13.13). These two examples notwithstanding, the coastal hazards and risks in this coastal segment are very low, due primarily to sparse development and limited access to the shoreline. However, potential coastal hazards related mainly to unpredictable seacliff retreat are locally serious for projects built near the cliff edge.

TUNITAS CREEK TO SAN GREGORIO–PESCADERO

The 6.2-mile-long linear coastline between Tunitas Creek on the north and Pescadero Creek on the south consists of high coastal bluffs and vertical seacliffs cut into high, rolling hills that are dissected by deep, narrow canyons and slightly broader valleys. Seasonal sand beaches extend along most of this coastline during the summer months. At the mouth of Pescadero Creek a permanent sand spit capped by dunes separates tidal lagoons and marshes from the open ocean.

This hilly coastal area is virtually undeveloped, with sheep and cattle grazing representing its primary land use. State Highway 1, which runs locally along the crest of high vertical seacliffs, traverses the area from north to south. The only permanent structures near the coastline are a few homes near the mouth of Tunitas Creek. Access roads from the eastern part of the county intersect the coastal highway near the mouths of San Gregorio and Pescadero creeks. Parking lots and paths to state recreational beaches are situated near the mouths of San Gregorio, Pomponio, and Pescadero creeks. Sandstones and mudstones form the seacliffs along this entire segment of coastline. Locally, a thin veneer of terrace sand and gravel overlies these older rocks.

The rocks that form the 400- to 500-foot-high irregular coastal bluffs in the northern part of this area, between Tunitas and San Gregorio creeks, are sheared and fractured by the San Gregorio Fault, leaving these bluffs unstable and scarred by deep gullies and large landslides. Movement on a landslide just south of Tunitas Creek damaged the coastal highway in the 1950s. Also just south of Tunitas Creek, the old rock-cut for the Ocean Shore Railroad grade (built in 1905–8 but never used) that traverses these bluffs about 100 feet above the beach is gradually being destroyed by gullying and small landslides caused by wave erosion at the base of the slope. Grade destruction was greatly accelerated by severe cliff erosion during the storms of January 1983.

The rocks exposed in the seacliffs between San Gregorio Creek and Pescadero Creek in the southern part of this coastal segment are fairly competent and are not highly deformed. Therefore, the seacliffs in this area are nearly vertical and locally overhang broad and shallow sea caves that form along widely spaced fractures in the rock. Seacliff retreat in this area is very slow and poses no immediate threat to the coastal highway. However, the occasional fall of large sandstone blocks from the vertical cliffs are a hazard to the increasing number of people who use the beaches below for recreational purposes.

The large sand spit capped by coastal dunes that separates the tidal lagoons and marshes of Pescadero Creek from the open ocean has changed little since the preparation of the first coastal survey in 1856, and it is probable that southerly littoral drift will continue to supply an adequate amount of sand for the spit to maintain its position relative to the marsh. However, if sea level continues to rise, and erosion slowly forces the cliffs that lie north and south of the creek to retreat, the spit-dune complex will slowly migrate inland, encroaching on the western edge of the marsh. In January 1983, heavy storm waves cut deeply into the beach and dune ridge on the spit that form the lagoon near the mouth of Pescadero Creek. It is probable that as the spit-dune complex continues to migrate slowly inland we will see continued erosion, and eventually Highway 1 will be threatened and/or destroyed.

In summary, there are minimal coastal hazards in this area because of the generally stable seacliffs and the lack of urban development. However, future development proposals should carefully consider the potential for slow erosional shoreline retreat and the continued landward migration of the Pescadero sand spit–dune complex, which may eventually destroy Highway 1.

PESCADERO CREEK TO FRANKLIN POINT

The highly irregular 11.2-mile-long coastline between Pescadero Creek on the north and Cascade Creek on the south consists of low, rocky cliffs and sparse small pocket beaches backed by low, terraced hills. Bedrock exposed in the cliff faces along most of this open coastline consists of sandstones, mudstones, and boulder conglomerates that are highly resistant to wave erosion. A thin layer of terrace sand and gravel overlies these older rocks and forms the upper part of the cliff face in some areas. At Franklin Point, near the southern end of this coastal segment, sand dunes now stabilized by vegetation overlie the terrace deposits.

Agriculture is the primary land use in this sparsely populated coastal area. In the central part of the area scattered homes are present west of State Highway 1, which traverses the region close to the shoreline.

Because the rocks exposed along the shoreline in this area are highly resistant to wave erosion, long-term rates of seacliff retreat are virtually nil. It is probable that along most of this shoreline there has been no significant erosion over the past several hundred years.

Although much of this shoreline is generally stable, several areas have eroded significantly over the past 25 years. During the severe storms in January 1983, large waves associated with very high tides and storm surge locally eroded weak terrace sands and gravels exposed in the upper part of the seacliffs, threatening a residence and a short section of the coastal highway between Pescadero Point and the mouth of Pescadero Creek. A second home south of Lake Lucerne was also threatened. Wave erosion during these storms also cut deeply into the sand dunes on the north shore of Franklin Point. Continued erosion, culminating in the large storms of the winter of 2002–3, undermined and damaged several portions of Highway 1 south of Pescadero Creek. Early in 2003 the roadbed for Highway 1 was moved inland and riprap revetments were constructed at three points to protect the highway from further damage (Figure 8.4). In a similar fashion, continued slow seacliff erosion now threatens several homes built along the low cliffs south of Pescadero Creek.

These minor problems notwithstanding, actual and potential coastal hazards and risks in this coastal segment are very low due to the relatively stable seacliffs and sparse development.

POINT AÑO NUEVO

Point Año Nuevo is a broad, low-lying headland extending 1.7 miles west from the foot of the Santa Cruz Mountains near the southern boundary of San Mateo County. The gently sloping to nearly flat surface that forms the point is a marine terrace, an uplifted former ocean floor that probably formed during two separate episodes of sea-level rise 105,000 and 83,000 years ago. The terrace surface is underlain by a 5- to 30-foot-thick layer of soft, easily eroded sands, gravels, and silts that overlie hard, erosion-resistant bedrock consisting of sandstone and mudstones. Historic and present erosion rates at the point are primarily controlled by which of these rock types is exposed to wave attack in the surf zone. The formerly active, 5,000-year-old dune field that covers the seaward half of this low headland has become stabilized and heavily vegetated over the past 110 years.

The area is sparsely populated, with agriculture and ranching the principal land uses for the past 200 years. However, in the 1970s Año Nuevo Island along with the western and southern parts of the headland became part of the California State Reserve System. Over the past 30 years the point has become one of the principal breeding areas for the Northern Elephant Seal, and access to the point is restricted, particularly during the breeding months of December through April. Recreational uses such as hiking and docent-led nature walks as well as elephant seal research are the primary uses of the state reserve. State Highway 1 lies inland along the base of the Santa Cruz Mountains and runs along the base of the seacliff south of Año Nuevo.

The 1.0-mile-long northern shoreline of Point Año Nuevo is a straight, sand-gravel beach backed by low bluffs cut mainly into terrace sands and stabilized dune sands. Bedrock exposed at the northern point of Año Nuevo and at Table Rock to the north forms small headlands at both ends of this beach. The 0.8-mile-long southwest-facing shoreline of Point Año Nuevo is a curved, sandy beach that forms a bight between two rocky points. Año Nuevo Island lies about 2,300 feet offshore from the southern point. The irregular 1.0-mile-long southern shoreline of Point Año Nuevo consists of low vertical to overhanging cliffs. Here, small coves, arches, and large sea caves are cut into resistant bedrock consisting of sandstones and siliceous mudstones. Offshore rocks, small sea stacks, and resistant headlands are common along this southeast-facing shoreline. The siliceous mudstones of the Monterey Formation exposed in the seacliff contain abundant nodules of chert (a glassy sedimentary rock that the Native Americans used for making arrowheads). A straight shoreline backed by high, vertical cliffs cut mainly into sandstones and mudstones and fronted by seasonal sandy beaches extends 1.2 miles southeastward from the base of Point Año Nuevo to the county line.

Because the shoreline of this coastal segment is completely undeveloped and will most likely remain that way, there are no major coastal hazards here. However, the shape of the coastline has changed rapidly and dramatically in the past several hundred years and will continue to do so. A few examples of these dramatic coastal changes along this remote coastline are worth mentioning.

The first written record of Point Año Nuevo occurs in the ship's log of Sebastian Vizcaino, who sailed past the point on January 3, 1603. In his log he makes no mention of an island off the headland; therefore, the present Año Nuevo Island was probably part of the mainland at that time (Figure 13.14). The first mention of an island is on the 1798 map prepared by Sgt. Pedro Amador. Consequently, it is probable that the island formed in the late seventeenth or early to mid-eighteenth century. Exactly how this breakthrough forming the island occurred is not known, but it is possibly related to crustal movements associated with prehistoric earthquakes along the San Gregorio Fault zone, which intersects the coastline at Point Año Nuevo. What is clear is that an enormous amount of erosion has occurred at the point since it was first sighted by Vizcaino in 1603.

Exceedingly rapid coastal retreat can be clearly demonstrated by comparing the United States Coastal Survey Map of 1853 with more recent maps and aerial photos (Figure 13.14). Although the channel is currently about 2,300 feet wide and the island inaccessible except by boat, in the mid-1800s it was possible to walk to the island at low tide along the sand spit. The

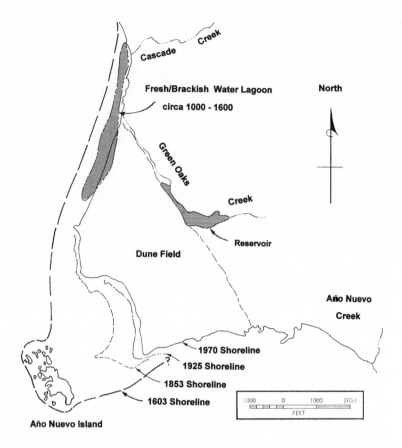

Figure 13.14 Map showing erosional changes at Point Año Nuevo over the past 400 years. The 1603 shoreline is extrapolated from the changes that occurred between 1853, 1925, and 1972. The presence of a broad north beach and a large, possibly intermittent fresh- to brackish-water lagoon is based on Late Holocene peat deposits interbedded in dune sands along the north shore of Point Año Nuevo. Illustration by Gerald E. Weber.

materials used for construction of the lighthouse keepers' houses on the island in the 1870s to 1890s were taken to the island by wagon, across the sand spit at low tide. The rapid retreat of the southern point and the bight shoreline (and to a lesser extent the erosion along the north and south shore since 1853) is consistent with the separation of the island from the point in the late seventeenth century (Figure 13.14).

The presence of the 6,000-year-old dune field at the point clearly indicates that Point Año Nuevo once formed a trap for the southward littoral drift of sand along the coastline. Because of this southward littoral drift, sand accumulated along the north shore of both Franklin Point and Point

Año Nuevo, forming wide beaches. Under natural conditions, the prevailing northwest winds would blow this sand across the low headland of Franklin Point to form migrating dunes that moved inland. At Point Año Nuevo the sand dunes moved across the point, eventually cascading over the southern cliffs and back into the ocean (Figure 13.15). Dune migration has virtually halted on both headlands within the past 70 years. The dune stabilization at Point Año Nuevo is due to a combination of (1) increases in the natural and introduced vegetation in the dune fields, (2) changes in the groundwater conditions due to agriculture, (3) sand quarrying operations that depleted a portion of one large dune in the 1950s, and (4) the disappearance of the broad beach along the north shore that supplied sand for the dune field. Clearly, the disappearance of the north shore beach was caused by the formation of the channel between the point and the island. This event allowed the sand on the beach to slowly erode away and be added to the littoral drift that moves along the Santa Cruz County coastline.

As a result of the reduction of sand in the littoral zone, beaches at the point have thinned drastically over the past 150 years, and accelerated erosion occurs along most of the Point Año Nuevo shoreline. Historical maps and photographs indicate that little if any erosion had occurred along the once-vegetated cliffs north of the mouth of Año Nuevo Creek during the 110- to 125-year period prior to 1972 (Figure 13.16). These cliffs are now bare of vegetation and have eroded back approximately 50 feet over a period of about 25 years (Figure 13.17). Similarly, rapid erosion occurs along the southeast-facing shoreline, the point, and the north shore. Large waves associated with the storms of 1983, 1995, and 2002 cut deeply into the unprotected dunes on the north shore of Año Nuevo, and storms of the late 1990s and early 2000s have carved large caves into the hard, cherty mudstones exposed along the south shore (Figure 13.16). The sudden roof collapse of a large sea cave along the south shore in 1971 may have been related to increased wave erosion due to the recent disappearance of beaches.

To summarize, the formation of the channel between Año Nuevo Point and the island resulted in the release of a large reservoir or point source of sand for littoral drift. Rough estimates suggest that as much as 12 to 18 million cubic yards of additional sand-sized sediment was made available to littoral drift over the past 200 to 330 years. During that time the normal flow of sediment in littoral drift along the Santa Cruz County coastline was supplemented by many tens of thousands of cubic yards of additional sand on a yearly basis. As a result, many beaches down drift (southeast) of the point widened, including the beaches in northern Monterey Bay. The beaches most affected by this event were partially sheltered or protected from the large northwest swells of winter storms. Consequently, behind many of

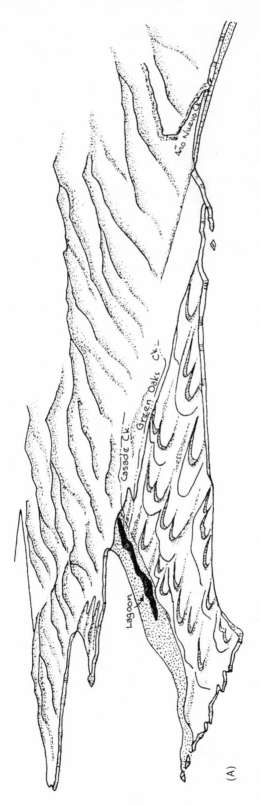

Figure 13.15 Interpretive drawings of the major erosional changes at Point Año Nuevo over the past 400 years. (A) 1603: A broad beach (possibly a broad sand island) lies along the north shore of the point that is separated from the shore by either a freshwater lagoon or brackish-water slough connected to the ocean. The dune field is active, with dunes forming along the shore from Cascade Creek southward to what is now Año Nuevo Island.

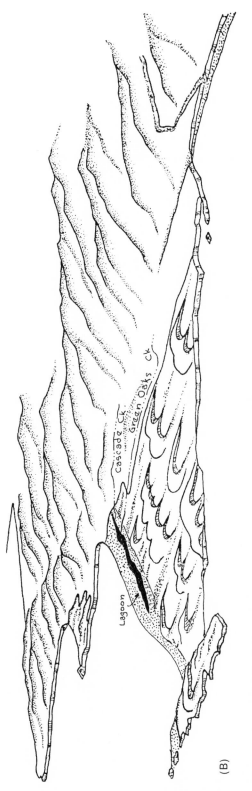

Labels within figure: Lagoon, Cascade Ck, Green Oaks Ck

(B)

Figure 13.15 (*Continued*)   (B) Early 1700s: Surface faulting and/or subsidence during a large earthquake have changed conditions sufficiently to initiate erosion of a channel between the present-day point and island. Following formation of the channel, the sand that has been trapped behind Point Año Nuevo (and eventually large portions of the sandy sediment that make up the point) are now available to the littoral drift through the newly formed channel.

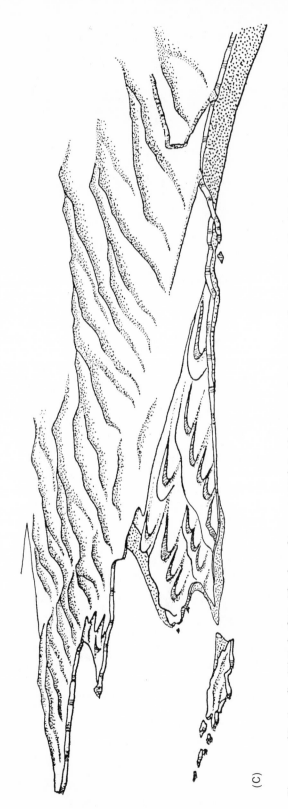

Figure 13.15 (*Continued*) (C) Early 1800s: The channel is over 1,000 feet wide, but a sand spit connects the point to the island during low tides. The north beach has eroded and thinned, but there is still sufficient sand to feed the dune field.

(C)

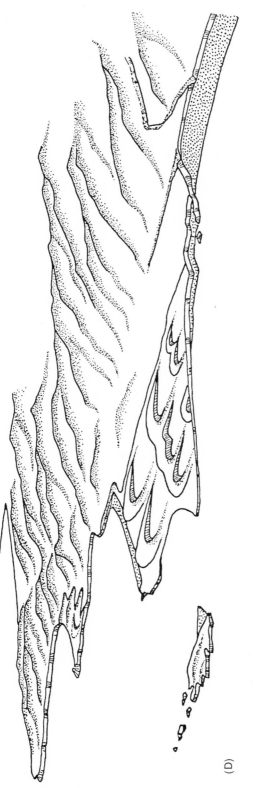

(D)

Figure 13.15 (*Continued*) (D) 1900: The channel is over 1,800 feet wide and large portions of the point have been incorporated into the littoral drift. The north beach has thinned to the point where there is insufficient sand to feed the dune field. Beaches at the point have thinned as the amount of sand available to littoral drift has declined rapidly.

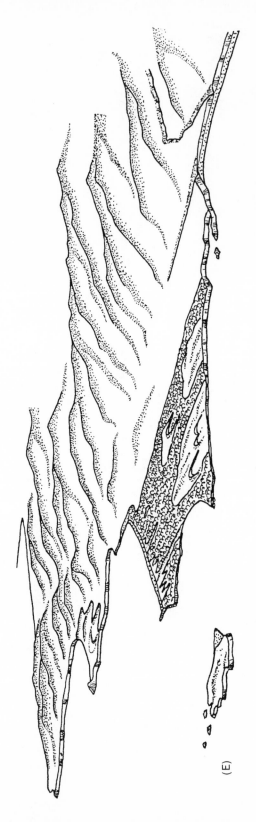

(E)

Figure 13.15 (Continued) (E) 2003: Active dune formation has essentially stopped along the north shore. Only a few thin longitudinal dunes are now forming. The broad beach that once supplied sediment to the dune field has essentially vanished, and the low seacliffs along the north beach are eroded on a yearly basis. Wave erosion is rapidly eroding the cliffs behind the southern beaches and along the south-facing seacliffs, an event not common for at least several hundred years. The large point source of sand that fed excess sand into the littoral drift for 200 to 300 years is gone, and the original equilibrium has been reestablished. The "tail end" of this large slug of sand is now slowly migrating down coast toward northern Monterey Bay. Illustration by Gerald E. Weber.

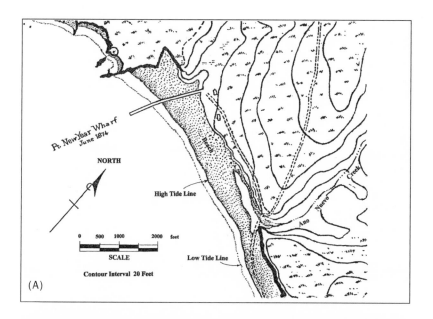

Figure 13.16 Erosional changes along the south shore of Point Año Nuevo during the past 150 years. (A) Portion of the 1853 shoreline map showing a straight cliff face east of the long, linear ridge of rock that extends southwesterly from the shoreline. (B) April 1972 aerial oblique of the same portion of seacliff showing two large coves eroded into the seacliff over the past 150 years. The eastern cove is approximately 200 feet across and over 100 feet deep. The smaller, western cove (about 75 to 100 feet across) formed in mid-April of 1971 in what must have been an essentially instantaneous sea cave collapse. During the ensuing 30 years the entire mass of collapsed rock has been removed by wave erosion. Compare the width and position of the beach on the 1853 map with the 1972 photo. The 1853 beach is more than twice as wide as the present beach. Illustration by Gerald E. Weber.

Figure 13.17   Erosional changes south of Point Año Nuevo. (A) August 1973. Note the position of the rock attached to the seacliff in the foreground, the vegetated seacliff, and the partially vegetated dune ramp several thousand feet down the beach. (B) Same view in February 2000. The lone rock is the same is in (A). The face of the seacliff is freshly eroded and the dune ramp has disappeared. During the 27 years between these photographs, the seacliff was eroded back 50 to 75 feet. In the 1973 photo a lone remaining piling from Waddell's Wharf (built in 1862) can barely be seen at the base of the seacliff. Illustration by Gerald E. Weber.

these beaches cliff erosion was either slowed or halted altogether. Because that large point source of sand is now gone, there is no addition of sediment to the littoral drift immediately south of Point Año Nuevo, and the tail end of the wave of extra sand is slowly migrating down coast. Figure 13.15 is an interpretive set of drawings of the probable changes in the shoreline at Point Año Nuevo over the past 400 years.

The most interesting conclusion that can be inferred from the erosional changes at Point Año Nuevo is that the volume of sand in littoral drift along the northern coastline of Santa Cruz County has been slowly decreasing over the past 25 years. The decrease has occurred because the large volume of sand created by the erosion of the channel at Point Año Nuevo over the past 250 years has ended. As the tail end of this point source of sand moves southward along the coast, the once-wide beaches will slowly thin (Figure 13.18). The eventual result may be accelerated erosion along portions of sea-cliffs in Northern Monterey Bay, where erosion has been minimal over the past two centuries or more.

The erosional changes at Point Año Nuevo is also a reminder that it is difficult to characterize or determine the long-term equilibrium of geologic processes based on short-term observations. It is probable that the amount of sand moving in littoral drift along the Santa Cruz County coastline over the past 200 to 250 years (since occupation by the Europeans) is greater than what had been moving along the coastline over the past 4,000 to 5,000 years. Hence, in Santa Cruz County we have been and still are living in a geologic instant when there is a temporary excess of sand.

SUMMARY AND CONCLUSIONS

The primary coastal hazards in San Mateo County include slope failures at Thornton Beach, Mussel Rock, and Devil's Slide. Upslope expansion of the large landslides at Thornton Beach, Mussel Rock, and Seal Cove has damaged and continues to threaten homes and streets at the top of the high coastal bluffs. Movement on the large landslide at Devil's Slide periodically closes State Highway 1. Wave erosion at the toe of the high, unstable bluffs originally caused and has subsequently maintained the slope instability in these four areas.

Wave erosion and inundation is a serious, long-term hazard at Sharp Park, Rockaway Beach, Seal Cove, and Half Moon Bay. Of these hazards, only the one at Half Moon Bay was caused by human activity, in this case the construction of the breakwater between 1959 and 1960. Large waves accompanying the heavy storms in the winters of 1982 and 1983 eroded many low-lying coastal bluffs that had not been threatened historically.

Figure 13.18 Erosional change at the site of Waddell's Wharf. (A) Photo of lone remaining piling taken in January 1977. Piling is within 3 feet of the base of the seacliff. (B) Photo taken in December 1977. Piling is about 18 feet from the cliff face. Cliff retreat during portions of two storm seasons was approximately 15 feet.

Most of this damage took place on the north side of headlands, indicating that the damaging waves came from the northwest; in the past most coastal damage was caused by waves from the southwest.

South of Miramontes Point coastal hazards are currently minimal, mainly because of the low population density and sparse development. However, between Pescadero Beach and Franklin Point the low, exposed coastline has few hazards, primarily because of the resistant nature of the rock exposed to wave erosion in the seacliffs.

Point Año Nuevo has historically experienced the greatest changes and most rapid erosion of any area in San Mateo County. The formation of the channel between the point and the island occurred only recently, approximately 300 to 350 years ago. Development of the channel changed the equilibrium (which may have existed for as long as 5,000 to 6,000 years) at the point, allowing the destruction of the large beach along the north shore that fed the dune field and protected the point from erosion. Channel formation was followed by a 300-year period of rapid erosion in which the soft terrace and dune sands were rapidly eroded, drastically changing the shape of the point. Erosion continues, but at a reduced rate, as the waves are now starting to attack the hard sandstones and mudstones of the bedrock that underlies the terrace deposits. There are no coastal hazards because of the absence of development along this stretch of coastline.

CHAPTER FOURTEEN

AÑO NUEVO TO THE MONTEREY PENINSULA

GARY GRIGGS AND KIKI PATSCH

HISTORY

The first inhabitants of the central California coast arrived well over 10,000 years ago. Archeological evidence, including radiocarbon dating, suggests that the initial inhabitants were hunters; however, about 7,500 years ago a strong dependence on shoreline resources and seed gathering developed. Ancient kitchen middens (shell mounds), which clearly record both dietary and waste disposal practices of the original inhabitants, are widespread from Año Nuevo south to Santa Cruz, in the mid-bay region, and on the Monterey Peninsula. In addition to access to a rich intertidal area for food, the Native peoples sought two of the things we still search for today: freshwater and protection from the wind.

The first Europeans to reach the area, a Spanish expedition led by the Portuguese explorer Cabrillo, entered Monterey Bay in 1543. Sixty years later, in 1602, Sebastian Vizcaino sailed into Monterey Bay and described it as "a harbor sheltered from all winds." Although it may have been somewhat protected from northwest winds during that voyage 380 years ago, the coastline of the bay has been repeatedly attacked and damaged over the years by severe storms from the west and southwest. Vizcaino passed Point Año Nuevo on January 3, 1603, and gave it its present name (Punta del Año Nuevo). The early marine and overland expeditions along this coastline provided clear evidence for the rate of coastal retreat and shoreline changes at Point Año Nuevo.

It seems apparent from the travel accounts of Father Crespi (1769), Father Palou (1776), and Captain Vancouver (1798) that what we know today as Año Nuevo Island was a peninsula in the not-too-distant past. As late as 1857 at least one detailed and official map shows no island, only a point connected by a sandbar. Now that the point here has been breached, erosion of the low terrace is proceeding rapidly, averaging about 9 feet per year for at least the past hundred years and leaving behind an island covered with thousands of seals and sea lions.

By the mid-1800s, a variety of settlers had been attracted to the central coast from Año Nuevo to Monterey by the region's natural resources: fertile, flat land and water for agriculture and grazing; redwood for lumber; and limestone for cement. Although these same natural resources are important parts of the coastal economy today, tourism and recreation and higher education have also become increasingly important to the local economy. By the late 1800s California's inland population was attracted to the cooler coastal climate as an escape from the summer heat in the inland valleys. Hotels and resorts were followed by more permanent residents who built cliff-top and ocean-view homes. Within the last 30 years, many large-scale oceanfront developments have been built along much of the inner Monterey Bay coastline as the outside world has discovered the quiet beach town. The early developments consisted of small vacation or second homes, but as demand for housing has increased, many of these part-time residences have become year-round homes. Almost all real estate in Santa Cruz, but particularly oceanfront or ocean-view homes, now commands astronomical prices. From 2003 to 2005, median home prices in the Santa Cruz area ranged between $550,000 and $750,000, among the highest in the nation; and these aren't all high-end houses. Homes on the beach or on the coastal cliffs with unobstructed ocean views typically have asking prices in the $2 to $10 million range! For people who can afford these prices, ocean views and beach access have been far more important than the hazards of winter wave attack or seacliff erosion. Most oceanfront property sells in the warm summer months, when the waves are small and the beaches wide. People pay a lot for their oceanfront property, and often they have to pay again to protect it when severe winter storms roll through.

AÑO NUEVO TO NATURAL BRIDGES STATE BEACH

PHYSICAL SETTING AND LAND USE

The 25-mile-long coastline between Año Nuevo and Natural Bridges State Beach is for the most part undeveloped. A sequence of up to five elevated marine terraces or benches, used principally for agriculture (brussels sprouts and artichokes are grown on the lowest terrace) and grazing, flank the coastline throughout much of this area. State Highway 1 follows the inner edge of the lowest and youngest (roughly 100,000 years old) of these uplifted benches (Figure 2.3). These nearly horizontal terraces were cut by wave action in the surf zone. The sequence of terraces exposed here and in

A. Highway One endangered by rock falls from steep bluffs prior to protection by debris fence. Storms in 1983 endangered highway fill leading to an additional 2000 feet of riprap to protect road.

B. Near vertical cliffs in Santa Cruz Mudstone nearly 150 feet in height with no permanent beach.

C. Greyhound Rock is a tombolo or large rock mass connected to the shoreline by a sand spit which is periodically washed over by waves.

D. The mouth of Scott Creek migrates across the beach in response to stream flow and littoral drift. The combined effects of high creek flow and winter wave erosion at high tides has led to the armoring of the road fill along the back beach with riprap.

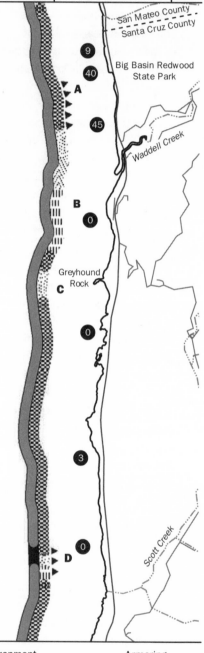

Hazard Level

Stable: low risk
Caution: moderate risk
Hazard: high risk

Erosion Rates

1 Inches per year

Shoreline Environment

Sandy beach
Beach backed by marsh
Beach backed by dunes
Cliff or bluff fronted by beach
Rocky coast with no beach

Armoring

▲ ▲ ▲ Riprap
Seawall
Other

many other places along the California coast result from a combination of the rise and fall of sea level over the past 500,000 years combined with the slow uplift of the coast (Chapter 2).

The small community of Davenport, which is built around a large cement plant, is the only town in the area. In 1867 Captain John Davenport, a whaler from New England, established a whaling and lumber shipping settlement at Davenport Landing, about a mile and a half north of the present town. A 450-foot-long wharf was soon constructed and was used as late as 1905. The sea, however, eventually claimed this pier, as well as the inner portion of another one constructed in 1934 at Davenport for shipping cement by sea. The first major cement company started out in 1906 as the Santa Cruz Portland Cement Company, and although it has changed hands several times, it is still a major producer of cement and the main employer in the community, a true company town. Today the cement is shipped by trucks and rail.

Most of the coastal land between the city of Santa Cruz and Scott Creek was owned for many years by ranching families or corporations and maintained in agriculture. Increasing property taxes and economic incentives for other land uses in the late 1960s and early 1970s led to two major proposals that would have changed the character of this area forever. The 7,000-acre Coast Lands and Dairy property was optioned to Pacific Gas and Electric Company for what was proposed as the world's largest nuclear power plant, and a firm that proposed the development of homes for 25,000 people and annexation to the city purchased the 2,000-acre Wilder Ranch. Local opposition developed, and after many years of studies and public meetings the properties were ultimately purchased with a combination of state bond money and funds from several private foundations. About 9,000 acres of coastal land extending from the Santa Cruz County line to north of Scott Creek are now set aside for permanent recreational and agricultural use.

GEOLOGICAL CONDITIONS

The bedrock exposed in the seacliffs, which range from 30 to nearly 200 feet in height along this reach of coast, is mudstone that is moderately resistant to erosion. More rapid erosion occurs along zones of weakness within the mudstone, such as faults and joints, as well as within the sandstone intrusions that are unique to this stretch of coast. This differential or preferential erosion has produced a variety of scenic arches, tunnels, and caves along the coastline. A close inspection of a particular stretch of cliff where an arch or tunnel exists will usually reveal the zone of rock weakness responsible for the differential erosion. The sandstone intrusions that are visible in the seacliffs north and south of Davenport resulted from the liquefaction of buried sand that was saturated with fluid (either water or oil), probably during major earthquakes.

Figure 14.1   Sandstone intrusions at Yellow Bank Beach. Two different series of sand-stone dikes have been squeezed up into the overlying Santa Cruz Mudstone. Photo by Gary Griggs and Kiki Patsch.

It is believed that the sand had its origin at greater depths and was then squeezed under pressure, much like toothpaste from a tube, into the overlying mudstone to create the distinct features visible in the cliffs (Figure 14.1).

A low mudstone bench or shore platform 50 to 100 feet wide extends seaward at the base of the cliff along much of this coastline (Figure 14.2). Most of the wave energy is expended against the outer edge of this resistant bench rather than against the seacliff. During major winter storms and high tides, however, breaking waves will reach across the platform and remove loose rock and debris. A shore platform is well developed north of Natural Bridges State Beach and provides not only excellent tide pooling, but also

Figure 14.2   A set of low shore platforms fronting the seacliff at Wilder Ranch State Park. Photo by Gary Griggs and Kiki Patsch.

important long-term protection to the cliffs fronting a mobile home park and a marine laboratory. The mast of a ship, the *La Feliz,* which went aground on the platform in 1924, provides evidence for this natural protection. The mast was removed from the ship, placed against the cliff, and then used with a block and tackle to help salvage the cargo. Eighty years later, the mast is still leaning against the cliff in the same location, providing clear evidence for the lack of cliff erosion in this period of time (Figure 14.3).

The only significant beaches along this stretch of coast occur where the small coastal streams have eroded their way through the marine terraces and near-vertical cliffs. The steep-sided canyons of these incised streams were major obstacles to travel along this coast for nearly a century. When the highway and railroad were first constructed up the coast in the early 1900s, many of the streams were confined to rock tunnels bored through the mudstone, which was less expensive than building large bridges over the canyons. Because of low summer flow in most of these streams, littoral drift often closes off the creeks so that lagoons form at their mouths.

CLIFF EROSION AND PROTECTION

Considerable wave energy is expended against the exposed stretch of seacliffs along the north coast of Santa Cruz County. Yet average long-term erosion rates in most places are moderately low (3 to 6 inches per year or less) because of the erosion resistance of the mudstone and the presence of

Figure 14.3 (A) The wreck of the *La Feliz* on the shore platform immediately north of Natural Bridges State Beach in 1924. (B) The mast of the *La Feliz* is still visible in 2003 leaning up against the cliff. Photo by Gary Griggs and Kiki Patsch.

Figure 14.4 (A) A natural arch at Wilder Ranch State Park, ca. 1900. (B) The same Wilder Ranch location in 2002. Photo by Gary Griggs and Kiki Patsch.

protective shore platforms in many places. Periodically an arch or sea cave collapses, producing rapid retreat (up to 75 feet in one case) nearly instantaneously (Figure 14.4).

Because the uplifted coastal terraces are used primarily for agriculture and recreation, the erosion of the seacliffs has produced no major hazards or damage to date. The structures facing the greatest threat from coastal erosion over the long term are Highway 1 and the railroad between Santa Cruz and

Figure 14.5 (A) Looking south along Waddell Bluffs from the Santa Cruz–San Mateo county line in 1954. Photo by U.S. Army Corps of Engineers. (B) Looking south at Waddell Bluffs in 1983, with riprap to protect road fill from wave erosion. Photo by Gary Griggs and Kiki Patsch.

Davenport. Combined stream runoff and wave action in 1973 undercut about 30 feet of railroad track just south of Davenport, leaving it dangling in midair. Riprap was subsequently emplaced and the track was replaced.

Another problem location exists between Waddell Creek and the San Mateo County line, where high, steep bluffs flank the inland side of Highway 1. These actively eroding bluffs mark the former coastline and historically formed a barrier to transportation between Santa Cruz and San Francisco. The Ocean Shore Railway opened in 1908 and was intended to connect San Francisco and Santa Cruz along a coastal route. The railroad was never able to bridge the Waddell Bluffs gap, however, due to constant failure of the bluffs. Travelers made the connection for years on a Stanley Steamer that crossed the beach below the cliffs at low tide. In 1947, the state removed about a million cubic yards of loose rock at the base of the cliffs and constructed the present highway. A 600-foot stretch of riprap was emplaced during construction to protect the section most exposed to wave action (Figure 14.5). During the 1983 El Niño storms, waves removed all of the loose rock and fill protecting about 2,000 feet of highway at the southern end of the bluffs. Twenty-four thousand tons of rocks were emplaced to save Highway 1 that year.

Weathering of Waddell Bluffs has led to continual sloughing of both small chips and large rocks. As the maintenance ditch at the base of the cliff filled, rocks falling onto the roadway at times became a hazard to motorists. A debris fence has been added to catch materials coming off the bluffs before they reach the highway.

NATURAL BRIDGES STATE BEACH TO NEW BRIGHTON BEACH

PHYSICAL SETTING AND LAND USE

The urbanized portion of Santa Cruz County begins at Natural Bridges State Beach and extends well into Monterey Bay. The 10-mile-long stretch of coastline between Natural Bridges and New Brighton Beach consists primarily of seacliffs that range in height from 25 to 75 feet and are often fronted by sandy beaches of varying widths. Wider pocket beaches exist at the mouths of streams and at natural headlands or where artificial barriers to littoral transport (jetties and groins) have been constructed (Figure 14.6).

Lighthouse Point forms the northern edge of Monterey Bay and provides protection for the beaches of the northern bay by sheltering them from the dominantly northwesterly waves. The wide, protected beaches of northern Monterey Bay combined with the excellent climate and proximity to the

A. The flat terrace at El Jarro Point was proposed for a large nuclear power plant in 1969 but local opposition and environmental concerns terminated the project.

B. Several piers have been built and destroyed over the years at Davenport. The remnants of a pier built to load cement onto a ship is still visible today, along with the large steel doors exposed on the cliff face across from the cement plant where a tunnel was constructed to allow cement to be carried by rail to the waiting ship.

C. In order to construct State Highway 1 and the railroad up the north coast, the stream valleys were bridged with fill, carried along a wooden trestle, and the streams were rerouted through hand-dug tunnels through the mudstone cliffs.

D. Yellow Bank or Panther beach, named for the yellowish colored cliffs or the apparent outline of a panther in the cliff, is the best exposure of large sandstone intrusions along the north coast. A large arch has been eroded along the south side of the beach.

E. Sand Hill Bluff is unique in that relict sand dunes atop the terrace have been preserved. These would have formed when sea level was much lower and the dunes were connected to a beach as a source of sand.

F. Highway 1 follows the back edge of the first marine terrace and the road cut at Majors intersects the old seacliff (~100,000 years old) that can be seen in the cut. Small asphaltic sandstone intrusions can be seen in this road cut and another a few hundred feet south.

G. This is one of the only locations where all five elevated marine terraces are exposed.

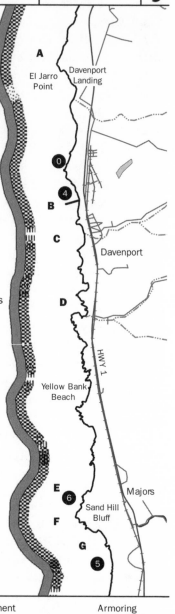

Hazard Level
- Stable: low risk
- Caution: moderate risk
- Hazard: high risk

Erosion Rates
- 1 Inches per year

Shoreline Environment
- Sandy beach
- Beach backed by marsh
- Beach backed by dunes
- Cliff or bluff fronted by beach
- Rocky coast with no beach

Armoring
- ▲▲▲ Riprap
- Seawall
- Other

Figure 14.6 The coastline of northern Monterey Bay from the boardwalk to the Santa Cruz Small Craft Harbor, showing beaches up coast from major littoral drift barriers, including San Lorenzo Point (center of photo, at the river mouth) and the west jetty of the harbor. Photo by Gary Griggs and Kiki Patsch.

greater Santa Clara Valley area, as well as one of the few remaining beach boardwalks, have attracted summer tourists to the Santa Cruz area for well over 100 years. Oceanfront hotels and motels, piers, casinos, and the board-walk attractions have come and gone with fires and storms over the years. Older homes with ocean views are being replaced by large, newer homes with prices in the multi-million-dollar range.

GEOLOGICAL CONDITIONS

The Santa Cruz Mudstone that makes up the cliffs along the north coast of the county extends a short distance into the city of Santa Cruz and underlies the West Cliff Drive area up to about Almar Avenue (about three-fourths of a mile north of Lighthouse Point). Here the mudstone dips beneath the waves, and the overlying younger rock, the Purisima Formation, a sedimentary rock consisting of sandstones and siltstones with occasional layers or lenses composed almost entirely of mollusk shells, makes up the low coastal cliffs.

Much of the city of Santa Cruz has been built on the lowest marine terrace, a flat benchland up to several miles wide. Where the terrace is well exposed along West Cliff and East Cliff drives, the flat bedrock bench contains the well-preserved shells of boring clams. Uplift of the terrace was not uniform, however. In the East Cliff Drive area between Corcoran Lagoon and Moran Lake, the bedrock of the terrace is nearly at sea level, leaving only the overlying loose sand and gravel deposits to resist wave action. As a result, seacliff erosion rates in this area are very high, and most of this coastline has been armored over the past 30 to 40 years. To the east the terrace climbs quickly, and from Opal Cliffs to Capitola the bedrock is elevated to a height of about 70 feet.

CLIFF EROSION AND PROTECTION

Many detailed studies have focused on the coastal area from Santa Cruz to New Brighton Beach because of the relatively rapid rate of erosion and the impact of cliff retreat on roads, sidewalks, and homes. Historical ground photographs of this area provide a good record of the changes that have taken place over the past century (Figure 14.7). Various factors control the rate at which the seacliffs erode in this and similar areas: the resistance of the cliff materials to wave impact as well as terrestrial processes (runoff, landsliding, and other types of slope failure); structural weaknesses within the bedrock, primarily jointing and its orientation (Figure 14.8); the seasonal presence or absence of a protective beach; and human activities such as placement of storm drains, and the direct and indirect effects of coastal engineering structures (seawall, riprap, and jetties or groins).

For the most part the long-term retreat rates of the mudstone cliffs extending along West Cliff Drive are relatively low (4 to 8 inches per year or less on the average). Photographs of some areas show no recognizable changes in over 75 years (Figure 2.2). When erosion does occur, however, it is usually episodic rather than gradual. For example, two of Natural Bridges' three arches collapsed during heavy storms (Figure 14.9). The major threats along West Cliff Drive are to the road and adjacent bicycle path, parking areas, and the lighthouse, which is now endangered by wave-eroded caves.

Figure 14.7   (A) Coastline immediately east of Lighthouse Point in Santa Cruz, also known as Steamer Lane, ca. 1900. (B) Photo of the same location as in part (A), taken in 2003. Photo by Gary Griggs and Kiki Patsch.

The first riprap was placed along West Cliff Drive around 1965. In subsequent years, and particularly during the last 25 years, many additional areas have been armored. During the winter of 1983, combined high tides and extreme waves struck the low bluffs with fierce intensity, resulting in considerable retreat and damage to both the bike path and roadway; this damage

Figure 14.8 Erosion along low coastal cliffs at Point Santa Cruz is strongly influenced by the orientation and spacing of joints in sedimentary rocks (the joints are designated with arrows). Photo by Gary Griggs and Kiki Patsch.

led to additional armoring. Today, much of West Cliff Drive is protected by various structures, primarily riprap.

The Purisima Formation presents far less resistance to wave attack than the mudstone. The indentation of the coastline that forms the northern edge of Monterey Bay is due primarily to the presence of this weaker rock type. A few resistant headlands in the Purisima do stand out, such as Lighthouse Point and Soquel or Pleasure Point, as a result of harder layers within the generally weak Purisima sandstones and silt-stones. Although inner Monterey Bay is more protected from dominant northwesterly waves than the exposed coast to the north, the cliffs cut into the Purisima erode considerably faster than those in the mudstone.

Figure 14.9 Progressive coastal cliff retreat at Natural Bridges State Beach has reduced the original three bridges to a single arch. (A) Pre-1900. (B) 1970. Photo by Gary Griggs and Kiki Patsch. (C) 2003. Photo by Gary Griggs and Kiki Patsch.

Because of the relatively rapid cliff erosion throughout this area over the years, and the almost complete urbanization of the cliff top from West Cliff to Capitola, considerable effort and expense currently and for the past 50 years have focused on preventing these cliffs from eroding. Of the 10 miles of coastline between Natural Bridges and Capitola, approximately 4.5 miles, or nearly 50 percent, are protected by some type of armor, usually riprap. Although the effectiveness of the protective measures varies considerably, it is clear that the armor has significantly reduced or at least temporarily halted the cliff erosion. Many of the erosion rates given in the accompanying maps represent conditions prior to the emplacement of any riprap or seawall.

The jetties of the Santa Cruz Small Craft Harbor were completed in 1965 and initially had a pronounced effect on the coastline, both immediately up coast and for a considerable distance down coast (Figure 14.6). The west jetty formed a major barrier to littoral drift moving eastward and as a result created a wide, stable beach (Seabright or Castle Beach) in an area that had previously undergone rapid erosion (Figure 14.10). A large portion of the approximately 300,000 cubic yards of annual littoral drift fills the harbor entrance and requires annual dredging, with a price tag of over $500,000 each year. In the first several years following harbor construction, before annual dredging was initiated and the sand flow returned, there was a sand deficit down coast, resulting in a decrease in beach and accelerated bluff erosion. Erosion rates in the weak Purisima Formation and along joints and fractures had already created an area of rapid cliff retreat, and harbor construction initially added to the problem. Most cliff-top homeowners have built seawalls or placed riprap to protect their homes, and additional projects are still being proposed. Cost and effectiveness have varied depending on the amount, structure, and size of rock used. Many homeowners had to replace or reinforce their protection after settlement and ultimate collapse of the structures. Following the 1983 storms, one property owner spent $100,000 on additional protective rock (Figure 14.11). This is a high price to pay for an unobstructed ocean view.

Within several years of harbor construction, the beach at Capitola disappeared, and the winter waves began to attack the oceanfront structures and parking area. To alleviate this situation and to provide a beach for the summer tourists, on which the community depends, Capitola eventually constructed a 250-foot-long groin at the down-coast end of their beach and brought in about 2,000 truckloads of local quarry sand to build a beach. Capitola's sandy beach was stabilized and provides a protective buffer except during periods of severe storm waves from the west or southwest, which occur during El Niño winters (Figure 3.4). In the El Niño winters of 1978

Figure 14.10 (A) No beach existed at Seabright in the winter prior to harbor construction. (B) Construction of the jetties at Santa Cruz harbor produced year-round widening of Seabright Beach. Photo by Gary Griggs and Kiki Patsch.

and 1983, waves washed debris over the low seawall and into the downtown village of Capitola.

Average annual erosion rates along East Cliff Drive between the harbor and New Brighton Beach prior to harbor construction and armor emplacement varied between several inches and 1 to 2 feet. Some unprotected areas are still retreating at rates as high as a foot per year, leading to road closure

Figure 14.11 Between 1950 and 1984, additional rock was emplaced to protect this property along East Cliff Drive 13 times, but the rock continues to sink into the sand and move seaward. Photo by Gary Griggs and Kiki Patsch.

and threats to houses, which in some cases results in relocation, and to city streets and utility lines. Particular problem areas include Pleasure Point, Opal Cliffs Drive between 41st Avenue and Capitola, and Depot Hill, between Capitola and New Brighton Beach. On Depot Hill, apartments were undermined during the 1989 Loma Prieta earthquake (Figure 14.12), roads fell in, and entire parcels of land disappeared. There is no such thing as a permanent protective beach, and the rock weaknesses favor the failure of large blocks of rock. Although wave attack and undermining is the most common failure mechanism, seismic shaking during the 1989 Loma Prieta earthquake produced major failure along the Depot Hill cliffs and led to the demolition of six apartment units (Figure 14.12). Many different cliff protection projects have been proposed over the years for the Depot Hill cliffs, but to date not one has been approved or carried out.

NEW BRIGHTON BEACH TO THE PAJARO RIVER

PHYSICAL SETTING AND LAND USE

The coast from New Brighton Beach to the Pajaro River constitutes southern Santa Cruz County and lies within the usually protected inner portion of Monterey Bay. Wide, sandy beaches border this stretch of coast and provide nearly permanent protection to the bluffs from wave attack. From New

Figure 14.12   These apartment buildings on Depot Hill were damaged by cliff failure during the 1989 Loma Prieta earthquake and were ultimately demolished. Photo by Gary Griggs and Kiki Patsch.

Brighton to La Selva Beach an uplifted marine terrace about 100 feet in height is the dominant coastal landform, with a steep cliff forming the seaward edge of the terrace. To the south, from Manresa Beach to the Pajaro River the terrace disappears, and recent and ancient sand dunes dominate the coastline.

A. Four-Mile beach is a popular surfing spot and has a well-developed shore platform along the northern side of the beach.

B. 2000 acres between Wilder Creek and Baldwin Creek (Four-Mile Beach) are now protected as Wilder Ranch State Park with scenic trails along the coastal cliffs.

C. Shore platform protects cliffs from Terrace Point to Natural Bridges State Beach. Mast of the *La Feliz* has been leaning against the cliff since 1924 indicating no erosio.

D. There were originally three arches at Natural Bridges State Beach. Two have collapsed over the past 30 years.

E. West Cliff Drive follows the outer edge of the marine terrace and is only about 30 feet above sea level. The bike path and some parking areas are frequently overtopped during intense winter wave activity.

F. Only a single house is sited on the ocean side of West Cliff and it is on a resistant point and protected with gabion baskets of rock.

G. The cliff is very low at Woodrow Avenue, where a small drainage discharges and winter waves frequently carry water and debris onto the bike path and roadway.

H. Steamer Lane is a world-famous surfing spot with very large winter waves that continue to erode the point. Two wave-eroded caves extend under the point toward the lighthouse.

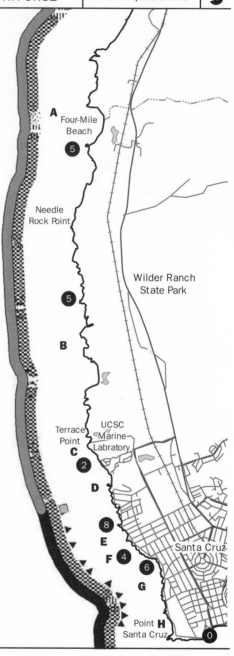

Four-Mile Beach

Needle Rock Point

Wilder Ranch State Park

Terrace Point

UCSC Marine Labratory

Santa Cruz

Point Santa Cruz

Hazard Level

☐ Stable: low risk
▨ Caution: moderate risk
■ Hazard: high risk

Erosion Rates

① Inches per year

Shoreline Environment

Sandy beach
Beach backed by marsh
Beach backed by dunes
Cliff or bluff fronted by beach
Rocky coast with no beach

Armoring

▲▲▲ Riprap
■■■ Seawall
⊡⊡⊡ Other

A. While cliffs were eroding prior to harbor construction due to winter wave attack, Seabright Beach is now about 600 feet wide as a result of sand impoundment against the west jetty of the Santa Cruz Small Craft Harbor.

B. Low cliffs in Purisma Formation between the harbor and Soquel Point are completely developed and historically subject to winter wave attack.

C. High tides and storm waves in recent winters have eroded the beach and brought logs and debris across East Cliff Drive and into Corcoran Lagoon.

D. East Cliff Drive at Moran Lake is very low and is periodically overtopped by severe winter waves during high tides. A house built on the beach was destroyed here in 1983 storms.

E. Bluffs between Soquel Point and 41st Avenue are eroding at about a foot/year and East Cliff Drive and the sewer and water lines are now threatened. A bluff stabilizaton project is being studied (2005), and some emergency repairs have been completed.

F. Two houses are perched precariously on the ocean side of East Cliff with cliffs covered with gunite.

G. Cliff erosion along joints in the underlying Purisima bedrock is becoming a threat to homes on ocean side of Opal Cliffs Drive. While some protective structures have been emplaced, lack of easy access has made the installation of any protective structure difficult.

H. Downtown Capitola is periodically flooded by high tides and storm waves. Oceanfront businesses are frequently damaged from wave impact. Low groin at south end of beach has stabilized the city beach.

I. Depot Hill is an area of active cliff erosion due to weak rocks and lack of a protective beach. Buildings have been demolished and moved and the cliff-top road is now gone.

J. Beachfront homes between New Brighton State Beach and Seacliff State Beach are periodically subject to wave attack and inundation. Bluffs failed during 1989 Loma Prieta earthquake cutting off access to the homes.

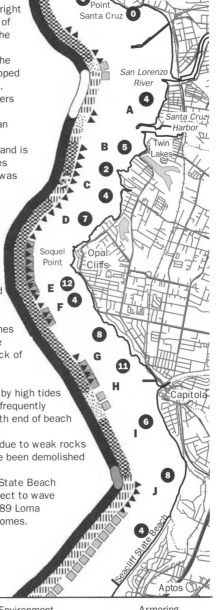

Hazard Level

- ☐ Stable: low risk
- ◼ (gray) Caution: moderate risk
- ◼ (black) Hazard: high risk

Erosion Rates

- **1** Inches per year

Shoreline Environment

- Sandy beach
- Beach backed by marsh
- Beach backed by dunes
- Cliff or bluff fronted by beach
- Rocky coast with no beach

Armoring

- ▲ ▲ ▲ Riprap
- ◼◼◼ Seawall
- ⊡⊡⊡ Other

Although both the seacliff and the beach have been heavily developed in the north (Rio Del Mar and Aptos Seascape, for example), much of the southern segment has historically been devoted to agriculture and parklands. Major portions of the beaches have been set aside as state parks (New Brighton, Seacliff, Manresa, and Sunset state beaches).

GEOLOGICAL CONDITIONS

The terrace and seacliff between New Brighton and Aptos Creek are composed of the moderately resistant sandstones and siltstones of the Purisima Formation. This is the same geological formation that makes up the cliffs in the Santa Cruz to Capitola area, but the inner-bay cliffs are more stable and vegetated because they are protected from wave action and erosion by wide, sandy beaches.

About a mile and a half south of Aptos Creek the material exposed in the bluffs changes to poorly consolidated ancient dune sand (the Aromas Sand). This material is far less stable than the bedrock to the north and can erode very quickly where it is subject to rainfall, stream flow, or wave action. It is prone to gullying, slumping, and sliding and therefore should be considered potentially unstable in any hillside or bluff area. The loose and unstable nature of this sand was not given much consideration in the development of the bluffs.

The Sunset Beach to Pajaro Dunes area forms the southern Santa Cruz County coastline. The dunes here are active, meaning they are still connected to their beach sand sources and undergo periodic erosion under severe storm conditions, followed by subsequent accretion or rebuilding.

COASTAL EROSION AND PROTECTION

The interior of northern Monterey Bay presents a clear example of the problems associated with constructing permanent structures on the beach. A wide, sandy beach, in equilibrium with the predominantly northwesterly waves, normally flanks this entire stretch of coast. The northwesterly waves undergo considerable refraction or bending as they enter the bay and thereby lose much of their energy, which is why the inner-bay beaches are generally quite safe and popular with locals and visitors. The historical record, however, shows repeated impacts of storm waves from the west or southwest characteristic of El Niño winters. Waves from the west or southwest reach the shoreline with little refraction or energy loss and as a result can quickly erode the beach, often depositing large logs and other debris carried down the rivers and creeks during the associated heavy rains. Beach sand and stranded driftwood logs at the base of a seacliff are good evidence for past storm wave inundation activity; however, they are not always recognized as such. The back beach area, now intensively developed around northern Monterey Bay, is equivalent to a river's floodplain. The question potential homeowners should ask themselves is not whether the back beach will be inundated, but how often and to what depth this inundation will occur, and how wave impact will affect their homes.

For a distance of nearly 3 miles from Pot Belly Beach (just southeast of New Brighton State Beach) to Aptos Seascape, extensive public and private development has taken place on the back beach area. Dozens of private homes in addition to a recreational vehicle campground (Seacliff State Beach), a roadway, restrooms, and a major sewer line have been built on or buried beneath the beach. Storm damage in 1978 and 1983 was extensive in this area. A look at the historical record, however, shows that these storms are not new to this portion of northern Monterey Bay.

The history of recurrent storm damage at Seacliff State Beach, for example, is well documented. Ten times in 58 years, or about once every 6 years on average, seawalls and bulkheads on the beach at Seacliff, which protect a parking and picnic area as well as a recreational vehicle campground, have been destroyed or heavily damaged. After extensive damage in 1939 and 1940, the timber seawall was rebuilt. Storms in the winter of 1941 destroyed it again. Following extensive damage to a piling and timber bulkhead in 1978 and again in 1980, a new bulkhead 2,600 feet long and identical to the previous structure was constructed along with the campground. This new structure was built at a cost of $1.7 million ($2.9 million in 2005 dollars) and was intended to last 20 years. In late January 1983, within two months of its completion and dedication, El Niño storm waves combined with very high tides overtopped the bulkhead (Figure 14.13). Large logs battered and loosened the timbers and pilings, and about 700 feet of the bulkhead were destroyed. The parking lot, recreational vehicle campsites, and restrooms were heavily damaged. Waves carried logs, sand, and debris over and through the battered bulkhead to the seacliff (Figure 7.11). Damage costs reached $1.5 million (in 2005 dollars), or half the cost of construction of the new bulkhead. The California State Parks Department, in its post-1983 reconstruction effort, placed "sacrificial" timbers on the seaward side of the bulkhead, hoping to reduce the impact of logs on the timber structure in the future.

Private development at Pot Belly Beach and Las Olas Drive immediately up coast of Seacliff, and houses along Beach Drive just down coast, suffered heavy damage in the same severe 1983 winter. Nearly 100 homes have been built over the years directly on the beach, some on pilings or concrete caissons, and some on slabs. Many of these homes have suffered serious damage, some repeatedly, particularly during severe winters of the past 25 years. Decks and stairways to the sand collapsed, and windows, doors, and entire

Figure 14.13   Storm waves overtopping a timber bulkhead at Seacliff State Beach during the 1983 El Niño. Notice the large logs and debris that have been washed over the bulkhead. Photo by Gary Griggs and Kiki Patsch.

walls facing the ocean were destroyed by wave action (Figure 6.2), leaving furniture afloat in seawater. In 1983, two houses on shallow pier foundations along Beach Drive collapsed onto the beach when 6 feet of sand was scoured away, undermining the piers (Figure 6.4).

At Aptos Seascape, just to the south, 21 homes were built on fill above the beach in 1969 and protected by riprap. Political pressure led to approval of the project despite objections from the county planning department about the difficulty of guaranteeing protection from wave action. The homes were initially required to be set back 20 feet from the top of the riprap. In time, the setback was reduced to 10 feet and ultimately eliminated altogether. During the January 1983 El Niño storms, which arrived at a time of very high tide, waves overtopped the protective riprap and damaged 19 of the 21 homes. In most cases the waves broke through the windows, doors, and house fronts facing the ocean and washed into the homes. One house partially collapsed (Figure 6.4). About half of the riprap was lost as sand was scoured from beneath it and the rock tumbled onto the beach. In June 1983, less than six months later, the county approved a $5.8 million (in 2005 dollars) 1,000-foot-long seawall. Insurance paid for the construction of the seawall, which cost about $3,000 per fronting foot or $100,000 per property owner.

There are clear risks associated with beachfront living in any area. Whether one likes history or not, the record of the risks and the costs are

A. Oceanfront homes along Beach Drive have been inundated and damaged during high tides and severe wave attack. Many previous seawalls have been damaged or destroyed. Present structures have survived since the late 1980's. Additional damage to homes built at the base of the bluff has occurred due to slope failure during intense rainfall or seismic shaking.

B. Homes along Via Gaviota extensively damaged in 1983. Large concrete seawall now protects this area.

C. Bluffs in this area failed during 1989 earthquake threatening bluff top homes. Retaining walls subsequently constructed.

D. Bluffs consist of weakly consolidated sands of the Aromas Formation that are prone to gullying from surface runoff.

E. Sandy bluffs above Place de Mer consist of ancient sand dunes that failed during Loma Prieta earthquake requiring demolition of one bluff top home. Wave runup and erosion at base of bluff has threatened some of the townhouses. Some protective rock has been emplaced.

F. Most of bluff top land to the south used primarily for agriculture.

G. At Sunset Beach area, shaking during Loma Prieta earthquake produced failure in the old dunes threatening some of the homes.

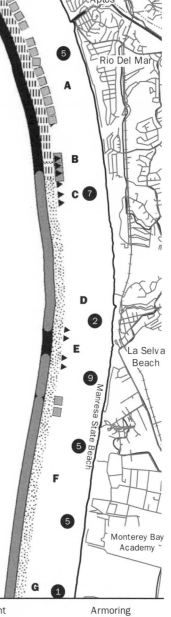

Hazard Level

☐ Stable: low risk
▨ Caution: moderate risk
■ Hazard: high risk

Erosion Rates

❶ Inches per year

Shoreline Environment

▨ Sandy beach
🌱 Beach backed by marsh
▨ Beach backed by dunes
⫿⫿ Cliff or bluff fronted by beach
▩ Rocky coast with no beach

Armoring

▲▲▲ Riprap
■■■ Seawall
⊡⊡⊡ Other

A. Condominiums and homes built directly on the frontal dunes were threatened with erosion from storm waves during high tides in 1969, 1978 and 1983. Emergency rock was emplaced that was replaced with a mile long revetment at a cost of ~$5 million.

B. Pelican Point condominiums at the mouth of the Pajaro River built very close to sea level without protection. Riprap installed following threat of inundation in the 1983 winter.

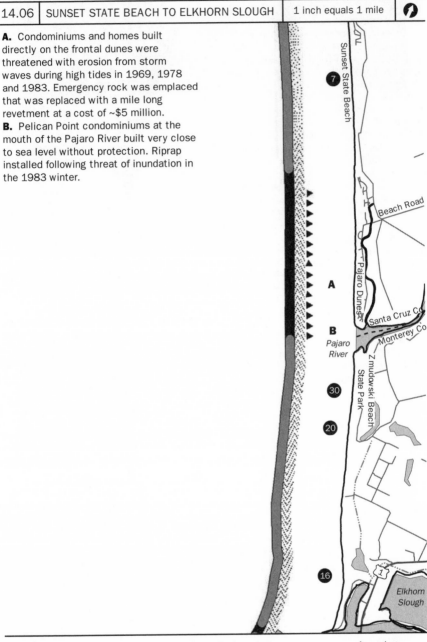

Hazard Level
- ☐ Stable: low risk
- ▨ Caution: moderate risk
- ■ Hazard: high risk

Erosion Rates
- ① Inches per year

Shoreline Environment
- ░ Sandy beach
- ⌇ Beach backed by marsh
- ▨ Beach backed by dunes
- ⦀ Cliff or bluff fronted by beach
- ▦ Rocky coast with no beach

Armoring
- ▲▲▲ Riprap
- ■■■ Seawall
- ⊡⊡⊡ Other

Figure 14.14   Coastal cliff failure in Rio Del Mar from prolonged and intense rainfall
in January 1982. The house at the base of the cliff was destroyed and subsequently
replaced with a triplex. Photo by Gary Griggs and Kiki Patsch.

clear. No one should naïvely purchase beachfront property or a home without carefully examining the impact of winter storms.

Most of the seacliffs backing the northern portion of the bay are protected
from wave impact and high tides by some combination of houses, riprap, and
seawalls. The bluffs are not immune to rainfall and earthquakes, however, as
homeowners have discovered in recent years. Prolonged, intense rainstorms
in early January 1982 generated widespread failure in the loose cliffs above
Beach Drive. Two homes at the base of the cliff were totally destroyed (Figure
14.14), and many others suffered heavy damage. Property owners atop the cliff
lost backyards, and some stood in danger of losing their homes.

Figure 14.15 Condominiums built on low dunes on a sand spit at the mouth of the Pajaro River. Photo by Gary Griggs and Kiki Patsch.

In October 1989 the magnitude 7.1 Loma Prieta earthquake hit Santa Cruz County, causing slope failures on the bluffs. Loose rock and debris slid down the steep bluffs, cutting off access to many beach-level homes and damaging the foundations of six apartments in Capitola and several houses on the bluff in Rio del Mar to the extent that they had to be demolished.

The sand dunes at the southern end of the Santa Cruz County coastline present a different set of concerns. The largest development in the area, Pajaro Dunes, consists of 396 condominiums, 24 townhouses, and 145 single-family dwellings. All units are built on the active sand dunes, with many of the houses, townhouses, and condominiums built directly on the foredune above the beach or on the beach itself (Figure 14.15). The pattern in this area over the past 50 to 75 years, which is evident in historical aerial photos, is one of dune erosion or removal during severe storms, followed by gradual build-up of sand, or accretion, during the subsequent calmer years. Thus, although there does not appear to be any significant net retreat of the shoreline, the advance and retreat of the dunes may move the shoreline 40 or 50 feet during a single winter. Unfortunately, the condominiums and homes do not shift with the dunes. Since development was initiated in 1969, four major El Niño winters (1978, 1980, 1982–83, and 1997–98) have brought large waves from the west and southwest, combined with elevated sea levels, and significantly eroded the dunes. The January 1983 storms cut back the dunes up to 40 feet and left a near-vertical cut measuring 15 to 18 feet that came right to

A. Head of Monterey Submarine Canyon lies directly offshore from harbor entrance and is the sink for littoral drift from the north.

B. Sand spit at Moss Landing subject to liquefaction during seismic shaking which destroyed the Moss Landing Marine Laboratories during the 1989 earthquake.

C. Major beach and dune erosion occurred in 1983 at the Monterey Dunes Colony site but adequate setback from shoreline prevented any damage to structures.

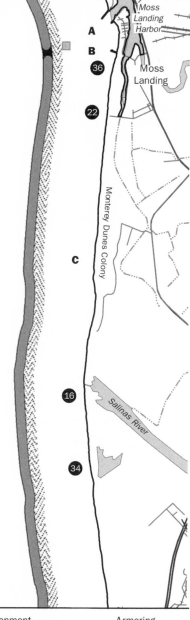

Moss Landing Harbor

Moss Landing

Monterey Dunes Colony

Salinas River

Hazard Level

☐ Stable: low risk
▨ Caution: moderate risk
■ Hazard: high risk

Erosion Rates

① Inches per year

Shoreline Environment

▦ Sandy beach
〰 Beach backed by marsh
▧ Beach backed by dunes
||||||| Cliff or bluff fronted by beach
▨ Rocky coast with no beach

Armoring

▲▲▲ Riprap
☐☐☐ Seawall
◉◉◉ Other

the foundations of many of the homes (Figure 4.12). Only the emergency emplacement of thousands of tons of rock saved these expensive homes from disaster. At the end of the storm season, a permanent revetment was built along the seaward frontage of this development at a cost of several million dollars (Figure 7.7). Although the revetment has provided some protection, by the time the 1997–98 El Niño hit Pajaro Dunes, much of the well-planned revetment was scattered across the beach (Figure 7.8). Any resemblance to the original, natural dune environment has disappeared.

PAJARO RIVER TO THE MONTEREY PENINSULA

PHYSICAL SETTING AND LAND USE

The coastline from the Pajaro River to the Monterey Peninsula consists of broad coastal lowlands fronted by wide, sandy beaches. A wide belt of dunes, commonly rising as high as 100 feet, flanks the beaches. The broad beaches and dunes are the result of large quantities of sand brought to Monterey Bay over thousands of years by the Salinas and Pajaro rivers. The combination of abundant sand, a broad low-lying area landward of the beach, and a dominant onshore wind direction has enabled a large dune field, extending from Sunset Beach to Monterey, to form and migrate.

Between the mouths of the Salinas and Pajaro rivers lie several lagoons or sloughs that are remnants of former drainages and estuaries. These brackish water bodies (particularly Elkhorn Slough) are important refuges for aquatic birds, mammals, and other wildlife. Rivers along this coast have very little if any runoff during the summer and fall, and consequently sandbars commonly block their mouths during these seasons.

During the early part of the century the mouth of the Salinas River flowed northward along the coast, paralleling the shoreline and separated from the sea by a narrow stretch of dunes. The river actually discharged into Elkhorn Slough until about 1910, when it broke through the dunes at approximately its present location. Ultimately, a dike was constructed to prevent the river's northward flow into its old channel.

Immediately offshore from the mouth of Elkhorn Slough lies the head of Monterey Submarine Canyon, one of the world's deepest and largest underwater canyons. The canyon is the major sink for the annual littoral drift of 300,000 cubic yards of sand moving southward along the shoreline of the northern bay, and it may intercept some sand moving northward as well.

The low oceanfront land from the Pajaro River mouth to Monterey is mostly undeveloped. Much of the land occurs in state beaches (Zmudowski State Beach, Salinas River State Beach, Moss Landing State Beach, Monterey State Beach, and now the coastal portion of the former Fort Ord Military

Reservation). As a result, many of the beaches and dunes are used for recreation; however, in recent years some development of the dunes has taken place. At Moss Landing, two marine research institutions have facilities along the shoreline. Just north of the Salinas River mouth, the Monterey Dunes Colony occupies a 1.25-mile-long strip of active coastal dunes, but the structures are set well back from the frontal dune. Farther south, a hotel and a large condominium complex occupy the oceanfront.

GEOLOGICAL CONDITIONS

Beaches and dunes form the dominant coastal geological features in this area, although the Monterey Submarine Canyon is an impressive and important feature as well. The oceanfront dunes are geologically quite young (less than about 3,000 to 5,000 years old) and many are still active, although the supply of sand is significantly reduced compared with the past. The migration of the dunes inland can be seen clearly along Highway 1 from Marina to Sand City, where drifting sand is regularly removed from the roadway. Farther landward, principally beneath the former Fort Ord area, older dunes extending 4 to 5 miles inland are now stabilized by vegetation and attest to the great volumes of sand that were transported to the coast by the Salinas River in the recent geological past.

Monterey Submarine Canyon (Figure 4.15), which is over 3,000 feet deep, is a key element in the littoral sediment transport system within the bay and is the largest sink for beach sand in this area. The head of the canyon extends inshore almost to the mouth of Elkhorn Slough and intercepts the sand moving south along the shoreline. The sand is carried down the canyon by sand flow, periodic slumps, and underwater turbidity currents or mud flows that move the sand many miles offshore, where it is deposited in a large submarine fan and lost forever to the beaches. It is believed that this huge underwater canyon owes its origin to an ancient river system that drained California's ancestral Central Valley many millions of years ago, and cutting by the large volumes of sand carried down the canyon by turbidity currents continues, much like flood flows carve river canyons on land.

Another major historical loss of sand in southern Monterey Bay was due to the sand mining in the Marina and Sand City areas that took place from the early 1900s until the late 1980s (Figure 4.13). It is believed that about 300,000 to 400,000 cubic yards of beach sand were removed annually for decades. The sand from the beaches and dunes of southern Monterey Bay is unique for its color, hardness, particle shape, and wide range of grain sizes. The sand has a high value due to a variety of industrial uses, including sandblasting and abrasives, water filtration, and stucco manufacture. As the erosion of the shoreline of southern Monterey Bay became more evident in the 1980s and oceanfront structures were threatened, people began to question the connection between beach sand mining and bluff retreat. Because of the apparent connection between the two, most beach sand mining was terminated in the late 1980s.

COASTAL EROSION AND PROTECTION

The generally smooth outline of inner Monterey Bay and its wide, sandy beaches and active dunes suggest an equilibrium coast, or one essentially in balance with the forces or processes acting upon it. Although this may be the long-term picture, historical and photographic records combined with recent observations indicate that the shoreline is as dynamic here as elsewhere. The beaches advance and retreat, and the dunes are cut back and subsequently rebuild. Of major concern for any coastal land use is the amount of change involved in these cycles or processes. How far does the shoreline migrate from year to year or during extreme events, and is long-term or net retreat of the shoreline going on as well? This information can be obtained from a careful analysis of historical aerial photographs and maps, but unfortunately, much of the coastal development in place today along the coastline of southern Monterey Bay did not have the benefit of this type of investigation.

Although aerial photographs from 1937 to 1980 show erosion alternating with accretion on the coast near Moss Landing, the overall trend was one of gradual beach accretion at an average rate of about 2 feet per year. Despite this overall pattern of build-up, however, severe erosion can occur in any individual year. An addition to the Moss Landing Marine Laboratories was proposed in 1982, to be constructed on the dunes. The geological consultants were convinced that the area was quite stable and that the structure would need no seawall or other protection. Immediately after the foundation was poured in December 1982, however, storm waves removed 17 feet of the dunes and threatened the foundation, requiring emergency riprap to be brought in. After the subsequent El Niño storms of early 1983, a concrete seawall was designed and emplaced for future protection.

The sand spit on which the original Moss Landing Marine Laboratories were built was also a site that experienced well-documented liquefaction and major ground failure during the 1906 San Francisco earthquake. This history was disregarded when the labs were built in the 1960s. Although the 1989 Loma Prieta earthquake was a smaller earthquake, its epicenter was far closer. The severe shaking again produced liquefaction in the Moss Landing area (Figure 14.16), which unfortunately led to the complete destruction of the laboratories. It took nearly a decade to obtain the funds and permits to rebuild these facilities a short distance inland on a stable site.

To the south, Marina State Beach and the former Fort Ord Military Reservation are areas experiencing rapid erosion, primarily because the

Figure 14.16 Liquefaction during the 1989 Loma Prieta earthquake led to the failure of Jetty Road near Moss Landing. Reproduced by permission of Jeffrey Marshall and Dan Orange.

material making up the sandy bluffs is unconsolidated or loose sand. Erosion rates in the Marina State Beach area between 1937 and 1983 averaged 5 to almost 7 feet per year. The Marina sewage treatment plant experienced significant wave damage during the 1982–83 winter when a 40-foot storm section of the emergency bypass outfall line was undercut and collapsed.

Erosion in the Fort Ord area is the most severe in the southern Monterey Bay because of wave refraction patterns. Bluff retreat is dramatically episodic and is well documented. Several large-diameter sewer outfall lines on the military base were placed atop massive concrete piers (set 35 feet apart) and

A. Active sand mining from the beach was terminated in the mid-1980s, although sand is still dredged from a large back-beach pond.

B. The sandy bluffs of the former Ft. Ord military base historically eroded faster than any other areas around Monterey Bay. Large volumes of rock and broken concrete were dumped over the bluffs repeatedly to try to protect the former Soldier's Club, which has become a peninsula. Property now belongs to State Parks Department. Soldier's Club (Stillwell Hall) was removed in 2004.

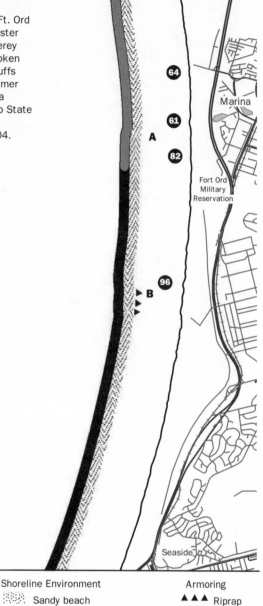

Hazard Level

Stable: low risk
Caution: moderate risk
Hazard: high risk

Erosion Rates

1 Inches per year

Shoreline Environment

Sandy beach
Beach backed by marsh
Beach backed by dunes
Cliff or bluff fronted by beach
Rocky coast with no beach

Armoring

▲▲▲ Riprap
■■■ Seawall
◨◨◨ Other

buried below beach level during building in 1962. When observed in 1983, the pipes were 20 feet above beach level, and 175 feet of retreat had occurred in 21 years, or 8.3 feet per year on average. This is a very high rate of coastal erosion.

The former Soldiers Club at Fort Ord, an entertainment center where many World War II soldiers spent their last evenings before being shipped overseas, was abandoned some years ago after being designated as too hazardous for continued use because of its precarious cliff-top location. The Army repeatedly dumped rock and broken concrete at the base of the sandy bluff in an attempt to slow the erosion and save this historic structure. Winter waves, however, broke over the rock and undermined and eroded the loose sand. By early 1984 the edge of the eroding bluff was within 15 to 20 feet of the structure. The Army at this point looked into different approaches for saving the structure and decided that complete relocation or reconstruction was far more cost-effective than trying to halt the ongoing bluff erosion. The base was subsequently closed, with the structure sitting on an eroding sandy peninsula. The California State Parks Department has taken over the property, and the structure has now been demolished.

Four miles south of Fort Ord, a hotel was constructed in 1968 right on the back beach, where long-term shoreline retreat rates have averaged about 2 to 3 feet per year (Figure 14.17A). To protect the building, a concrete panel seawall was built completely across the front and along both sides of the hotel. Wave overtopping and surge during the severe 1983 storms penetrated the joints between some of the concrete panels, leading to some loss of fill behind the wall. Large waves combined with high tides in 2002 destroyed the south end of the seawall and began to erode a parking lot before the emergency emplacement of rock could be achieved (Figure 14.17B). The dune edge is now about 50 feet landward of the front of the seawall, so that the hotel is becoming a peninsula and during high tides there is no beach access in front of the structure.

A mile south of the hotel, a large, 172-unit apartment complex was built on the dunes between 1972 and 1974, where long-term bluff erosion rates were 1 to 2 feet per year. The oceanfront units were first seriously threatened by the large waves, high tides, and elevated sea levels of the 1982–83 El Niño event. Wave erosion broke the water line, which had to be rerouted, and threatened the sewer and electrical lines. Continuing dune erosion by January 1984 had come to within 14 feet of the shallow pilings supporting the apartments. Five thousand tons of rock was brought in to provide emergency protection for the oceanfront apartments, but because the apartments were built right to the property line, the rock had to be placed on beach belonging to the city of Monterey (Figure 14.18A). Ultimately, the city required that the emergency rock be removed. The frontal units were resupported by a series of 50- to 55-foot-deep concrete piers connected by grade

Figure 14.17 (A) The Monterey Beach Hotel is now surrounded on three sides by beach as the shoreline continues to erode. Photo © 2002–2004 Kenneth and Gabrielle Adelman, California Coastal Records Project, www.Californiacoastline.org. (B) Wave erosion during December 2002 undermined and led to the partial collapse of seawall protecting the Monterey Beach Hotel, after which emergency riprap was emplaced. Photo by Gary Griggs and Kiki Patsch.

beams. In addition, the apartment units were converted to condominiums and sold to many different owners. Although the shoreline has advanced and retreated seasonally, the overall pattern has been one of continuing retreat of the bluff edge at an average rate of nearly 2 feet per year. The 1997–98 El Niño caused additional retreat (Figure 14.18B), and by 2002 more

Figure 14.18 (A) The Ocean Harbor House condominiums in 1984, following emplacement of emergency rock for protection against wave attack. (B) Ocean Harbor House in 1998, with supporting concrete caissons installed following removal of emergency riprap.

Figure 14.18 *(Continued)* (C) Continued dune retreat and undermining led again to emplacement of emergency riprap at Ocean Harbor House in 2002. Photos by Gary Griggs and Kiki Patsch.

emergency rock had to be emplaced to prevent undermining of additional units (Figure 14.18C). As of 2004, property owners and their consultants had considered many different proposals for the threatened structures and evaluated the environmental impacts of each. The proposal that moved forward was a concrete seawall, although it was recognized that with continued shoreline retreat the beach in front of the site would gradually be lost along with beach access.

SUMMARY

The sandy beaches and dunes of the Monterey Bay area have been popular summer recreational areas for nearly a century, but the coastline is a far different place during the winter months. Although the beaches, dunes, cliffs, and bluffs are regularly attacked by winter storm waves and high tides, these same environments have also been intensively developed, particularly since the late 1960s. The hazards of oceanfront living are now more clearly recognized and understood, as are the high costs of protecting or attempting to protect such property. The damage and erosion caused by the frequent El Niño storms between 1978 and 1998, in addition to longer-term, historical

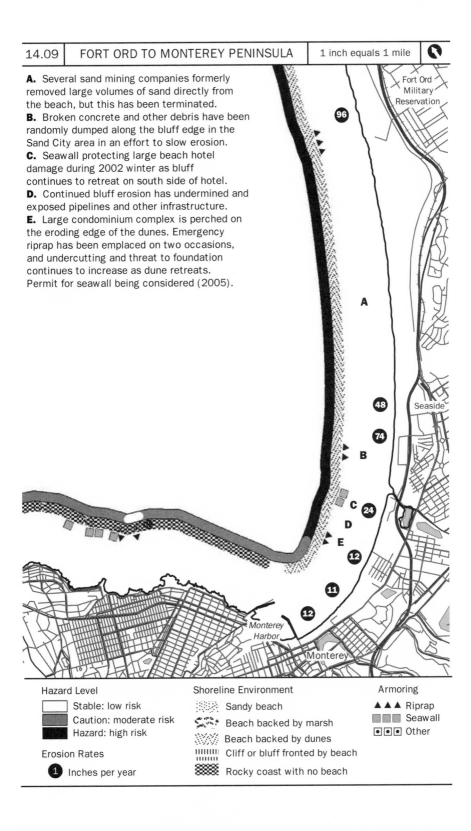

A. Several sand mining companies formerly removed large volumes of sand directly from the beach, but this has been terminated.
B. Broken concrete and other debris have been randomly dumped along the bluff edge in the Sand City area in an effort to slow erosion.
C. Seawall protecting large beach hotel damage during 2002 winter as bluff continues to retreat on south side of hotel.
D. Continued bluff erosion has undermined and exposed pipelines and other infrastructure.
E. Large condominium complex is perched on the eroding edge of the dunes. Emergency riprap has been emplaced on two occasions, and undercutting and threat to foundation continues to increase as dune retreats. Permit for seawall being considered (2005).

Fort Ord Military Reservation

Seaside

Monterey Harbor

Monterey

Hazard Level
- ☐ Stable: low risk
- ☐ Caution: moderate risk
- ■ Hazard: high risk

Erosion Rates
- ① Inches per year

Shoreline Environment
- Sandy beach
- Beach backed by marsh
- Beach backed by dunes
- Cliff or bluff fronted by beach
- Rocky coast with no beach

Armoring
- ▲▲▲ Riprap
- ☐☐☐ Seawall
- ⊡⊡⊡ Other

coastline changes, must be carefully evaluated prior to approval of any additional oceanfront development. We face significant challenges in dealing with the hundreds of developed oceanfront parcels on the beach, dunes, and bluffs of this region when the historical record and all indications for the future point to a continuing rise in sea level, resulting in the continuing erosion of the coastline.

CHAPTER FIFTEEN

THE MONTEREY PENINSULA TO MORRO BAY

CHERYL HAPKE

PHYSICAL SETTING

The central coast of California from the Monterey Peninsula south to Morro Bay includes some of the most spectacular coastline in the United States. Along much of this section of the coast, the Santa Lucia Mountains, part of the Coast Ranges geomorphic province, rise steeply from the waters of the Pacific Ocean to elevations of nearly 1 mile within an inland distance of just 3 miles (Figure 15.1). The geology varies dramatically along the coast as different types of rocks are juxtaposed along numerous faults and shear zones associated with the San Andreas Fault. Because of the variable geology, the coastline is very irregular where weaker rocks erode more rapidly than adjacent stronger rocks. Elevated marine terraces that end abruptly at the coast with near-vertical cliffs characterize this section, especially in the south.

Erosion hazards and processes are highly variable along the central California coast, primarily because of the diverse geology of the region. High-energy waves and saturated ground from winter rains combine to create erosion hazards ranging from surficial block falls along the marine terrace cliffs, to slumping of highly weathered bedrock, to large, deep-seated landslides in the steep mountainous terrain. In addition to the threat of damage to homes and other structures from these erosional processes, large landslides along the narrow Highway 1 on the Big Sur coast frequently block or damage the road (Figure 15.2). This road is the only paved route for the small communities established there, and is itself a major tourist attraction, providing visitors with vistas of the spectacular scenery and access to many recreational sites.

For most of the Big Sur coast, small pocket beaches occur between more resistant rocky headlands, and narrow beaches form at the base of eroding cliffs or at river mouths. Wide, sandy beaches are uncommon along this

Figure 15.1 View looking north toward the Santa Lucia Mountains and the Big Creek bridge. Within 3 miles of the coast, these mountains reach heights of nearly a mile. Photo by John Duffy, Caltrans.

section of the coast. Exceptions include several beach and dune systems, in some cases forming a tombolo (a causeway of sand connecting the mainland to a small offshore island) such as Point Sur (Figure 15.3).

In 1992 the Monterey Bay National Marine Sanctuary was established by the National Oceanic and Atmospheric Association (NOAA). The waters immediately adjacent to the Big Sur coastline are included as the central part of the marine sanctuary. In addition to banning oil and gas exploration in the pristine waters of this region, the sanctuary provides protection for the diverse ecosystems of the area, including animals ranging from abalone to sea otters.

HUMAN HISTORY

The first known occupants of the region, dating back 8,400 years, were aboriginal peoples in the area of San Luis Obispo County near the present-day town of Cambria. Farther north along the more rugged Big Sur coastline, the oldest evidence dates the first inhabitants at 4,600 years before the present. Several Native American peoples reside in the area, including the Ohlone, the Esselen, and the Salinan.

Juan Cabrillo was the first European to describe the Big Sur and adjacent coast during a sailing expedition in 1542: "There are mountains which seem to reach the heavens, and the sea beats on them." Two hundred years passed

Figure 15.2   J. P. Burns landslide in 1983. Removal of landslide debris and slope stabilization measures resulted in the closing of State Highway 1 for over a year. Landslide material is covering the highway in the central portion of the photo, and a few vehicles can be
seen parked near trees to the right of the landslide. Photo by Lynn Harrison, Caltrans.

before Europeans "discovered" and settled in the Monterey and Carmel
areas. By the early 1770s, Monterey was a well-established coastal town and
port, serving as the capital of Alta (upper) California.

Early settlement along the Big Sur coast south of Monterey and Carmel
was due to land grants given by the Mexican government in the mid-1800s,

Figure 15.3 Looking south at Point Sur and the tombolo that connects the point to the mainland. Photo by Cheryl Hapke.

but these extended no further south than Point Sur. In the period immediately following the discovery of gold at Sutter's Mill (1848), prospectors and settlers began to flood California, and some made their way to the central coast, where they found most of the land around Monterey already privately owned. A small number of settlers migrated down coast into the rugged and isolated Big Sur area, where they established small homesteads. Many ridges and canyons in the area carry the names of the settling families, and many of the residents in the region are descendents of the original immigrants.

Because of the remote and rugged nature of much of this region, there was little development along the coast prior to the completion of the Carmel–San Simeon Road, now known as coastal Highway 1, in 1937. The opening of this road brought an influx of tourists and new tourist facilities and opened the gate for development of the previously inaccessible coastal areas.

MONTEREY PENINSULA TO
THE CARMEL HIGHLANDS

The rocks of the Monterey Peninsula to the Carmel Highlands are very erosion-resistant, coarse-grained granites that outcrop at numerous locations throughout the peninsula and along the coast. At the southern end of Carmel Bay, at Point Lobos State Reserve, the muddy sandstones and shales of the Tertiary Carmelo Formation overlie the granite. In the central part of Carmel Bay, softer sediments deposited by the ancestral Carmel River are exposed along the coastline. The granitic rocks form low-relief cliffs throughout the peninsula and increase in height in the Carmel Highlands area south of Point Lobos. The coarse-grained granitic rocks are very resistant to the battering of the waves but eventually provide the sand for the beautiful beaches common to the Monterey Peninsula. The spectacular irregular, crenulated coast owes its existence to well-developed joints within the granitic rocks that erode out much faster than the solid rock around the joints.

The Monterey Peninsula is heavily developed and includes the city of Monterey; the towns of Pacific Grove, Carmel-by-the-Sea, Carmel Highlands, and Pebble Beach; and a number of well-known golf courses. Many expensive houses have been built in this area, and it has remained an exclusive community since the 1880s, when wealthy vacationers came to stay at the Hotel Del Monte, which still stands today as part of the Naval Postgraduate School complex. Del Monte Properties developed Seventeen Mile Drive in the 1920s, and today is a gated community owned and run by the Pebble Beach Company.

Although the granitic rocks of this section of coast erode quite slowly, hazards still exist in the form of wave-flung debris over the low-relief cliffs, and the occasional roof collapse of caves that are common in the highly jointed granite. Boulders, logs, and other debris can be tossed by large storm waves, damaging roads and other facilities. A severe storm in 1931 threw the mast of a ship, wrecked 35 years earlier, some 50 feet across Seventeen Mile Drive.

Asilomar State Beach is known as a good winter surf spot, with waves often breaking at 8 feet or more. At the southern end of Asilomar is Spanish Bay, which in the 1920s housed a boardwalk and bathhouse operated by the Del Monte Hotel. Storm waves caused damage to these structures on several occasions, and wave run-up during one particular storm was documented as extending hundreds of feet inland. A luxury resort, the Inn at Spanish Bay, was built here in the early 1990s, despite the hazard of structural damage from debris and flooding.

The low terrace (<20 feet) can be inundated during heavy storms, carrying debris such as logs and boulders onto the terrace. In the 1930s, floating debris from several large winter storms caused damage to the Del Monte Hotel and nearly destroyed a municipal wharf. During the 1997–98 El Niño winter, many golf courses were damaged from waves overwashing the low

Figure 15.4 Development of Carmel Highlands atop the granitic cliffs. Erosion hazards are low here and are primarily related to the occasional collapse of a sea cave or failure along the well-developed joint system. Photo by Cheryl Hapke.

terrace. A major golf tournament was postponed at Pebble Beach because of a washed-out road in front of the 15th tee and a large sinkhole that developed in front of the 8th green. At the exclusive Cypress Point Club, only 10 of the 18 holes were playable after the intense early February storms.

The coastal cliffs of inner Carmel Bay are composed of softer sedimentary rocks that overlie the resistant granitic rocks. As a result, the erosion hazards here are higher, and numerous seawalls and shore protection structures were built as early as the 1920s. Farther south, the higher cliffs of Carmel Highlands provide homeowners with spectacular views of the Pacific Ocean and granitic sea arches and sea stacks just offshore (Figure 15.4). The primary hazards here are failure of the cliffs along the well-developed joints and the occasional collapse of one of the numerous sea caves.

THE BIG SUR COAST

The rocks of the Big Sur coast can be broadly divided into two complexes: hard, resistant granitic rocks similar to those described for the Monterey Peninsula; and the highly variable, sheared and faulted rocks of the Late Jurassic to Miocene Franciscan complex. Thick deposits of alluvial fan and older landslide deposits overlie the bedrock in much of the region. The

A. Moderate risk areas are primarily where the road is very close to the coastline and differential erosion of granite leads to undercutting of the road. Areas of low hazard may be subjected to wave-thrown debris during large storms.
B. Moderate hazard areas due to potential of wave-thrown debris during large storms and potential undercutting of road.

Hazard Level

- ☐ Stable: low risk
- ▨ Caution: moderate risk
- ▪ Hazard: high risk

Erosion Rates

- ① Inches per year

Shoreline Environment

- Sandy beach
- Beach backed by marsh
- Beach backed by dunes
- Cliff or bluff fronted by beach
- Rocky coast with no beach

Armoring

- ▲▲▲ Riprap
- ▪▪▪ Seawall
- ⊡⊡⊡ Other

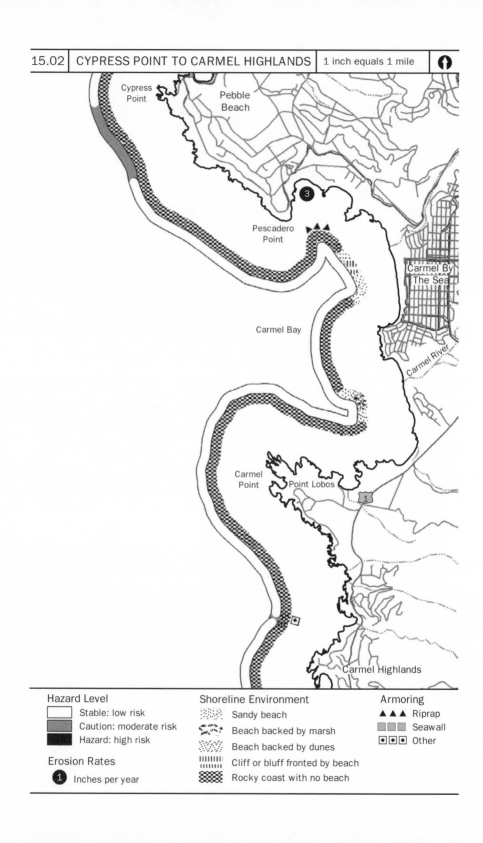

Cypress
Point

Pebble
Beach

Pescadero
Point

Carmel By
The Sea

Carmel Bay

Carmel River

Carmel
Point   Point Lobos

Carmel Highlands

Hazard Level

☐ Stable: low risk
▨ Caution: moderate risk
■ Hazard: high risk

Erosion Rates

① Inches per year

Shoreline Environment

Sandy beach
Beach backed by marsh
Beach backed by dunes
Cliff or bluff fronted by beach
Rocky coast with no beach

Armoring

▲▲▲ Riprap
▨▨▨ Seawall
⊡⊡⊡ Other

granitic rocks are the predominant rock type along the northern third of this 75-mile stretch of coast, and the Franciscan complex dominates along the southern two-thirds. The landscape is dramatic, with the Santa Lucia Mountains rising steeply from the Pacific Ocean. Although the development along this section of coast is sparse, a few scattered communities are concentrated in the north. The greatest hazard along the Big Sur coast are large landslides on the steep slopes that damage or block Highway 1, the only paved transportation route for the area. During El Niño winters, Highway 1 is often blocked or undermined by slope failure from intense rainfall and storm waves. In 1983 the highway was blocked by 42 different landslides. The Julia Pfeiffer Burns (J. P. Burns) landslide closed the highway for over a year (Figure 15.2) as the California Department of Transportation (Caltrans) rebuilt the road and stabilized the hillside to prevent future slides, the largest land-moving operation in Caltrans history.

Prior to the opening of Highway 1 in 1937, the only access to the rugged Big Sur area was a wagon road that connected the present-day village of Big Sur to Monterey. In the 1920s, many original landholdings from the homesteaders of the 1800s were acquired by wealthy filmmakers and actors for use as vacation retreats. Development proceeded sporadically throughout the area and included exclusive gated communities, resorts, retreats, hermitages, and several small towns such as Lucia and Gorda. Although the local economy relies heavily on the tourist dollar, the people who inhabit the Big Sur coast heavily guard their privacy, and NO TRESPASSING signs are common.

Erosion hazards are fairly low along the section from Carmel Highlands to the village of Big Sur. The average cliff erosion rate for this area is approximately 6 inches per year. The town of Big Sur is situated substantially inland from the coast, and most developed stretches north of Big Sur are built on low terraces of erosion-resistant granitic rock. Though uncommon along this part of the coast, a large, deep-seated landslide at Hurricane Point closed the highway for several weeks in 1999. In addition, where the coarse-grained granitic rocks are deeply weathered, the loose weathered material is easily mobilized as debris flows during periods of heavy rainfall. Debris flows from the steep coastal mountains nearly buried the village of Big Sur in the winter of 1978 following a major summer brushfire and prolonged, intense winter rainfall.

South of the town of Big Sur, numerous private homes and exclusive resorts are perched on top of high cliffs of both granite and the siltstones and sandstones of the Franciscan complex. This area is very susceptible to slumping, especially during major winter storms. Just south of Grimes Point to McWay Canyon, granitic rocks are again the dominant rock type along the

A. Granitic rocks are very resistant to erosion and this stretch of coast has little development. Marine terraces have development potential but also have a history of landsliding which poses a moderate hazard.

B. This is the site of the Hurricane Point landslide which resulted in the closure of Highway 1 for three months during the winter of 1999. The area has been mapped as a location of ancient (Holocene) landsliding. Much of the hill slope around Hurricane Point has been stabilized.

C. Coastal Highway 1 is significantly inland here but some homes are built near top of ridge. Ocean facing slope has numerous fresh scarps and studies show evidence of large, episodic slumps.

D. Moderate hazard zone is area of recent landslide activity. In 1983 the Sycamore Draw landslide closed Highway 1 for approximately eight months.

E. Northern high hazard area is the location of the JP Burns landslide which closed Highway 1 for over a year in the winter of 1983. The upper slope is now stabilized but loose material on the seaward side of road continues to erode rapidly.

F. Landslides during the 1998 El Niño destroyed the famous Esalen Institute hot tubs. They were subsequently rebuilt much lower on the slope.

G. Hot Spring Creek is an area of chronic landsliding. The highway was closed in February 1987 for several days and again in the winter of 1995 for approximately 1 week.

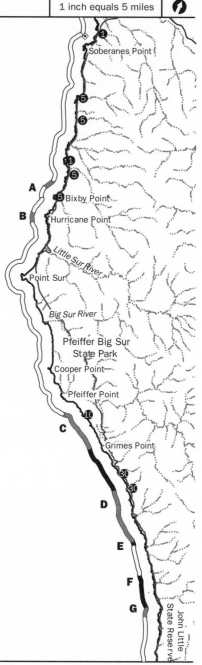

Hazard Level
- ☐ Stable: low risk
- ☐ Caution: moderate risk
- ☐ Hazard: high risk

Erosion Rates
- **①** Inches per year

Armoring
- ▲ ▲ Riprap
- ☐ ☐ Seawall
- ⊡ ⊡ Other

coast. This section is very rugged and remote with little development. Landslide hazard is fairly low, although the J. P. Burns landslide (mentioned earlier) occurred within the granitic rocks along this stretch of coast. The extreme rainfall associated with the 1982–83 El Niño is thought to have induced this large slide. Prior to this landslide, a waterfall here poured directly into the Pacific Ocean (Figure 15.5A). As a result of the deposition of landslide material at the base of the slope, a beach formed and the waterfall to this day empties onto the beach rather than into the water (Figure 15.5B). The average cliff erosion rate here (about 9 inches per year) is higher than in similar granitic rocks to the north.

From McWay Canyon to the south end of the Big Sur coast, the weak and unstable rocks of the Franciscan complex form the coastal slope. Although there is little development from McWay Canyon to Lopez Point, a few developed sites exist, perched on top of coastal terraces (Figure 15.6). The steep seaward slopes of these terraces are highly susceptible to slumps and landslides. At an area locally known as Cow Cliffs near the University of California Big Creek Reserve, the hillslope above the road is prone to frequent rockfalls and small landslides. Photographs of the road during construction in 1934 (Figure 15.7A), when completed in 1937 (Figure 15.7B), and as it is today (Figure 15.7C) show that the slope here is still quite active. Cliff erosion rates along this section average about 8 inches per year.

At Lopez Point, the coast takes a sharp, nearly 90° turn. This change in direction suggests that the rocks originally eroded at a much faster rate than those to the north. Even today, this portion of the coast, especially from Lopez Point to Rockland Landing, is extremely landslide prone. The highway along this stretch is very uneven, testimony to the ever-moving slopes. The town of Lucia (a restaurant, gift shop, and 10 cabins) is perched on the high cliffs overlooking the Pacific Ocean here. Although the view is spectacular, the hazard from landslides is very high. The roof peaks on the small rental cottages were at one time vertically aligned with each other. Today, they are visibly offset as the ground beneath them slowly shifts downslope.

South of Rockland Landing to Pacific Valley, the only developments on the coast are several state parks. Landslide hazard along this stretch is high and average long-term cliff erosion rates are about 8 inches per year. Due to the ruggedness of the terrain, however, little developable land exists. At Pacific Valley, the land immediately adjacent to the coast flattens out along a well-developed marine terrace. Although the coastal cliffs are eroding, most land immediately adjacent to the coast is parkland or is used for grazing, and all developed properties are on the inland side of the highway.

Figure 15.5　J. P. Burns waterfall, showing the beach that formed as a result of the sediment added to the system from a large landslide. (A) Photo taken July 10, 1963, shows waterfall pouring directly into the ocean. Photo from Caltrans photo archives. (B) Photo taken June 10, 2003, shows water now pouring onto a beach that formed after the 1983 landslide. Photo by Cheryl Hapke.

Figure 15.6 Development on the narrow, high marine terrace north of Lopez Point. Photo © 2002–2004 Kenneth and Gabrielle Adelman, California Coastal Records Project, www.Californiacoastline.org.

(A)

Figure 15.7 "Cow Cliffs," just north of the Big Creek Reserve. Historical photos show that this slope has been active since the time of road construction. (A) May 1934. Photo from Caltrans photo archives.

Figure 15.7 *(Continued)* (B) August 1937. Photo from Caltrans photo archives. (C) June 2003. Note the rock net (shown by arrow) installed to protect the road and cars from falling rock and debris. Photo by Cheryl Hapke.

Figure 15.8 The town of Gorda in the southern portion of the Big Sur coast. The town itself is built within an ancient landslide, and portions of the slide have been reactivated. Note in the middle of the photograph the retaining wall built by Caltrans in 1995 to stabilize the road. Photo © 2002–2004 Kenneth and Gabrielle Adelman, California Coastal Records Project, www.Californiacoastline.org.

From Pacific Valley to the southern end of Big Sur at San Carporforo Creek, the terrain again becomes very steep and rugged. This entire section contains some of the weakest rock (serpentinite) of the Franciscan complex and is very prone to landsliding. Some development, single-family homes and small ranches, exists on the steep slopes west of Highway 1, even though the risk from landslides is quite high. The town of Gorda is built within a large ancient landslide complex (Figure 15.8). The interior portion of this large slide continues to slip on smaller, active slide planes, resulting in very high landslide risks. A large retaining wall and dewatering system were installed along the seaward edge of the highway in 1995 to stabilize the highway after winter storms resulted in significant damage (Figure 15.8).

The Big Sur coast continues to draw millions of tourists each year, who come to drive Highway 1 and observe the offshore marine life in the Monterey Bay National Marine Sanctuary. Caltrans is responsible for maintaining the highway and keeping it safe for tourists and travelers, and is working with the sanctuary on a plan to ensure that any material generated from road failure and repair does not cause damage to the sanctuary's coastal ecosystems.

A. "Cow Cliffs," an area of very active, unstable slopes. Landslides closed HYW 1 here in January 1983.

B. Landslides closed the highway at Big Creek for approximately one week in March 2000.

C. Cliff erosion rates are highly variable due to different lithologies exposed at the base of the slope. This is an area of high landslide occurrence with very unstable slopes. The "Grandpa's Elbow" landslide displaced 20 million cubic yards of material in the winter of 2000.

D. The "Pitkins Curve" landslide area has been a recurring area of road closures since at least 1973. As recently as March 2000, the highway was closed for three and a half months, and cost an estimated $3 million for repairs and stabilization. Caltrans is ultimately considering a bridge across this area due to its extreme instability.

E. The cliff edge along the well-developed marine terrace has high erosion rates, but no development currently exists on the seaward side of the road.

F. Sand Dollar Beach is one of the largest and most accessible beaches along the Big Sur coastline.

G. Very steep and landslide-prone stretch of coast. The town of Gorda sits in an ancient landslide and portions of it have been active as recently as 1995 when a retaining wall was constructed downslope of HWY 1. In January 1997 a landslide at Gorda closed the highway for a month. Repairs cost an estimated $5.5 million. A slide at Cape San Martin closed the highway for ten days in January 1965.

H. Several damaging mud slides have occurred near Redwood Gulch and resulted in road closures both in 1974 and 1982. In March 1986 landslides closed the highway for over two months at Redwood Gulch.

I. In Northern San Luis Obispo County, the steep mountains of Big Sur give way to rolling hills and well-developed marine terraces.

Hazard Level
- ☐ Stable: low risk
- ▨ Caution: moderate risk
- ■ Hazard: high risk

Erosion Rates
- **1** Inches per year

Armoring
- ▲▲ Riprap
- ■■ Seawall
- ⊡⊡ Other

THE NORTHERN SAN LUIS OBISPO COUNTY COAST

The Big Sur coast ends at San Carporforo Creek, where the steep, dramatic slopes of the Santa Lucia Mountains transition to more gently rolling hills and a wide marine terrace. The coastal landscape here is dominated by open terraces and low coastal bluffs, with a scattering of offshore sea stacks. This portion of the central California coast is very sparsely developed and is dominated by grazing land. The Hearst Ranch owns much of the land, and long stretches of the accessible coastal areas are established as state beaches (William R. Hearst Memorial State Beach and San Simeon State Beach). Several coastal access points near Piedras Blancas allow steady streams of tourists to view the breeding and molting of elephant seals. Narrow beaches front the actively eroding low bluff in this area. The erosion in places undermines parking lots and coastal Highway 1, such as the area just north of the Piedras Blancas Motel (Figure 15.9).

The first developed community in the northern San Luis Obispo coast area, south of Big Sur, is the small town of San Simeon, which was named by Padre Fermin Francisco de Lasuen in 1797. The town was a thriving whaling village from 1852 to 1880, and at that time all the buildings of the town were on San Simeon Point at the northern end of San Simeon Bay. When the whaling industry in the region died out, the town was abandoned except for Sebastian's General Store, which was moved from the point to its present-day location along the northern portion of San Simeon Bay. In 1865 George Hearst purchased 40,000 acres of land surrounding San Simeon Bay. In the 1920s his son, William Randolph Hearst, built a large compound, La Cuesta Encantada, now known as the Hearst Castle, in the hills overlooking San Simeon Bay. A Spanish-style warehouse and employee residences were built along the northern portion of the bay by Hearst. The northern portion of the bay is now part of the William Randolph Hearst Memorial State Beach. In 1998 the California Coastal Commission denied the Hearst Corporation a permit to develop a 356-acre luxury hotel and golf course just north of the memorial park.

The inner part of San Simeon Bay is a sandy beach backed by low bluffs that continue along the coast to the south. The bluffs are composed of relatively weak rocks of the Franciscan complex and are capped by Pleistocene marine terrace deposits. Active erosion of the weak marine terrace deposits is especially common in the winter, when the ground is weakened from rains and large waves reach the base of the bluffs. The inner portion of the bay is protected from large waves by San Simeon Point, unless the waves arrive from the west or southwest. A 1,000-foot-long fishing pier was built

Figure 15.9   Riprap protecting coastal Highway 1 just north of the Piedras Blancas Motel in northern San Luis Obispo County. Photo by Cheryl Hapke.

in 1957; however, this is the third pier to be built at this location, evidence that destructive waves occur within this protected bay. The original pier was built in 1869 but was destroyed by a storm just a few years later. The next pier, built in 1878, was destroyed during a storm in the 1940s.

Farther to the south is San Simeon Acres, a commercial development that exists primarily to serve the tourists visiting Hearst Castle and the southern entrance to the Big Sur coast. The coast between San Simeon and San Simeon Acres is a series of small pocket beaches backed by low bluffs. In San Simeon Acres, several motels and homes are built on the west side of Highway 1 on the actively eroding marine terrace. Some properties are set back from the coast, and those that sit directly on the bluff edge are at high risk from undercutting and damage.

CAMBRIA TO MORRO BAY

The coast southward to Cambria is largely undeveloped; much of it is part of San Simeon Beach State Park. Cambria is separated into unincorporated east and west villages, and most of the residential areas are concentrated on the coast side of Highway 1. The oceanfront houses are built on actively eroding bluffs composed of marine terrace deposits overlying the faulted

A. Road is in danger of being undercut anywhere it runs close to the edge of the actively eroding marine terrace.

B. Higher hazard area where soft rocks of the Franciscan Complex undermine support for the fishing pier within San Simeon Bay.

C. Small community of homes and condos built at the edge of a marine terrace. Cliffs are being undercut and are now protected with riprap and a seawall.

D. All developed coastal areas of Cambria are in high hazard zones and nearly every individual parcel has either a seawall or riprap for protection at the cliff base.

E. Coastal area is undeveloped but bluffs composed of Franciscan Formation rocks are actively eroding.

F. Seawalls and riprap protect nearly each individual parcel along developed sections of Cayucos except where small promontories of more resistant rock occur.

G. Wide sandy beach and dune system provide protection for development north of Morro Rock. Hazards are related to possible flooding if dunes are overwashed during a large storm.

Hazard Level
- Stable: low risk
- Caution: moderate risk
- Hazard: high risk

Erosion Rates
 Inches per year

Armoring
- ▲ ▲ Riprap
- Seawall
- Other

Figure 15.10   Houses in Cambria are perched at the edge of an eroding cliff composed of sheared and fractured Franciscan complex rocks. A narrow, seasonal beach provides little protection for the houses from winter wave attack. Photo by Cheryl Hapke.

and sheared sandstones and siltstones of the Franciscan complex. This is an area of high erosion hazard, and many houses on the cliff top are in danger of being undercut and damaged during large waves and high tides (Figure 15.10).

Farther north, the community along Moonstone Drive, which runs along the top of the cliff, is a mile-long stretch of coastal land developed in the mid-1960s and 1970s, primarily as motels, bed-and-breakfast establishments, and restaurants. The 20-foot-high cliffs are formed of marine sandstones and siltstones that are moderately resistant to erosion, although deep gullies have undermined the cliff-top trail in many areas. Continued erosion threatens Moonstone Drive in some locations. The narrow beach here has little sand and is composed primarily of rounded pebbles. In the winter, storm waves break directly on the cliffs.

South of Cambria, the highway turns inland, and the coast consists of relatively undeveloped ranchland to the town of Cayucos, which is the northernmost community on Estero Bay. Just south of Cayucos Point, the material forming the coastal bluffs changes from the sandstones and siltstones of Cambria to very weak mélange of the Franciscan complex. This mélange is characterized by highly sheared and fractured blocks, some of

Figure 15.11   Construction continues on the northern Cayucos cliffs even though erosion rates are high. Riprap has been emplaced to protect two houses south of the one under construction. Photo by Cheryl Hapke.

which are composed of easily eroded serpentinite. Differential wave erosion along the numerous shear zones and fractures creates an area of high erosion hazard for cliff-top homes (Figure 15.11).

Along the northern portion of Estero Bay, near Cayucos Pier, there are no bluffs; homes and other buildings are built directly on the beach and the sediments of Cayucos Creek. To the south, bluffs increase in elevation and are completely developed. Many homes and businesses are built near the edge of these eroding bluffs and are thus at high risk of being damaged during severe winter storms. Armoring is common and provides evidence of the high erosion hazard along most of this coast. Where the bluffs are composed of stronger units within the Franciscan complex, rapid erosion of the weak marine terrace deposits that cap the bluffs can undercut building foundations.

The coastline along the southern stretch of Cayucos is the beginning of Morro Strand State beach, which was once divided into Morro Strand (north) and Atascadero (south) state beaches. The two stretches of beach are connected along an undeveloped portion of coast, where riprap at the cliff base protects the road, suggesting that erosion is a continual problem. The bluffs above the northern stretch of Morro Strand State Beach in

Figure 15.12 The developed cliffs of Cayucos are protected by a variety of structures (seawalls and riprap). Photo by Cheryl Hapke.

Cayucos are heavily developed (Figure 15.12). Although the wide beach provides some buffer from waves, larger winter storm waves reach the bluff base and rapidly erode the weak rocks.

The southern portion of Morro Strand State Beach, in the community of Morro Bay, is wider than the northern beach and is backed for most of the park by a large, vegetated dune field. The marine terrace in this area is depressed and the dunes lap onto the low terrace. There are a few houses at the northernmost end of this portion of the state beach; seawalls and riprap at the base of the cliff protect many of these homes. Although there is a hazard from wave inundation during extreme storm events, such as those that occurred during the 1982–83 and 1997–98 El Niños, the wide beach and dunes along the remainder of the park provide protection for the houses and other buildings constructed along the coast in this area.

Sand accumulates along the southern portion of Morro Strand State Beach. Near Morro Bay High School, the beach increased in width nearly 500 feet over a 60-year period. The beach accretion is directly related to the trapping of sand up coast from the Morro Bay harbor breakwater and the Morro Rock causeway. Morro Rock, the dominant geological feature of

Morro Bay, is a remnant of a volcanic plug (the inner core of an ancient volcano). Morro Rock once stood 1,000 feet offshore until a causeway, constructed from material quarried from the rock itself, was built during the 1930s and 1940s. Sand accumulated in front of the causeway, eventually forming a tombolo.

CHAPTER SIXTEEN

MORRO BAY TO POINT CONCEPTION

ANTONY R. ORME

The south-central California coast extends 100 miles from Point Estero to Point Conception, covering the coastal portions of southern San Luis Obispo County and western Santa Barbara County. In general terms, the coast consists of six natural units: (1) the 20-mile-long, west-facing coast of Estero Bay, dominated by Morro Rock and the barrier beach enclosing Morro Bay; (2) the 13-mile-long, southwest-facing coast fronting the San Luis Range; (3) the 15-mile-long, west-facing coast of the Santa Maria Valley fronting San Luis Obispo Bay; (4) the 8-mile-long, mostly southwest-facing coast of the Point Sal Ridge; (5) the 18-mile-long, west-facing, embayed coast of the San Antonio and Santa Ynez valleys; and (6) the 18-mile-long, mostly southwest-facing coast at the western end of the Santa Ynez Mountains (Figure 16.1). There is a certain repetition to these units: The west-facing bays are backed by sandy beaches, extensive dune fields, and alluvial valleys; and the intervening southwest-facing coasts are the seaward edges of resistant rock structures that form prominent cliffs or low rocky shores backed by marine terraces.

Human impacts on this coast are variable in both time and space. The rocky promontories have largely escaped development, whereas the bays have seen light to moderate impacts. The most developed segments occur around Morro Bay and northern San Luis Obispo Bay, the latter notable for the "five cities" around Pismo Beach. Land uses in this area range from single-family homes and apartments to hotels and recreational facilities, the whole backed by farmland and pastures. The coast south of Pismo Beach is much less developed owing to the extensive land holdings of the state and county park systems, and of the U.S. Department of Defense at

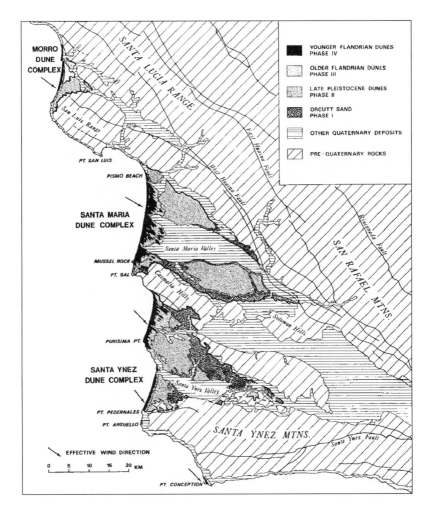

Figure 16.1   Coastline between Morro Bay and Point Conception, showing dune systems and major topographic features. Illustration by Antony Orme. From Quaternary dunes of the Pacific coast of the Californias. In *Aeolian Geomorphology*, ed. W. G. Nickling. New York: Allen & Unwin, 1986.

Vandenberg Air Force Base. Elsewhere, farms devoted to a range of agricultural crops and ranches engaged in rearing livestock complete the landholding pattern.

Nowhere here does the development density approach the scale of the San Francisco Bay area or southern California. Nevertheless, the expansion of residential properties and the infrastructure necessary to support a growing population, not only at the coast but also in the immediate hinterland around the small cities of San Luis Obispo (population 42,000) and Santa Maria (62,000), poses serious challenges to the present and future management of this coastal region. Along the immediate coast, many such developments have occurred with scant regard to natural hazards or to the problems posed by an expanding infrastructure.

PHYSICAL SETTING

The alternation of sandy bays and rocky headlands along the south-central California coast reflects the geological evolution of the region over Late Cenozoic time, and especially the processes that have shaped coastal cliffs, marine terraces, alluvial valleys, sand dunes, and beaches over the past 125,000 years.

In essence, collisional plate tectonics over the past 30 million years have fashioned volcanic, sedimentary, and meta-sedimentary rocks into a series of folded and faulted structures that reach the coast as prominent ridges separated by basins. The main ridges make up the Santa Lucia Range, a major component of the Coast Ranges, which parallel the coast north of Estero Bay; the smaller San Luis Range and Point Sal Ridge; and in the south the Santa Ynez Mountains, which are the westernmost part of the Transverse Ranges. The principal intervening basins contain the Los Osos Valley, the Santa Maria Valley, the San Antonio Valley, and the Santa Ynez Valley. In general terms, these valleys, together with the San Luis Range and Point Sal Ridge, are all part of the Santa Maria Basin, a broad, triangular structure constrained between the Coast Ranges to the northeast, the Transverse Ranges to the south, and the active Hosgri fault zone offshore to the west. This basin began opening about 12 million years ago, when it became the locus of marine deposition.

The ridges have been the focus of intermittent uplift during the last million or so years. This is shown by sequences of marine terraces that rise 650 feet above sea level north of Point Estero, 810 feet in the San Luis Range, 870 feet east of Point Sal, and 1,080 feet east of Point Conception. The shoreline, cut by interglacial seas roughly 125,000 years ago, when sea level was about 20 feet higher than at present, has since been raised from 40 to over 100 feet above present sea level.

This alternation of rising ridges and subsiding basins strongly influences the character of the present coast. The ridges form rocky shores whose subtle changes in rock type and structure are etched by waves and provide valuable habitats for suitable plant and animal life. The basins provide for sandy beaches, muddy lagoons, and wetlands whose character reflects the relative importance of fluvial, marine, and aeolian processes.

Sediment for the sandy beaches and dunes is derived mostly from coastal watersheds, introduced to the coastal zone by floods, moved onto and along beaches by waves and currents, and then reworked by wind from the beaches to form coastal dunes. There are two major conveyors of sediment, the Santa Maria and Santa Ynez rivers, but both of these have been dammed and therefore deliver less sediment than in historic times. Smaller streams such as San Luis Obispo Creek, Arroyo Grande, San Antonio Creek, and Jalama Creek are locally important. Even so, it is difficult to explain the vast amount of sediment stored in ancient dunes onshore without reference to past climates and tectonics. The wetter climates and floods of Pleistocene time certainly yielded far more sediment than today's regimes. More sediment reached Morro Bay, for example, before Pleistocene tectonic deformation of Los Osos Valley curtailed sediment delivery.

About 30 to 50 percent of storm waves and swells approach this coast from the northwest, with the remainder mostly from the west-northwest or west. Thus, most of this coast is fully exposed to, and indeed reflects, these predominant swells and storm waves. The prevailing longshore current and littoral drift are directed from north to south, while the winds that trigger these swells are reflected in the northwest–southeast axes of the coastal dunes. In addition, southerly and southwesterly swells generated in late summer by tropical cyclones off western Mexico and distant Southern Hemisphere storms may cause local reversals of the littoral circulation, notably along rocky, southwest-facing shores. These shores may also suffer significant impacts from winter cyclonic storms passing eastward along more southerly tracks. Mean tidal range along this coast is around 3.6 feet, and the range between mean lower low water (MLLW) and mean higher high water (MHHW) is 5.2 feet. The extreme range during spring tides may reach almost 9 feet. Thus, tidal range is not a major factor in shaping the open coast but is important in the low-gradient estuary of Morro Bay. However, if astronomical high tides are superimposed onto superelevated ocean levels, which occurred during the 1982–83 El Niño winter, serious beach erosion can occur throughout this coastal area.

DEVELOPMENT BACKGROUND AND PLANNING ISSUES

Middens, hearths, and stone artifacts buried beneath coastal dunes and alluvium provide abundant evidence for early occupancy of this coast, with people thinly present in Early Holocene time and more numerous after 4000 years B.P. (before present). The first Spanish overland foray, the Portola expedition of 1769, encountered many small Chumash villages, fish camps, and semi-nomadic groups along the coast. The Chumash people, relatively late immigrants to the region, extended their culture from Estero Bay to Santa Monica. The founding by the Franciscans of the Mission San Luis Obispo de Tolosa in 1772 led eventually to the decline in population of indigenous culture and the introduction of widespread grazing. The Spanish and Mexican interludes probably impacted the coastal dunes through livestock grazing, especially during the severe drought of 1828–30, when pasturage inland was severely limited. These interludes have also left a legacy of place names that tell much about the early historical environments of the region—for instance, *alamo* (cottonwood) and *oso* (bear).

The transfer of California to American administration in 1850 saw little immediate change in the region, other than in disputed titles regarding landholding. Cattle grazing flourished, except during the drought of 1862–64, and such infrastructure as existed catered mostly to livestock. Wharf construction, which began at Port Harford (later Port San Luis) in 1868, at Morro Bay in 1872, and at Spooner's Landing in 1890, was designed initially for the export of hides and live cattle, and later of dairy products and grain, as well as for timber imports from northern California. A coastal steamship operator then opened a narrow-gauge railway from Port Harford to San Luis Obispo in 1876. This railway was extended piecemeal over an additional 76 miles through Arroyo Grande (1881), Central City (later Santa Maria, 1882), and Los Alamos (1883) to Los Olivos in 1887. Meanwhile, the Southern Pacific Railroad, which had reached San Luis Obispo from the north in 1884, completed its coastal route from Santa Barbara westward around Point Conception to San Luis Obispo in 1901. Deprived of passengers and through freight, the Pacific Coast Railway, which had linked with the Southern Pacific at Guadalupe in 1909, depended on the local movement of farm produce and construction aggregate from gravel pits in the Sisquoc valley. It closed in 1941.

Although an oil boom and urban growth followed the discovery of petroleum beneath the Santa Maria Valley in the early 1900s, the region remained predominantly agricultural and relatively isolated until the mid-twentieth century. Then, with improved highways, car ownership, and paid vacations, the coast was "discovered" by the expanding urban populations of southern California and the Central Valley for recreation and retirement homes. Viticulture also expanded inland. The subsequent establishment of state and county parks and preserves, the presence of the U.S. Department of Defense at Vandenberg AFB, and the persistence of private ranches in the San Luis Range, the Point Sal Ridge, and around Point Conception have focused coastal development issues mostly on areas around Morro Bay and San Luis Obispo Bay.

ESTERO BAY AND MORRO BAY

Development around the broad sweep of Estero Bay focuses mostly on Morro Bay (population 10,000), with smaller communities to the north at Cayucos and to the south at Baywood Park (3,000) and Los Osos (8,000). These are essentially recreational, residential, and retirement communities into which an oil-fired power plant and oil storage tanks intrude at Morro Bay. The development of Morro Bay itself began with the construction in 1872 of a wharf for the import of timber and export of cattle products. A local fishing industry soon emerged, but significant growth began only after World War II.

Morro Bay originated as an estuary around 5,000 years ago, toward the close of the Holocene marine transgression. The seas flooded the seaward end of a subsiding structural trough. Gradually the estuary became separated from the open ocean first by the accumulation of submarine bars and emergent islets south of Morro Rock, and then by the growth of an impressive barrier beach and dune complex, which has in turn migrated eastward over former estuarine deposits. The barrier is now 1,000 to 2,000 feet wide and extends for 4 miles across the seaward front of Morro Bay. Behind it the sheltered bay has been transformed into an estuary that traps most of the sediment entering from Chorro Creek and Los Osos Creek, allowing the formation of intertidal mudflats and marshes. It also traps sand blown inland from the dunes. Sedimentation has significantly reduced the lagoon's size.

Sand dunes are a major feature of this coast, both as active dunes on the barrier beach and mainland shore north of Morro Rock, and as paleodunes blanketing the coastal hills and rocky coast south of Morro Bay (Figure 16.1). These dunes reflect prolonged episodes during which sand moved onshore with the prevailing winds from the northwest. The oldest paleodunes form a ramp 50 to 150 feet thick rising from below present sea level and feathering out around 1,000 feet against the San Luis Range. These overlie the last interglacial marine terrace and were deposited during the last glaciation, when sea level was lower than today. Radiocarbon dating of buried soils and Indian midden refuse indicates that a later sequence of dunes formed during and after the last marine transgression, became stabilized, and then partly eroded by gullying and wave action. These dunes are interbedded with numerous middens in the 4,000- to 2,000-year range, indicating a healthy presence of people along the coast at that time. The most recent dune phase is represented by active dunes on the Morro barrier,

A. Morro rock is the dacite neck of a former volcano, the westernmost of a series of Oligocene volcanoes extending eastward to San Luis Obispo. Until 1910, it was an offshore island linked to the mainland by sand at low tide.

B. In 1910, the north entrance to Morro Bay was closed by a revetment that later decayed but was rebuilt and extended after 1935.

C. The present entrance channel to Morro Bay began as a wartime emergency measure with the construction of the north breakwater in 1942. Shoaling problems were countered by the subsequent construction of the south breakwater and other engineered changes to the channel. Because these structures have modified littoral drift, periodic dredging of the channel has been necessary since 1949, rendering this a high hazard zone.

D. Morro Bay is a triangular lagoon, 3 to 5 miles to a side, that is separated from the Pacific Ocean by a late Holocene barrier beach and dune complex that began forming about 5000 years ago. This complex has been subject to reshaping by both natural processes and human activities.

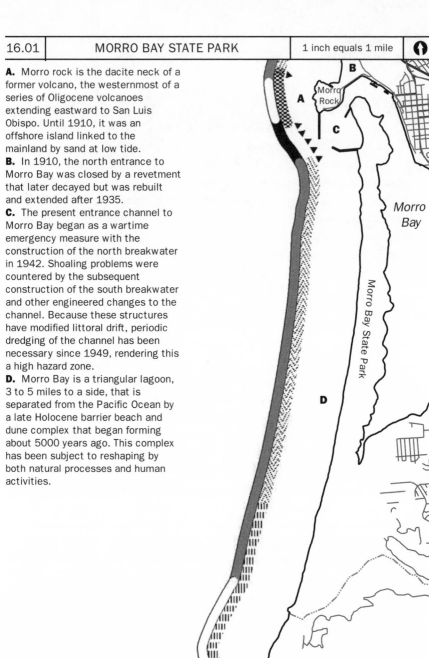

Morro Bay

Morro Bay State Park

Hazard Level
☐ Stable: low risk
▦ Caution: moderate risk
■ Hazard: high risk

Erosion Rates
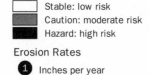 Inches per year

Shoreline Environment
▦ Sandy beach
▦ Beach backed by marsh
▦ Beach backed by dunes
▦ Cliff or bluff fronted by beach
▦ Rocky coast with no beach

Armoring
▲▲▲ Riprap
■■■ Seawall
⊡⊡⊡ Other

Figure 16.2   Morro Rock in 1904, before any coastal engineering structures were built. Photo courtesy of the U.S. Geological Survey.

fed from the modern beach, from reworking of dredge spoil, and by desta-bilization of older dunes. These dunes and their stabilizing vegetation have been much affected over the past century by fire, grazing livestock, military activities, and off-road vehicles, but are now largely protected within Montaña de Oro State Park.

Morro Bay's limitations as a port, however, have challenged sailors and coastal engineers for more than a century. The port is sheltered in the lee of Morro Rock, a vast volcanic edifice rising to 578 feet at the seaward end of a dozen or more volcanic necks extending southeastward to beyond San Luis Obispo. Describing Morro Rock in 1769, Juan Crespi "saw a great rock in the form of a round morro, which at high tide is isolated and separated from the coast by little less than a gunshot" (Figure 16.2). Throughout the nine-teenth century, Morro Rock remained an island tied to the mainland at low tide by shifting sand and flanked by unstable tidal channels both north and south of the rock. The shoaling of one entrance channel commonly accom-panied the deepening of the other. As early as 1872 Army engineers sug-gested closing the south channel, but in 1910 it was the north channel that was closed by a small revetment built by the San Francisco Bridge Company. This revetment decayed for want of maintenance after 1913, was rebuilt and extended to 1,700 feet in length and 12 feet above MLLW in 1935–36, and was further reinforced in 1943, by which time accretion against the north side of Morro Rock had eliminated the north channel (Figure 16.3). The diversion of Morro Creek directly to the sea north of Morro Rock and later dumping

Figure 16.3   Aerial photograph of Morro Rock and Morro Bay on August 19, 1938, with part of the town of Morro Bay visible in the lower left corner. Photo courtesy of the U.S. Army Corps of Engineers.

of dredged spoil has favored accretion, causing the beach to widen as much as 500 feet and support a small dune field within Morro Strand State Beach.

Closure of the north channel focused attention on the navigation problems posed by the south channel. In 1942 the 1,900-foot-long north breakwater was built southward from Morro Rock to provide a safe haven for naval patrol boats. Because this was a hurried wartime project, neither wave analysis nor a model study was conducted prior to construction. Wave refraction around the breakwater soon caused shoaling leeward, and the south breakwater was subsequently built in an attempt to restrict northward

Figure 16.4   Morro Bay, 1947, showing north and south breakwaters and sand spit with dunes to the south. Spence Collection. Reproduced by permission of the Department of Geography, University of California, Los Angeles.

littoral drift from moving into the entrance channel (Figure 16.4). The combined effect of these structures has been less than satisfactory. North of Morro Rock there is a net southward littoral drift of over 18,000 cubic yards of sand annually, but south of the rock the net annual drift of 32,000 cubic yards is directed northward. It seems that construction of the north breakwater has augmented the natural impact of Morro Rock on reversing longshore currents, thereby accentuating northward drift at the expense of the eroding barrier beach farther south.

The entrance channel between these breakwaters remains a troublesome sand trap, necessitating frequent dredging in order to maintain an adequate navigation channel. Since dredging began in 1949, about 115,000 cubic yards of sediment have accumulated annually in the entrance channel. Between 1949 and 1987, nearly 5 million cubic yards of sediment was removed. However, placement of the slurried dredged spoil has generated further problems. At one time or another, spoil has been placed onshore north of Morro Rock (where it led to beach widening and blowing sand), offshore (but this sand soon returned to the entrance), at the north end of the barrier opposite Morro Bay wharf (but this blew into the navigation channel), and as a slurry on the barrier south of the south breakwater (but pipeline construction and bulldozing have made for unstable foredunes). The

A. In the vicinity of Montaña de Oro, a sequence of late Pleistocene and early Holocene dunes dominates the coast. This sequence overlies a shore platform from the last interglacial which is cut across sedimentary rocks.

B. The rocky coast between Montaña de Oro and Point San Luis is characterized by low cliffs fronting distinctive marine terraces related to high interglacial seas around 125,000 and 80,000 years ago. Prominent sea stacks lie on these terraces. Higher, older marine terraces rise inland to 800 feet above sea level against the San Luis Ridge.

C. Diablo Canyon Nuclear Power Station lies on a marine terrace.

D. This rocky coast is relatively stable over the short term. Over the longer term, however, marine terrace and fault evidence indicate that the local mountains are rising at a rate of up to 0.75 feet per 1000 years.

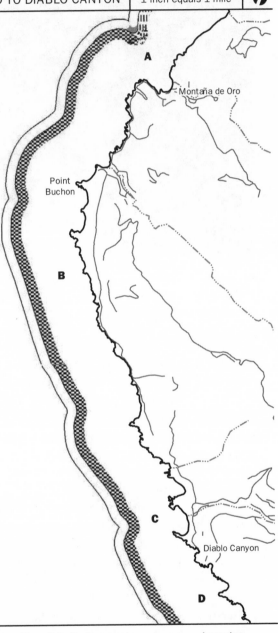

Hazard Level

	Stable: low risk
	Caution: moderate risk
	Hazard: high risk

Erosion Rates

 Inches per year

Shoreline Environment

Sandy beach

Beach backed by marsh

Beach backed by dunes

Cliff or bluff fronted by beach

Rocky coast with no beach

Armoring

▲ ▲ ▲ Riprap

Seawall

Other

A. This rocky, relatively inaccessible coast is noteworthy for its many sea stacks offshore and for abandoned Pleistocene sea stacks on the elevated marine terraces.

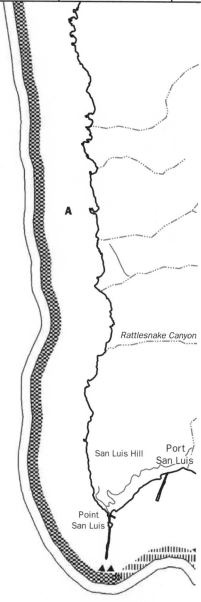

Rattlesnake Canyon

San Luis Hill

Port San Luis

Point San Luis

Hazard Level

☐ Stable: low risk
▓ Caution: moderate risk
■ Hazard: high risk

Erosion Rates

 Inches per year

Shoreline Environment

▒ Sandy beach
〰 Beach backed by marsh
〰 Beach backed by dunes
▥ Cliff or bluff fronted by beach
▨ Rocky coast with no beach

Armoring

▲▲▲ Riprap
■■■ Seawall
⊡⊡⊡ Other

Figure 16.5 Uplifted marine terraces of the San Luis coast, 1947. Spence Collection. Reproduced by permission of the Department of Geography, University of California, Los Angeles.

entrance to Morro Bay thus affords interesting lessons in the disruption of a natural system and attempts to manage an altered system.

SAN LUIS RANGE

The 12 miles of rocky coast from Montaña de Oro State Park to Point San Luis are backed by undeveloped ranchland and chaparral, currently inaccessible to the general public. A notable intrusion here is the Diablo Canyon Nuclear Power Plant, a controversial presence along a coast of known tectonic instability. In other respects, this is a remarkably picturesque coastal segment of rocky points, sea stacks, and pocket beaches that recaptures the way much of the coast may have appeared two centuries ago. The remaining 8 miles of rocky coast from Point San Luis to Pismo Beach harbor a range of uses—from the small boat harbor at Port San Luis, through the small community of Avila, to some unwise housing developments atop eroding seacliffs at Shell Beach.

The San Luis Range is formed mostly from Franciscan and younger marine sedimentary and volcanic rocks that have been uplifted over the past 15 million years into a series of west–northwest-trending fold and fault

Figure 16.6   Avila Beach, 1921. Spence Collection. Reproduced by permission of the Department of Geography, University of California, Los Angeles.

structures by compression within the broader tectonic environment of the San Andreas Fault system. Continuing uplift during the past 2 million years is demonstrated by a series of marine terraces that rise over 800 feet above sea level against the coastal hills (Figure 16.5).

Pocket beaches, sea stacks, and active seacliffs 30 to 200 feet high and backed by unstable hillslopes developed in Franciscan and younger rocks characterize the more developed coast from Point San Luis to Pismo Beach. The seacliffs, which reach 200 feet in height at Fossil Point and Pirates Cove, are prone to frequent rockfalls and occasional deep-seated landslides.

Port San Luis was founded as Port Harford around a modest wharf, warehouse, and horse tramway between 1868 and 1873. It began to grow, together with the nearby seaside community planned by the Avila brothers (Figure 16.6), when an 8-mile railway was completed to San Luis Obispo in 1876. Today Port San Luis is a commercial fishing port and recreational boat harbor, sheltered by Point San Luis from the predominant northwesterly swells but exposed to southwesterly storms. Accordingly, the present harbor authority, formed in 1954, maintains a protective rubble-mound breakwater 2,300 feet long. The exposure of this coast was well demonstrated during the El Niño winter of 1983, when southerly storm waves severely damaged the recreational

A. Protected by Point San Luis and its 2300 foot long breakwater, Port San Luis provides a relativelysafe anchorage, except from winter storms passing eastward along more southerly tracks. Under the latter conditions, and especially during El Niño events such as 1983 and 1998, both marine erosion and coastal landslides pose significant hazards.

B. Petroleum storage tanks.

C. Many homes are located on a arrow marine terrace near the cliff edge and require protection by seawalls, bulkheads, sandbags, and riprap. Rates of seacliff retreat range from 4 to 8 inches per year, usually from catastrophic rockfalls.

D. Narrow marine terrace subject to episodic seacliff retreat, endangering cliff-top roads and property.

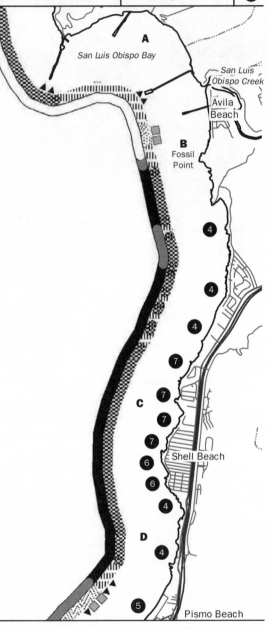

San Luis Obispo Bay

San Luis Obispo Creek

Avila Beach

B Fossil Point

A

4

4

4

7

7 **C**

7

7

6

6 Shell Beach

4

D

4

5 Pismo Beach

Hazard Level

- ☐ Stable: low risk
- ▨ Caution: moderate risk
- ■ Hazard: high risk

Erosion Rates

 Inches per year

Shoreline Environment

- Sandy beach
- Beach backed by marsh
- Beach backed by dunes
- Cliff or bluff fronted by beach
- Rocky coast with no beach

Armoring

- ▲▲▲ Riprap
- ■■■ Seawall
- ⊡⊡⊡ Other

pier at Avila Beach, led to seawall collapse at the coastline, and undermined seacliffs near the oil storage facility farther east. The accompanying heavy rains saturated coastal slopes, causing damage to roads and houses at Avila Beach, and flooded properties along San Luis Obispo Creek.

Farther east, the narrow marine terrace at Shell Beach supports many homes, apartments, and hotels built unwisely close to actively eroding seacliffs, which retreated up to 15 to 25 feet locally between 1965 and 1985. Since 1978, some cliff segments have retreated as much as 4 to 7 inches per year. Should these erosion rates continue, portions of U.S. Highway 101 will be undermined. Attempts to counter this erosion hazard involve sandbags, wood and steel bulkheads, and concrete seawalls, but in reality these properties should never have been built in harm's way.

THE SANTA MARIA VALLEY COAST

A sweeping sandy beach backed by extensive dune fields extending 30 miles inland dominates the 15-mile-long, low-lying coast from Pismo Beach to Mussel Rock. The principal source of this sand is the Santa Maria River, which drains a 1,880-square-mile basin extending seaward from the San Andreas Fault zone in the east through a series of tectonically active ridges and debris-choked valleys within the southern Coast Ranges. The Santa Maria River is formed by the confluence of the Cuyama and Sisquoc rivers only 22 miles inland from the coast. Below the confluence, the river has frequently shifted its course across a broad floodplain—for example, via Oso Flaco Lake—until being constrained by levees in recent decades. Much of the sediment brought by this river to the exposed continental shelf during the lowered sea levels of Pleistocene glacial stages was exposed to onshore winds that created the vast dune fields seen today.

Since construction of the Twitchell Dam in the lower Cuyama Valley just 27 miles from the coast in 1958, the lower reaches of the Santa Maria River have largely atrophied. Only when major floods in the uncontrolled Sisquoc River augment managed discharges from the reservoir does the river break through the coastal dunes to the shoreline. Sediment delivery to the coast has also been affected by sand and gravel extraction from the floodplain. The river mouth is thus inherently unstable, often more or less blocked by shifting beach and dune sands but sometimes breached by major floods. Nevertheless, there is an abundance of relict sand within this coastal cell that, driven south by the predominant swells, may still generate large volumes of littoral drift. Much of this sediment budget is transported onshore, where it is reworked by wind into backshore dunes, there perhaps to be stabilized by vegetation or further reworked by storm waves, winds, and river floods.

A. Wide recreational beach but backshore bluffs are prone to erosion, necessitating construction of seawalls and riprap protection.
B. Coastal management south of Pismo Beach is complicated by designation of beach and dunes for use by off-road vehicles.
C. Riprap used at the mouth of Arroyo Grande Creek to protect dwellings. Storm-wave impacts damage property and erode backshore dunes.
D. Coast between Pismo Beach and Santa Maria River comprises a broad sandy beach backed by a sequence of active, late Holocene dunes, and more stable, early Holocene and late Pleistocene dunes. Dune stability and vegetation cover are compromised by petroleum extraction facilities and off-road vehicle use, although the latter is now restricted.

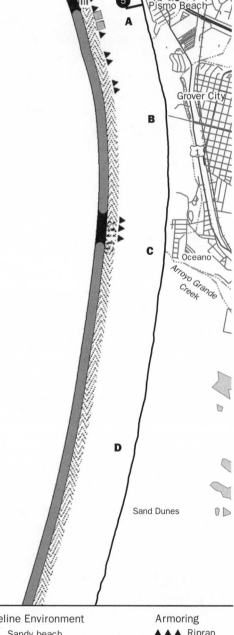

Pismo Beach

Grover City

Oceano

Arroyo Grande Creek

Sand Dunes

Hazard Level
- Stable: low risk
- Caution: moderate risk
- Hazard: high risk

Erosion Rates
- ① Inches per year

Shoreline Environment
- Sandy beach
- Beach backed by marsh
- Beach backed by dunes
- Cliff or bluff fronted by beach
- Rocky coast with no beach

Armoring
- ▲▲▲ Riprap
- Seawall
- Other

The Santa Maria dune complex consists of a series of older, inactive dunes (paleodunes) that extend for considerable distances inland, and the extensive modern, active dunes that extend from Pismo Beach to the Santa Maria River. Freshwater ponds and wetlands lie within the axes of these dunes. Some of the older dunes that have been reactivated directly or indirectly by human disturbances lie as much as three miles inland. The ancient and modern dunes are similar in orientation, structure, and mean sand grain size, indicating the persistence of similar dune-forming conditions along this coast over the past several hundred thousand years.

The Santa Maria dune complex has been much affected by human activity over the past 150 years—first by grazing, which disrupted the existing native vegetation, then by introduced plants and animals, sand mining, recreation, movie production, and petroleum extraction. As early as the 1920s, European dune grass was introduced here in an attempt to stop sand from drifting onto the railroad tracks. Oil companies acquired significant holdings in the dunes north of the Santa Maria River in 1938, and oil production from underlying Miocene traps began in 1948. The companies involved later became sensitive to the needs of habitat management. In 1962, a nuclear power station was proposed for Oso Flaco Lake, but the project was abandoned.

At one time, off-road vehicles had almost unrestricted access to the dunes behind Pismo Beach, but from 1983 onward these were progressively restricted to one designated area, the Oceano Dunes (Pismo Dunes) State Vehicular Recreation Area (Figure 16.7). Other active dunes are now largely protected within various state, county, and private parks and preserves. The paleodunes lie mostly in private hands.

The "five cities" of Shell Beach, Pismo Beach (population 8,000), Oceano (6,200), Grover Beach (12,000), and Arroyo Grande (15,000) have a major impact on this coast, both as growing residential communities and as recreational venues supported by hotels, rental properties, restaurants, campsites, and support infrastructure. Cliff-top development continues in Shell Beach, with many new, large residences being constructed nearly at the cliff's edge (Figure 16.8). Concrete seawalls and gunite are ubiquitous attempts to halt or slow the rate of cliff retreat. The Cliffs Hotel is a good example of both the issues associated with cliff-top construction and the challenges of permitting agencies in dealing with differing assessments of coastal erosion hazards (see Chapter 8 and Figure 8.1 for a more complete description of the Cliffs Hotel issue).

Pismo Beach is where most people and vehicles gain access to the shore and dunes, and this, together with exposure to storm waves, has necessitated a succession of unsightly shore protection structures. This is also the home

Figure 16.7   Beach access ramp, Pismo Beach, destroyed during March 1983 storms. Photo by Kevin Mulligan.

Figure 16.8   New cliff-top construction and concrete seawalls in the coastline of Shell Beach. Photo © 2002–2004 Kenneth and Gabrielle Adelman, California Coastal Records Project, www.Californiacoastline.org.

A. Dune stability within the Guadelupe Dunes has been compromised by petroleum extraction activities.

B. The Santa Maria River drains a 1880 square mile watershed, outflow from which has been controlled since 1958 by the Twitchell Reservoir. Nevertheless, unusually rainy years still see the river breach the beach-dune barrier at its mouth, generating a hazardous river-mouth environment.

C. The coast from Mussel Point to near Point Sal is underlain by Miocene bedrock overlain by massive accumulations of dune sand, forming coastal bluffs but now mostly stabilized farther inland.

D. Point Sal Ridge is prone to mass movement and seacliff erosion at rates of 5 to 12 inches per year. One large landslide pushed far into the surf zone in the 1930s.

E. The slopes extending southward from the Point Sal Ridge to the Santa Ynez River comprise a staircase of marine terraces mantled, especially along the coast and near the river, with aeolian dunes. This region lies within the limits of Vandenberg Air Force Base.

F. Military activities within Vandenberg Air Force Base impact coastal dune terrain. Coastal stability is favored by rock outcrops at the shore.

G. The flow and sediment production of the Santa Ynez River is now regulated by two dams upstream, but unusually wet winters will still generate significant discharge through the river mouth, rendering it a high hazard environment. Summer months see a barrier reform across the river mouth.

H. The coast from the Santa Ynez River southward to Point Pedernales comprises a sandy beach backed by a rising ramp of sand dunes which form bluffs at the shore. From Surf to near Jalama, this coast lies within Point Arguello U.S. Naval Missile Facility.

Hazard Level

- ☐ Stable: low risk
- ▨ Caution: moderate risk
- ■ Hazard: high risk

Erosion Rates

 Inches per year

Armoring

- ▲▲ Riprap
- ▨▨ Seawall
- ⊡⊡ Other

A. The coast between Point Pedernales and Espada Bluff is composed of low rocky cliffs and sea stacks fronting a relatively narrow marine terrace sequence. Landslides are common around Point Arguello and Rocky Point.

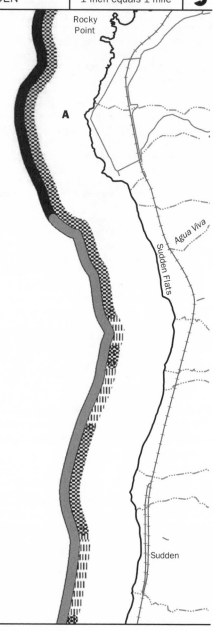

Rocky Point

A

Agua Viva

Sudden Flats

Sudden

Hazard Level
 Stable: low risk
Caution: moderate risk
Hazard: high risk

Erosion Rates
 Inches per year

Shoreline Environment
 Sandy beach
Beach backed by marsh
Beach backed by dunes
Cliff or bluff fronted by beach
Rocky coast with no beach

Armoring
 ▲ ▲ ▲ Riprap
 Seawall
 Other

site of the Pismo clam, a gastronomic delicacy that has become endangered from decades of unrestricted harvesting and the depredations of sea otters, although it ranges from central California to Baja California. At one time, wagonloads of Pismo clams were harvested from intertidal sand flats by teams of horse-drawn plows, but harvesting is now restricted. Farther south, various agencies have begun to achieve some success in recent decades with the enclosure and restoration of selected dune and wetland habitats. Officially established in August 2000, the Guadalupe-Nipomo Dunes National Wildlife Refuge is currently maintained by the U.S. Fish and Wildlife Service.

POINT SAL RIDGE

The beach-dune complex fronting the Santa Maria Valley ends at Mussel Rock, to be replaced by 8 miles of rugged rocky coast flanking Point Sal Ridge. A flight of up to nine marine terraces rising to 870 feet above sea level against the Casmalia Hills indicates continued uplift during the last several million years. Paleodunes rise to over 1,200 feet on the Point Sal Ridge, in part because they were blown upslope by the prevailing winds and in part because the ridge has been uplifted significantly since the dunes formed. Radiocarbon dating of paleodunes indicates massive dune deposition between 30,000 and 20,000 years ago, as sea level fell during the approach of the last glacial maximum and large quantities of sand were exposed on the emergent continental shelf.

Most of Point Sal Ridge remains private ranch land, where overgrazing may have been responsible for soil degradation, gullying, and dune reactivation. However, a dirt road provides access to the small Point Sal State Beach on the south side of the ridge, from which are visible a variety of rocky shore habitats and a massive landslide that slumped into the surf zone during the 1930s and has yet to be removed by wave action.

THE SANTA YNEZ VALLEY COAST

South of Point Sal Ridge, coastal cliffs decline in height and the coast is again dominated by mostly sandy beaches backed by extensive but degraded dune fields at the seaward ends of San Antonio Creek and the Santa Ynez River valley. Low cliffs occur around Purisima Point and rise farther south toward Point Pedernales and Point Arguello. The local rivers are again the principal source of sand for the beaches and dunes along this coast, but, as is the case farther north, sediment supplies have been curtailed in recent decades by the construction of dams on the upper Santa Ynez River and by floodplain levees downstream. Gibraltar Dam was built near the river's

A. Espada Bluff is the seaward edge of a 200-foot-high landslide, protected at its base by a seawall.

B. Bluffs fronting the marine terraces are subject to frequent rock falls and landslides.

C. Jalama Creek enters the sea between low dunes but the area has been much modified by recreational activities.

D. Cliffs of Miocene shale are subject to frequent rockfalls, occasional landslides, and marine erosion at rates of 12 to 20 inches per year.

E. Seacliff erosion and mass movement threaten the railroad and necessitate seawall protection.

F. Bedrock cliffs and dune bluffs are subject to frequent marine erosion.

G. Point Conception is a prominent headland formed in Miocene bedrock that is thoroughly exposed to marine erosion throughout the year.

Espada Bluff

A

B

Jalama Beach County Park

C

D Jalama

12
20

E

F

G
Point
Conception

Hazard Level
- ☐ Stable: low risk
- ▧ Caution: moderate risk
- ■ Hazard: high risk

Erosion Rates
- **1** Inches per year

Shoreline Environment
- Sandy beach
- Beach backed by marsh
- Beach backed by dunes
- Cliff or bluff fronted by beach
- Rocky coast with no beach

Armoring
- ▲▲▲ Riprap
- ■■■ Seawall
- ⊡⊡⊡ Other

Figure 16.9 Point Conception and Jalama Bay, 1931. Spence Collection. Reproduced by permission of the Department of Geography, University of California, Los Angeles.

headwaters in the 1930s, and farther downstream the Bradbury Dam, completed in 1953, forms Lake Cachuma.

The presence of Vandenberg Air Force Base, including the Pacific Missile range, ensures that most of this coast is inaccessible to the general public, except by special arrangement. Thus the main management challenge is to ensure that the military authorities take appropriate responsibility during their stewardship. The coast may be reached at Ocean Beach County Park at the mouth of the Santa Ynez River, and visual access may be obtained from passenger trains that pass close to the shore en route between Santa Barbara and San Luis Obispo. Near Point Pedernales, the railroad line is sufficiently threatened by cliff retreat to be protected by a concrete seawall and rubble revetment.

THE WESTERNMOST SANTA YNEZ MOUNTAINS

The Santa Ynez Mountains, the westernmost of the east–west-trending Transverse Ranges, reach the coast between the Santa Ynez River and Point Conception (Figure 16.9). Miocene marine shales and siltstones form the unstable and crumbling coastal cliffs, which are backed by up to 15 marine terraces, again indicating uplift during the past several million years. Deformed terraces related to sea levels around 125,000, 105,000, and 80,000 years ago extend around Point Conception and continue at intervals along the south-facing coast of the mountains toward Santa Barbara.

This is a rocky coast with numerous sea stacks, crumbling seacliffs, and small pocket beaches. Sand dunes are narrowly present against and atop these cliffs, and a more extensive dune field occurs where Point Conception interrupted prevailing northwesterly winds moving across the previously exposed shelf. Short-term rates of cliff retreat on the order of 12 to 20 inches per year have been recorded near Jalama Beach. This rocky coast, a dangerous lee shore for navigators seeking to pass around Point Arguello and Point Conception, is noteworthy for its numerous historical shipwrecks.

Vandenberg Air Force Base extends as far south as Jalama Beach County Park, the only public access to this coast, south of which the coast is private ranch land. Railroad construction across the lowest marine terrace has again been subject to cliff erosion and landslides, requiring seawalls and rubble revetments at Espada Bluff and between Jalama Beach and Point Conception. With so much coastal land under military, railroad, or ranch control, management issues are minimal at present. However, as in the San Luis Range and Point Sal Ridge, these sometimes controversial forms of land stewardship have saved much of this coast from more radical development projects.

CHAPTER SEVENTEEN

POINT CONCEPTION TO RINCON POINT

ROBERT M. NORRIS AND KIKI PATSCH

POINT CONCEPTION TO GAVIOTA STATE BEACH

Point Conception (Figure 17.1) marks a dramatic change in the orientation of California's coastline—from north–south to nearly east–west. Most of the shoreline from Point Conception to Gaviota State Beach is contained in the Hollister and Bixby ranches, which originated as Mexican land grants during the early nineteenth century. Hollister Ranch was sold in the early 1970s, and approximately 2 square miles were separated as the proposed site for a nuclear power plant, which was never built. The same site was subsequently proposed for a liquefied natural gas facility, which also met with opposition and was never approved. The remainder of this large ranch was subdivided into small ranches of about 100 acres each. Present development is light, with most structures set considerably back from the coastline. Most of the land is now used for grazing cattle and horses. To the west, Bixby Ranch, which extends from Cañada del Cojo to Point Conception, is virtually undeveloped except for dry farming and cattle grazing. There are no structures associated with the ranch that are endangered by seacliff retreat.

The Union Pacific Railroad tracks nearly parallel the coast through this entire 17-mile distance. The tracks were laid at the turn of the twentieth century, with the first trains running by 1901. Coastal erosion was apparently recognized fairly early on because the first seawalls designed to protect the tracks in this area were constructed in 1909. The dates of seawall construction are engraved in the ends of the concrete structures. Thus, the individual seawalls preserve a good record of seacliff erosion over different time periods in this area. Unfortunately, the seawalls were apparently constructed only where tracks were specifically endangered. Thus, control is excellent in the eastern part of the study area but virtually absent to the west.

Figure 17.1   Point Conception, where the California coastline makes a nearly 90°
change in orientation from north–south to east–west. Photo © 2002–2004 Kenneth and
Gabrielle Adelman, California Coastal Records Project, www.Californiacoastline.org.

PHYSICAL SETTING

The stretch of coast between Government Point and Gaviota Beach State
Park is relatively straight, with narrow, sand-deficient beaches and low to
moderately high (5 to 60 feet on average) seacliffs fronting a marine terrace
that is usually quite wide. The sand volume on the beach fluctuates from
year to year and season to season. During the spring and summer of 1980,
long stretches (up to several miles) of this coast were completely denuded
of sand cover, leaving only a bare rock platform. The presence of marine
sand deposits on the raised Pleistocene marine terrace in this area suggests
that this sand-deficient condition is probably not the norm. Much of the
sand historically supplied to the beaches as far south as Point Arguello by
up-coast rivers is now trapped in reservoirs behind dams. Dams now block
about 61 percent of the drainage basin of the Santa Maria River, for example, causing a 68 percent reduction in the sand supply from this river.
Similarly, dams have reduced the sand supply from the Santa Ynez River by
53 percent. A sand deficiency or a decrease in beach width along any coastline can be expected to increase coastal erosion or seacliff retreat, simply
because beaches are the principal buffer zones to wave attack. The effects of
sand reduction on beach widths along this coast are just now being studied,
however.

Historical cliff erosion data from Point Conception to Gaviota is sparse and incomplete. The seacliff and uplifted marine terraces that form the coastline throughout this area are eroded into sedimentary rocks, predominantly shales and mudstones of the Sisquoc and Monterey formations, which contain very little sand-sized material. The rocks are distinctly layered or well bedded for the most part, and are tilted moderately steeply seaward. These factors facilitate cliff erosion as individual layers are undercut at the beach and the overlying material fails with the removal of this support.

SEACLIFF EROSION

The amount of seacliff retreat in the Point Conception to Gaviota area was determined by directly measuring the distance from the ends of the seawalls protecting the railroad to the present seacliff. This method is based on the assumption that the wingwalls of each seawall were initially attached to the seacliff, as substantiated by bedrock fragments cemented into the end of the seawall where the wall turns toward the seacliff. In some cases, however, the part of the seawall indicating this retreat is an addition to the original structure. The geometry of the original seawall, for example, does not allow for a precise determination of the amount of erosion in the time span between installations of the two wall segments. High erosion rates over short time periods, interspersed with longer periods characterized by low rates, demonstrate once again that seacliff erosion is a highly episodic process.

At the end of one seawall west of Alegria Canyon on Hollister Ranch, 6.5 feet of erosion occurred between 1909 and 1915, when the original wall and an addition were built, respectively. Between 1915 and 1980, however, only 5.5 feet of additional erosion occurred. The average long-term rate since 1909 is only 2 inches per year, the 1909 to 1915 average rate is 13 inches per year, and the 1915 to 1980 average rate is only 1 inch per year. Similarly, at a seawall west of Agua Caliente Canyon, about 1 mile east of Alegria Canyon, the average retreat rate is 3 inches per year since 1928, but an addition in 1937 indicates that 6.2 feet of the total 12.8 feet occurred between 1928 and 1937, yielding a short-term average rate of 8 inches per year. The post-1937 average rate is only about 2 inches per year. Similar erosion patterns are indicated by data from other seawalls. From this data, it appears that a large part of the erosion along this section occurred over short time periods and perhaps during single seasons as a result of abnormally large storms or severe winters.

GAVIOTA TO RINCON POINT

In marked contrast to most of the west coast of North America, southern Santa Barbara County lies in the western Transverse Range province and as

A. Weak siltstone and shale. Erosion rates not determined due to lack of development. Beach sand often absent during winter.

B. Uninhabited coastline. Future development should be set back from the coast.

C. Due to lack of coastal development, railroad tracks are the only structures with potential problems. Short segments of seawalls have been used to protect railroad trestles.

D. Gaviota Beach State Park: Parking lot destroyed during 1983 storms.

E. Narrow boulder or pebble beach with numerous rocky headlands.

F. Refugio Beach State Park: Park damaged by storm debris during winter, 1983.

G. Sandy beach and boulder delta at El Capitan Beach State Park with campground on terrace.

H. Narrow to very narrow beach. Pebbles and boulders lie at the base of the cliff, with some sand seaward of the cliff. Occasional headlands of Monterey Shale. Steel rail pilings spaced 3 to 4 feet apart to protect the railway. Minor, eroded riprap behind pilings.

I. Pier serves boats that carry crew and equipment to offshore oil platforms. Large hotel/resort complex built on marine terrace directly adjacent to actively eroding cliff with no beach. Golf course on top of terrace and deteriorating timber seawall ~1.5 miles long follows base of cliff.

J. Beach widens as littoral drift is trapped behind Coal Oil Point with development of dunes on back beach.

Hazard Level		Erosion Rates	Armoring	
□	Stable: low risk	❶ Inches per year	▲▲	Riprap
▨	Caution: moderate risk		▨▨	Seawall
■	Hazard: high risk		⊡⊡	Other

a result has an east–west-trending coast, with the steep-fronted Santa Ynez Range dominating this coast. All rocks and geological structures in this area have a strong east–west trend. The coastal plain is no more than 2.5 miles wide at any point in this region. In fact, west of Dos Pueblos Canyon the coastal plain is almost completely absent. Locally, the coastal plain is interrupted on its shoreward side by a few short, elongated hills, some with elevations as much as 600 feet.

Much of southern Santa Barbara County consists of cliffs with heights averaging about 50 feet; however, some cliffs are over 200 feet high. These cliffs have typically been cut into uplifted marine terraces by wave action and are cut into sedimentary rocks that are extensively deformed. Because none of the rocks forming the cliffs are particularly resistant to erosion, these cliffs undergo fairly rapid retreat.

Three segments of this coast, however, lack cliffs or bluffs and are instead marked by low sandbars enclosing low-lying coastal plains and salt marshes. The largest of these is the relatively undisturbed salt marsh at Carpinteria, which is due in part to the presence of an offshore rocky reef exposed only at the very lowest tides. The wave shadow of this reef has produced a sandy, cuspate headland, Sand Point. This sandy headland encloses Carpinteria Salt Marsh.

The second segment of low-lying coast is at Montecito, which lies mainly to the west of Fernald Point. This part of the coast lies at the foot of a gently sloping alluvial plain formed from flood deposits left by the various short, intermittent streams that drain the south face of the Santa Ynez Range in the vicinity of Montecito. Bluffs, where present, are only a few feet high.

Prior to the urbanization that began about 170 years ago, the coast at the city of Santa Barbara was also low and marshy. It remains low, but little remains of the former marsh apart from the Andree Clark Bird Refuge. This stretch of coastline has been greatly modified by urban development and by the construction of Santa Barbara Harbor and breakwater in the late 1920s, which will be discussed in greater detail.

About 10 miles west of Santa Barbara is Goleta Slough, once a large salt marsh. This feature is almost entirely cut off from the sea by coastal hills and by a short sand spit blocking most of the entrance (Figure 17.2). Only a small amount of the original salt marsh remains. Much of it was filled during the development of Santa Barbara Airport in the early 1940s. Goleta Slough, like Carpinteria Salt Marsh, has been reduced in area by periodic severe floods that have brought large amounts of sediment into the marsh from the nearby mountains. The large areas of open water that characterized these marshes in the eighteenth and nineteenth centuries have been substantially reduced by both natural and human-related causes.

Figure 17.2   Goleta Slough and Goleta Beach County Park, where beach erosion has become a management problem. Photo © 2002–2004 Kenneth and Gabrielle Adelman, California Coastal Records Project, www.Californiacoastline.org.

A very small salt marsh, known locally as Devereux Lagoon, lies just west of Goleta Slough at Coal Oil Point. This feature is now flooded by occasional high tides chiefly during the winter and spring months, but is dry most of the year and supports very little salt marsh vegetation.

COASTAL HEADLANDS

The southeastern portion of the Santa Barbara County coastline is fairly straight, particularly from Point Conception eastward to El Capitan Beach, but beginning there and continuing beyond Rincon Point, the coastline is interrupted by a number of small headlands. At El Capitan Beach a small, cuspate boulder delta has been formed by deposition from Cañada del Capitan Creek (Figure 17.3). A similar but much smaller feature occurs at the mouth of Gato Canyon. More prominent headlands occur near Goleta and Santa Barbara, as does the broadest coastal plain in the Santa Barbara region. At Rincon Point, on the Ventura–Santa Barbara county line, another prominent deltaic headland has been produced by Rincon Creek. This feature is similar to the one at El Capitan. Several other small deltaic headlands occur along the Santa Barbara coast, but because the streams that supply them are short and intermittent, the deltas are quite small and have only minor effects on the general trend of the coast. As a rule, where boulder

Figure 17.3   Boulder delta at El Capitan Beach. Photo by Robert Norris.

deltas have developed, the tributary streams are longer, have relatively larger drainage basins that extend up to the high ridges, and have either permanent or nearly permanent flows.

The two prominent headlands, Goleta Point on the east and Coal Oil Point on the west, represent the seaward face of a block raised along the More Ranch fault and on which the UC Santa Barbara campus and the community of Isla Vista are located. Both of these headlands are formed from the Sisquoc Formation, a marine sedimentary rock slightly younger than the Monterey shale that makes up most of the seacliffs along the south coast of Santa Barbara County. Although the Sisquoc Formation is not a particularly resistant rock unit, here it has been armored to some extent by tar from large sea floor seeps off Coal Oil Point.

A prominent but low headland occurs at the mouth of Carpinteria salt marsh, although the natural shoreline is formed of loose, sandy beach deposits. The cause of this feature is a prominent rock reef lying just offshore from the salt marsh inlet. The reef is exposed only during the very lowest tides, perhaps two or three times a year, but is nonetheless an impediment to approaching waves. The reef acts as an offshore submerged breakwater, refracting and reducing the inshore height of the waves. As a result, both the longshore current and littoral drift are reduced at this location, leading to the deposition of sand and forming a feature akin to a tombolo

Figure 17.4 The entrance to the Carpinteria salt marsh has now been completely lined with riprap to protect the homes that line the sand spit enclosing the marsh. Photo © 2002–2004 Kenneth and Gabrielle Adelman, California Coastal Records Project, www.Californiacoastline.org.

(an offshore island tied to the shore by a sand spit) of sorts. The nearly complete development of the sand spit enclosing the Carpinteria salt marsh with homes, however, and the armoring of the entire sand spit (Figure 17.4) with riprap has greatly changed this environment.

ROCKS OF THE SEACLIFFS

Apart from the narrow canyon mouths, the entire coastline from Point Conception east to Goleta is cliffed and the bedrock exposed in the cliff face belongs mostly to the Monterey Formation. As noted, some areas near Goleta have cliffs cut in the Sisquoc Formation, and east of Santa Barbara near Summerland, some cliffs consist of the sandy non-marine Casitas Formation. The bedrock composing all of these cliffs is relatively weak and easily eroded by wave action. Numerous small rockslides occur where marine erosion has undercut the base of the cliffs or where the tilt of bedding is favorable to sliding (Figure 17.5). As a consequence of rock weakness, a prominent wave-cut platform has developed in front of the cliffed portions of the Santa Barbara coast, despite the partial barrier to waves from the open Pacific formed by the offshore Channel Islands. The wave-cut platform extends seaward perhaps as much as a mile in some places, but generally less than that. In winter, particularly after strong storms have stripped

Figure 17.5   Wave-cut bench in steeply dipping Monterey shales at Gaviota State Beach. Photo by Robert Norris.

most of the sand off the beaches, the inner edge of this wave-cut platform is well exposed.

East of Goleta Beach—at More Mesa, for example—the prominent 100-foot-high cliffs are composed of massive sandstones and siltstones belonging to the marine Santa Barbara Formation. These cliffs are eroding more rapidly than any others between Isla Vista and Santa Barbara. From More Mesa through Hope Ranch and eastward as far as Santa Barbara Point, the cliffs are again composed of siliceous shale of the Monterey Formation. From Santa Barbara to the western side of the Carpinteria lowland, the low cliffs are mostly cut into bouldery stream deposits except in the Summerland area, where a seacliff, nearly 200-feet high in a few locations, is made up of the sandy, non-marine Casitas Formation. Eastward from Carpinteria to Rincon Point, seacliffs are once again composed of the Monterey Formation.

Although most of the seacliffs are cut into the geological formations just mentioned, the uppermost portion often consists of nearly horizontal terrace and dune sands, typically 10 to 30 feet thick. The basal part of these nearly horizontal sediments contains some solitary corals and many fossil mollusks, including numerous rock-boring clams that have bored into the underlying bedrock. The existence of these clams well above sea level is a result of continued uplift of the coastline and an oscillating sea level. The clams bored into the wave-cut platform at a time when sea level was high, similar to conditions today; when sea level began to fall but the land continued to be uplifted, these clams were raised above the intertidal zone, died, and were preserved as fossils. The presence of these clams in their burrows suggests that some of the uplift was abrupt, probably accompanying strong earthquakes.

BEACHES

For the most part, beaches along the Santa Barbara coast are now thin and narrow, and from historical accounts and old aerial photographs, at least prior to human intervention, the beaches from Gaviota to Santa Barbara, east of Santa Barbara to Carpinteria, and along the Rincon area were always thin. These beaches are subject to erosion each winter during storm wave attack, and they then undergo gradual recovery or accretion during the late spring, summer, and fall months due to a more gentle wave climate. This seasonal fluctuation is certainly not unique to the Santa Barbara area but is particularly prominent here because of the normally small sand supply that can be removed quickly in winter, exposing the rocky platform fronting the seacliffs. These seasonal beach changes vary in duration and magnitude from year to year as a result of the combined effects of storm waves and tidal heights or sea level conditions. Some years show little change, whereas others, such as the abnormally stormy El Niño winter of 1982–83, produced the most severe beach erosion in many decades. The seasonal cycles are further modulated by other long-term changes, such as the rate of sand supply to the coast.

The large volumes of sand historically delivered to the coast of this area by streams such as the Santa Maria and Santa Ynez rivers appear to be largely blocked from reaching the south coast beaches by points Arguello and Conception. The Santa Ynez and Santa Maria rivers have also been dammed extensively, reducing their delivery of beach sand to the shoreline.

On the other hand, estimates of sand supplied to south coast beaches by cliff erosion and streams east of Point Conception seem inadequate to account for the annual influx of sand into Santa Barbara Harbor. The total amount of sand in transit along the coast has been measured at the Santa Barbara Harbor as a result of the biannual to annual dredging that has gone on since 1933. The dredging required to keep the entrance channel open averages about 300,000 cubic yards or approximately 30,000 dumptruck loads per year. This is equivalent to a dumptruck full of sand passing along the harbor breakwater every 17 minutes, 24 hours a day, 365 days per year.

However, the lack of any appreciable beaches at Point Conception (Figure 17.1), the very thin beaches as far east as Gaviota, and the differences in heavy mineral content of the beaches north of Point Conception from those to the east suggest that this point is a significant barrier, and therefore that the estimates of sand supply derived from the south coast are too low, particularly in view of the almost total absence of any independent evidence of sand transport around the headland at Point Conception.

Where might this missing sand come from, if not from north of Point Conception? Certainly not from the erosion of the seacliffs that are cut into the Monterey and Sisquoc formations, neither of which contains appreciable sand. The overlying terrace deposits average only 10 to 30 feet in thickness and, although sandy, probably are insufficient to make up the deficit. The contributions from the steep streams draining the south flank of the Santa Ynez Mountains are not known but may collectively provide significant additional sand. Very few of the smaller tributary streams draining the Santa Ynez Range are dammed or channelized, although many debris basins were constructed to trap the sediment eroded by winter rains following major brushfires.

Rainfall in this region of Mediterranean climate, however, produces a highly erratic pattern of stream flow. Higher-than-normal rainfall, that is, in excess of the usual 18 inches at the coast and 35 inches at the mountain crest, is no assurance that stream flow will be adequate to transport appreciable amounts of new sand to the beaches; the rains must be closely spaced and of sufficient magnitude and intensity to produce a sustained and vigorous runoff. Normally, the wetter years will produce more runoff and therefore more sand supply than drier years, but this is not always true. Many years ago a study by the Army Corps of Engineers showed a tendency for beach widening to correlate best with rainfall totals for two-year periods. Because sand delivered to stream mouths takes time to be carried along shore by littoral drift, the sand may not show up immediately on the beaches. The Corps reported a 3.5-year lag between heavy rainfall in the

Santa Ynez Mountains and a subsequent influx of sand to the Santa Barbara Harbor.

No significant barriers to the eastward longshore movement of sand exist between Point Conception and Santa Barbara, although several small headlands are present. A minor exception occurs at Coal Oil Point, where there is a short segment of west-facing beach. Because prevailing winds in the area come from the west, some beach sand is driven inshore, forming a very local area of sand dunes. Dune formation removes a small amount of sand from the littoral transport system.

At Santa Barbara, however, an L-shaped breakwater, open to the east, was constructed in the late 1920s and early 1930s to provide a protected anchorage for boats. This large structure has seriously interfered with the sand supply to down-coast beaches at least as far as Rincon Point. Because of the impoundment of nearly all of the littoral drift by the breakwater, the beaches between Santa Barbara and Rincon Creek were denied their normal sand supplies during most of the 1930s following the construction of the breakwater (Figure 4.9). The only sand reaching those down-coast beaches during that period came from small streams discharging to the coast east of Santa Barbara or from the immediate offshore area as a result of normal seasonal on- and offshore sand transport. Serious beach erosion occurred as a result. The beaches did not recover until the 1940s, when the sand discharged during the dredging of the harbor entrance was able to reestablish its transit down coast and again nourish these beaches. Further beach erosion and property damage have been prevented by the more or less regular sand bypassing at the harbor since then, as well as by the construction of seawalls and placement of riprap down coast. To the observant reader, it will now be evident that sand transport along the southern Santa Barbara County coast is dominantly from west to east (Figures 4.8 and 4.9). This direction of transport comes about because most waves enter the Santa Barbara Channel from the west, breaking at an angle to the shoreline and therefore driving sand eastward.

Because of the nearly continuous chain of islands on the south side of the Santa Barbara Channel, very few waves approaching directly from the south ever reach the Santa Barbara coast. The 10-mile gap between Anacapa Island, at the eastern end of the Channel Islands, and the mainland at Point Hueneme does allow some waves to enter the channel from the southeast. Such waves are produced, typically, by tropical storms off Mexico and Central America and are most apt to occur in late summer and early autumn. Some winters are also characterized by southeasters. Both of these situations may briefly reverse the normal direction of sand movement alongshore. The number of days during which some beach sand moves

A. Moderately wide beach fronting low seacliff. The marine terrace is undeveloped.

B. Beach narrows approaching developed portion of Isla Vista. Waves reach base of cliff even during summer months such that cliff is being undercut and is actively eroding. Two timber seawalls have been installed but cliffs continue to fail and have undermined the foundations of a number of cliff-edge apartments.

C. Goleta Point and cliffs immediately northeast protected by riprap.

D. Base of cliff beneath UCSB marine laboratory buildings armored with riprap.

E. Goleta Beach County Park was badly eroded during winter of 1982-83 and this trend has continued in 1995, 1998, 1999, and 2000. Beach, lawn area and parking lots have been threatened and damaged leading to installation of temporary riprap at west end and beach nourishment in 2000.

F. Steep cliffs are actively eroding and fronted by narrow sandy beach. Resistant rocks and natural tar seeps on beach.

G. Steep eroding cliffs fronted by sandy beach. Homes developed on large parcels on cliff top. Most of houses set well back from cliff edge. Winter waves typically reach base of cliff.

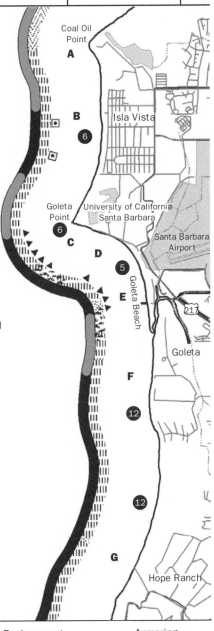

Hazard Level
Stable: low risk
Caution: moderate risk
Hazard: high risk

Erosion Rates
 Inches per year

Shoreline Environment
Sandy beach
Beach backed by marsh
Beach backed by dunes
Cliff or bluff fronted by beach
Rocky coast with no beach

Armoring
▲▲▲ Riprap
Seawall
Other

toward the west or up coast varies, being higher in El Niño years, when storms commonly come out of the southwest, but rarely exceeds 10 days in any single year. Despite such occasional reversals, beach and harbor maintenance within the Santa Barbara cell is almost entirely dependent on the west-to-east littoral transport direction.

THE IMMEDIATE OFFSHORE ENVIRONMENT

As noted earlier, the Santa Barbara south coast is dominantly a cliffed one, fronted by a rocky, wave-cut platform eroded from the same rocks that form the cliffs. The width of the platform varies but averages from about half a mile to a mile at most places. The outer margin of this rock platform is marked in many places by a band of kelp attached, for the most part, to bare rock. Between the kelp band and the beach, the bottom is generally sandy, as this is the zone where most wave energy is concentrated and where most of the littoral transport of sand takes place.

An artificial submerged sand ridge off East Beach in Santa Barbara is of special interest in this regard because it provides clear evidence of the water depths along the Santa Barbara coast where the available wave energy is incapable of moving significant quantities of sand. It also represents an early, unfortunate attempt to replenish the badly eroded beaches to the east of the harbor. In 1933, sand began to pass around the outer leg of the new Santa Barbara breakwater and accumulate in the harbor entrance. It became evident that sand had to be removed if the harbor was going to continue to function, and dredging was initiated in 1935 using a hopper dredge. The dredge required a clearance below the hull in order to discharge the load of sand. A disposal site was chosen off East Beach in about 20 feet of water, where about 200,000 cubic yards of sand was dumped. This ridge of sand was expected to be moved inshore by the waves and provide the east-flowing littoral drift with new supplies of sand to nourish the depleted beaches farther east. Unfortunately, this didn't happen, at least not to an appreciable degree. The pile of sand, about 2,000 feet long, remained nearly unaltered from 1935 to at least 1972 and appears plainly on the navigation chart of the latter date. But by the time the 1996 edition of the navigation chart was published, the sand ridge had been eroded or scattered so that it was not evident on the newer chart. This example shows that wave activity at Santa Barbara is generally too weak to move sand shoreward if it is in water more than about 15 feet deep. What happened to this ridge between 1972 and 1996 is not known, but it is very likely that most of the erosion or dispersion of the sand took place in the very stormy El Niño winter of 1982–83.

Subsequent dredging, in recognition of this problem, has utilized suction dredges at the end of the breakwater, and the sand has been pumped onto the face of East Beach near the foot of Stearns Wharf. This effort has stabilized, if not fully restored, beaches lying to the east.

HISTORY OF SANTA BARBARA HARBOR

Until the late 1920s there was no harbor at Santa Barbara, only a wharf. As a result, there was no safe anchorage for fishing, commercial, or recreational craft, and all vessels calling at Santa Barbara were at the mercy of the weather. Local interests on five different occasions had submitted requests for federal advice on the feasibility of building a harbor at Santa Barbara. The first of these requests was made in 1873 and the last in 1921. Each time, the Army Corps of Engineers advised against any new harbor construction. Nevertheless, local interests and desires ultimately prevailed, and an L-shaped breakwater parallel to the beach was constructed in 1928. Initially, the breakwater had a gap between the beach at Point Castillo and the short arm of the L, with the hope that the longshore currents would carry the littoral sand through the new harbor to the beaches to the east (Figure 17.6). What was then not sufficiently appreciated was that the longshore current and associated littoral transport of sand required the wave energy that the breakwater was designed to block. Unfortunately, this same lesson has had to be relearned at other places, such as at Santa Monica, where the same mistake was made only five years later.

The open L-shaped breakwater was completed early in 1929, and by the fall of that year shoaling inside the harbor due to the wave shadow had become so serious that the short end of the L was extended to the beach at Point Castillo, a distance of about 600 feet (Figure 17.7). This was followed by rapid deposition of sand on the west or up-coast side of the breakwater. By 1933, a triangular area west of the breakwater had been completely filled and the shoreline had realigned itself. Ultimately, Santa Barbara City College built a college stadium, several large parking lots, park facilities, and other structures on this new area of reclaimed sea floor. After filling the area up coast from the harbor, the sand moved along the long arm of the breakwater and eventually into the harbor entrance. While beaches immediately west of the breakwater built up and widened from 1928 to 1933, beaches as far east as Carpinteria, 13 miles to the east, suffered varying degrees of sand deprivation and eroded (Figure 4.9). Some previously sandy beaches were stripped of most of their sand, leaving only a narrow band of cobbles and boulders. It is curious that the full effects of impounding sand up coast of the breakwater were not appreciated because the general west-to-east drift

Figure 17.6   Aerial view of Santa Barbara breakwater off Point Castillo under construction in 1928. Courtesy of the Fairchild Aerial Photography Collection, Whittier College.

Figure 17.7   Closing the gap between the breakwater and Point Castillo in October 1929. Note the sand deposit inside the harbor that developed in 1928 and 1929. Courtesy of the Fairchild Aerial Photography Collection, Whittier College.

of sand had long been recognized in the area. A load of 2-foot-diameter boulders dumped on the beach just west of Summerland in 1917, for example, moved as much as 2,500 feet toward the east by 1938. They traveled mainly in the zone between high and low water and were observed to move more rapidly after 1938, when sand depletion became severe at that beach. A similar event took place in 1931, when a heavy boiler broke through a wharf and fell onto Summerland beach. It was deliberately filled with loose rock to prevent its movement, but despite this effort, it had moved 900 feet to the east by 1938.

Sand movement along the outer edge of the Santa Barbara breakwater and into the harbor entrance starting in 1933 had, by 1935, almost entirely blocked the harbor entrance. This happened many times in subsequent years, most dramatically in 1983, when the harbor was blocked for about a month after the abnormally stormy El Niño winter. The 1935 hopper dredging did almost nothing to check the erosion of beaches east of Santa Barbara because the spoil was not put on the feeder beach. Moreover, because the breakwater caused accretion on its up-coast side and because no significant amount of sand was bypassed to the feeder beach until 1938, the eastern beaches were subjected to sand starvation for about 10 years, resulting in severe beach erosion.

By 1942, however, the beaches for about 3 miles east of Santa Barbara had been restored almost to their pre-1928 condition because sand was now being dredged and pumped onto the immediate down-coast beach on a regular basis, but such was not the case farther to the east. In January 1940 the beach at Sand Point near Carpinteria eroded landward 245 feet and caused about $2 million in property damages. By 1941, however, with continued harbor dredging and sand bypassing, sand began to reappear on rocky beaches as far east as Fernald and Loon points in the Summerland area. Until the mid-1960s, dredging was generally performed on a biannual basis, but beginning in 1959 dredging became an annual or nearly continuous operation. As a result, beaches in southern Santa Barbara County have remained generally stable. However, owing to the vagaries of rainfall and winter storms, erosion is notably more severe in some winters than in others. The winter of 1982–83 is a case in point; beaches were severely eroded along the entire Santa Barbara County coastline and along virtually the entire coast of California. Although some substantial recovery was evident by early summer of 1983, many beaches 20 years later had not fully recovered to their pre-1983 width.

Various schemes for continuous sand bypassing at Santa Barbara Harbor have been discussed since the breakwater was constructed, but no permanent pumping plant has ever been installed. Although such a plant would have many advantages over periodic and irregular dredging, it is probable that any system, no matter how well designed, would miss some of the incoming sand, perhaps even a larger amount than the existing movable dredge system. This sand would then pass around the fixed intake and be deposited in the harbor entrance, where it would become a navigation hazard. In recent years annual dredging costs have ranged from as low as $350,000 to more than $1.5 million to move as little as 125,000 to more than 600,000 cubic yards of sand. Annual dredging volume now averages about 300,000 cubic yards.

Amusing suggestions for bypassing sand without dredging have been offered in local newspapers. Many of these were tried at other places and found wanting. A recurring suggestion involves some variation of a detached breakwater that is supposed to allow the sand to be transported along the foreshore through the harbor. All such suggestions overlook the fact that sand transport along the shoreline is driven by breaking waves, and breakwaters are designed to reduce wave energy (and therefore reduce littoral transport as well). This was the original, unsuccessful approach used when the Santa Barbara breakwater was constructed in 1928 and 1929.

Nature, of course, has her own way of taking care of the problem, and if allowed to work her will would create a new shoreline alignment altogether. This would come about, as has nearly occurred several times in the past, by the deposition of a curving sandbar from the tip of the breakwater to East Beach near the foot of Stearns Wharf. Once this bar or new beach was in place, sand could move uninterruptedly along the outer edge of the breakwater, across the harbor entrance, and eastward down the coast. The obvious problem with this solution is that it would eliminate Santa Barbara Harbor as a useful boat anchorage and convert it to a shallow lagoon.

It is clear at Santa Barbara, and at many other small-craft harbors and marinas along the coast of California, that the accumulation of sand and the annual dredging needed to maintain these harbors is primarily a result of harbor location within a littoral cell plus the rate of littoral drift at that location in the cell, rather than the configuration of the breakwater or jetties (Figure 17.8). Santa Cruz, Santa Barbara, Ventura, Channel Islands, and Oceanside harbors, for example, were built in the down-coast end of a littoral cell with high littoral drift rates. No matter how the harbor entrances were constructed, a large volume of littoral sand was trapped and would then have to be dredged out on a regular basis. On the other hand, Moss Landing, Monterey, Hueneme, Redondo-King, Newport, and Dana Point harbors were built between cells, at or just down coast from a submarine canyon where the littoral drift has been funneled off into deep water. As a result, these harbors have required either no annual dredging or only minimal maintenance dredging. There is now a considerable harbor dredging history and a reasonably

Figure 17.8   Sand spit accumulating at the end of the breakwater of Santa Barbara Harbor. This spit must be annually dredged to maintain a navigable channel into the harbor. Photo © 2002–2004 Kenneth and Gabrielle Adelman, California Coastal Records Project, www.Californiacoastline.org.

clear understanding of the boundaries and rates of littoral drift along California's 1,100-mile coastline, so that there is no excuse for planning a new harbor or marina today without full consideration of these issues.

CLIFF FAILURE AND COASTAL RETREAT

Probably the most dramatic example of shoreline erosion in the Santa Barbara area is the loss of a 245-foot-wide beach at Sand Point near Carpinteria in 1940. Somewhat less conspicuous is the steady if unspectac-ular erosion of the seacliffs. Rates of seacliff retreat in the Santa Barbara and Goleta area are reasonably well documented and probably are typical of the south coast because rock resistance appears to be similar throughout the area. The coastline has about the same trend and exposure to wave activity from Gaviota to Rincon Point.

Over the last 80 years a number of bench marks and triangulation sta-tion monuments have been placed along the top of the seacliff in the Santa Barbara area by such agencies as the U.S. Coast and Geodetic Survey and the U.S. Geological Survey. Because the original descriptions of these monu-ments usually include rather precise measurements to the upper edge of the seacliff, they can be very useful for determining rates of retreat. Other dated or datable structures such as storm drains, streets, sidewalks, and houses

A. Steep eroding cliffs fronted by sandy beach. Homes developed on large parcels on cliff top. Most of houses set well back from cliff edge. Winter waves typically reach base of cliff. Steep cliffs are unstable and frequently fail through landsliding.

B. Steep eroding cliff with nearly continuous development along cliff edge. Beach is very narrow, rocky in places.

C. Landslide in February 1978 along cliff edge.

D. Wide sandy beach formed following construction of Santa Barbara breakwater in 1928. Beach area now occupied by college parking lots and stadium, roads and park and harbor commercial areas. Severe beach erosion and damage to ocean front facilities during 1983 El Niño.

E. On average 300,000 cubic yards of sand is dredged each year from the Santa Barbara harbor and discharged on East Beach.

F. Broad sandy beach; waves and high tides reached to Cabrillo Boulevard in 1983.

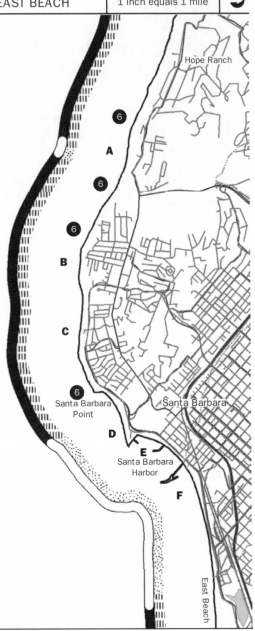

Hope Ranch

Santa Barbara

Santa Barbara Point

Santa Barbara Harbor

East Beach

Hazard Level
- ☐ Stable: low risk
- ▨ Caution: moderate risk
- ■ Hazard: high risk

Erosion Rates
- ① Inches per year

Shoreline Environment
- Sandy beach
- Beach backed by marsh
- Beach backed by dunes
- Cliff or bluff fronted by beach
- Rocky coast with no beach

Armoring
- ▲▲▲ Riprap
- ■■■ Seawall
- ⊡⊡⊡ Other

may also provide useful data. Stereo aerial photographs can also be very useful, and although they may not show small changes over short periods as clearly as do bench marks and triangulation stations, they are the most commonly used and widely available references for any long-term coastal change determinations. Plot maps of subdivisions, lot splits, parcel maps, or other property boundaries may also be very useful; dimensions of parcels are usually shown very precisely and for coastal bluff properties will often include exact measurements from the street to the cliff or bluff edge.

Many other records, including historical ground photographs, demonstrate the reality of cliff retreat, but they do not generally allow for the determination of erosion rates. Mature trees at the cliff edge obviously were not planted in these precarious positions, and where some of their root systems are exposed, erosion is clearly demonstrated. In the Santa Barbara area, as elsewhere, it is often assumed that cliff retreat is largely or entirely due to wave activity, and protective measures are designed to correct only this part of the problem. There is considerable evidence that non-marine processes such as earthquakes, landsliding, runoff, salt weathering, and groundwater seepage can also be very important and together may account for a significant amount of the cliff retreat documented. In the Santa Barbara area, non-marine processes are by far the most important cause of cliff retreat. Not infrequently, marine processes set the stage for the non-marine processes. Wave erosion at the base of the cliff may undermine the cliff and remove support for the overlying rocks. Heavy rains often trigger landslides in the upper part of the cliff well above the reach of the waves. Numerous examples of landsliding occur along the Santa Barbara coast, including some long-inactive examples as well as currently active ones.

One of the more dramatic slides in recent years occurred in the city of Santa Barbara in February 1978 following heavy winter rains in the preceding weeks. This slide destroyed or damaged several houses, garages, and properties in a few hours time, beginning in the evening. By noon the next day, most of the motion had stopped, as equilibrium had been reached. Waves will continue to gradually erode the toe of any landslide mass or rockfall, thus causing some renewal of movement. This particular slide caused a new, nearly vertical cliff to be established about 150 feet inshore from the previous cliff edge. The failure was a typical rotational slump on a curved failure surface that was nearly vertical at its head but flattened seaward (Figures 17.9 and 17.10).

Landslides initiated by burning tar seeps occur at several sites along the Santa Barbara coast as well as the coastline east of Rincon Point in Ventura County and on the north coast of Santa Cruz Island just offshore. The Monterey Formation in particular contains fractures and voids filled with

Figure 17.9   Vertical view of the February 1978 landslide at Camino de la Luz, Mesa area, Santa Barbara. Photo by D. W. Weaver.

Figure 17.10   Destroyed houses and garages, Camino de la Luz landslide, February 1978. Photo by Robert Norris.

Figure 17.11 Massive but inactive tar seep from Santa Barbara Formation. More Mesa Beach near Goleta, January 1978. Photo by Robert Norris.

natural asphaltum. For various reasons these deposits may ignite and burn or smolder for long periods, often producing smoke and steam, and baking or even fusing the enclosing rock. In some cases the heating seems to have induced fracturing, but in others the asphaltum had acted as a binder or cement in an already pervasively fractured rock. Once this material is burned, the rock crumbles readily and produces rockfalls or small slides.

Other examples of this process occur just east of Rincon Point in Ventura County, where the railroad company built electric fences at the cliff base to warn trains of rock slides blocking the tracks. These slides have been ignited several times. Another example of a burning tar seep creating problems for cliff-top development occurred west of Coal Oil Point at the time the Sandpiper Golf Course was being developed in the early 1960s. This seep had been smoldering, and workers extinguished it by excavating into the seacliff with a bulldozer and allowing seawater to enter. During the bull-dozing, the smoldering abruptly changed to open flames, from which the bulldozer operator narrowly escaped. The seawater eventually put out the fire and produced a number of temporary steam vents in the golf course area as water reached the burning asphaltum. In some cases, tar seeps have protected parts of the seacliff from erosion by producing a tough, con-glomeratic "rock." An area of currently inactive tar seeps just east of the mouth of Goleta Slough is a good example (Figure 17.11). Cliffs composed

A. High bluffs below cemetery subject to failure. A concrete seawall that formerly existed at east end of bluffs but has been destroyed. Rocky beach present at base of bluffs.

B. Narrow sandy beach backed by low riprap and older concrete seawall protecting roadway and Coral Casino. High tides and large storm waves in 1983 overtopped wall and covered road with debris in Biltmore area.

C. Miramar Cove-houses built on back beach subject to wave damage during high tide and wave conditions; extensive damage in 1983 El Niño event.

D. Fernald Point fronted with large slabs of broken concrete.

E. Long concrete seawall on back beach protecting railroad on low unstable bluff above beach.

F. Houses on bluff above wide sandy beach. Scattered riprap at base of bluff. One area near east end of Summerland Beach where large mass of broken concrete has been dumped at base of bluff.

G. Concrete seawall at base of steep unstable slope below railroad grade.

H. Loon Point: steep rocky cliffs undergoing failure fronted by a rocky beach.

I. Houses on beach with continuous riprap protecting homes and railroad. Large new developments. Extensive damage during 1983 storms to houses on Padaro Lane.

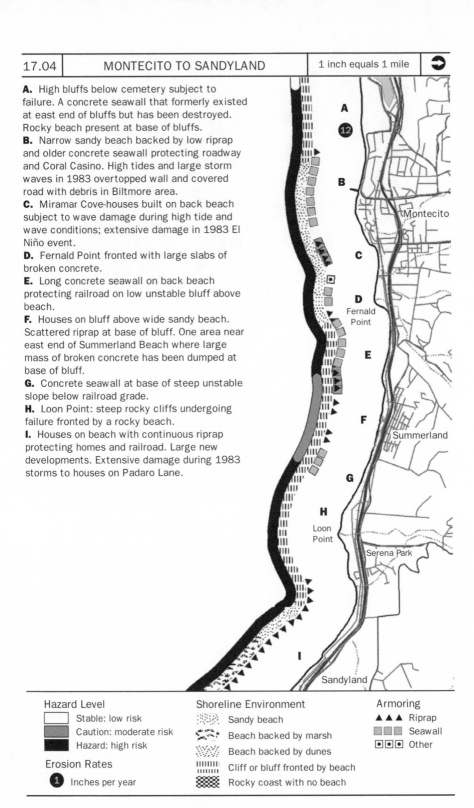

Hazard Level

Stable: low risk
Caution: moderate risk
Hazard: high risk

Erosion Rates

1 Inches per year

Shoreline Environment

Sandy beach
Beach backed by marsh
Beach backed by dunes
Cliff or bluff fronted by beach
Rocky coast with no beach

Armoring

▲▲▲ Riprap
Seawall
Other

of tar-cemented rubble are found at the eastern side of Carpinteria Beach State Park, and at one place, a small, active tar seep forms a fan-shaped mass of asphaltum on the beach.

COASTAL HAZARDS

Coastal hazards in the Santa Barbara area, as elsewhere, are usually the result of human occupancy or construction within an active geological environment. Where the coastline is undeveloped or is preserved as parklands or open space, there are fewer concerns with geological change, at least with respect to hazards. Goleta County Beach, however, just east of the UC Santa Barbara campus, has experienced significant wave erosion, land loss, and threats to park facilities in recent years, which have led to public outcry for protection but also concerns about armoring the shoreline. In 1971 only 3.2 percent, or 3.5 miles, of the Santa Barbara County shoreline were armored. However, by 1998 coastal armoring lined 12 miles, or 11.2 percent, of the shoreline (Figures 17.12 and 17.13), representing nearly a 350 percent increase in shoreline armoring over 27 years.

The progress of cliff erosion is notoriously erratic or episodic. Several years may pass with little or no visible erosion, but during El Niño years or periods when large storm waves coincide with very high tides, we may see slab failures or rockfalls with several feet of retreat during a single storm. The average of 3 to 12 inches of annual retreat in the Santa Barbara area is based on 50 to 60 years of experience and historical aerial photo analysis. However, these erosion rates were developed in the late 1970s during an unusually quiet period before the full and devastating effects of the El Niño–dominated winters of the last 25 years were felt. Thus, a 100-year average, which may result in an appreciably higher long-term erosion rate, is a much better indicator of long-term cliff erosion trends. In the Santa Barbara area all seacliffs are eroded into relatively weak sedimentary rocks, and all can be expected to erode landward well into the future, especially if sea level continues to rise.

The rates of erosion along the seacliffs of the Santa Barbara coast are not high relative to areas of rapid erosion elsewhere in California, but are still high enough that a substantial setback should be required for any permanent construction or significant improvements. Numerous examples of inadequate setbacks can be seen in the urbanized clifftops of Santa Barbara (Figure 17.14) and in examples of expensive and ineffective protective structures. It is difficult to generalize as to an appropriate setback because the rate of cliff retreat is influenced by variation in rock structure and stability, cliff height, susceptibility to landsliding, and the presence or absence of a

Figure 17.12   A large volume of broken concrete was dumped on the beach in Summerland to protect the bluff below the railroad from erosion. This was done prior to 1972 and would no longer be permitted. Photo © 2002–2004 Kenneth and Gabrielle Adelman, California Coastal Records Project, www.Californiacoastline.org.

Figure 17.13   Casablanca, a new development at the north end of the Carpinteria salt marsh, is protected by a large revetment that is continuous for over a mile along this stretch of coastline. Photo © 2002–2004 Kenneth and Gabrielle Adelman, California Coastal Records Project, www.Californiacoastline.org.

Figure 17.14 The Bacara Resort, immediately north of the Sand Piper Golf Course in Goleta. Note the very narrow setback between these structures and the edge of a cliff that is actively eroding. Photo © 2002–2004 Kenneth and Gabrielle Adelman, California Coastal Records Project, www.Californiacoastline.org.

protective beach. Moreover, the setback should be based on a specific time period. Is 100 years appropriate, or is 50 years more reasonable? Anything less than 50 years is certain to create problems because such a time span is too close to the useful lifetime of many buildings.

Long-term public interests as opposed to short-term private interests would seem to call for conservative setbacks based on at least a 50-year period of safety for any cliff- or bluff-top site (e.g., a setback equal to 50 years multiplied by the average long-term average annual erosion rate plus some safety factor). Many bluff- or cliff-top areas that are eroding at high rates would be much better utilized for parkland and open space than for development, where protection or removal issues will ultimately have to be faced. Shoreline Park in the Mesa area of the city of Santa Barbara is a good example of this approach; this beautiful park is an asset to the entire city and arguably of far greater public value than a row of privately owned homes. Regrettably, there are many houses and other buildings on cliff-top properties in the Santa Barbara area that are within 100 feet of the cliff edge, and a considerable number are within 25 feet or less. As long as these buildings remain, requests for cliff protection will be heard, and at least some public money will inevitably be required to protect private investments.

Figure 17.15 Isla Vista area, 2002. Student apartments are perched precariously on the edge of an actively eroding bluff. Photo © 2002–2004 Kenneth and Gabrielle Adelman, California Coastal Records Project, www.Californiacoastline.org.

Figure 17.16 (A) Seacliff behind existing timber seawall. Isla Vista area, February 1987.

The Isla Vista area of Goleta is the erosion hotspot for this reach of coast. Currently, apartments occupied by students attending UC Santa Barbara are perched precariously above severely eroding seacliffs (Figure 17.15). Although a few low-budget cliff protection structures have been installed, erosion continues to take place along the entire Isla Vista shoreline (Figure

(B)

(C)

Figure 17.16   *(Continued)* (B) Progress of cliff erosion behind timber seawall. Isla Vista area, February 1994. (C) Further erosion of cliff and damage to timber seawall. Isla Vista area, February 2002. Photos by Robert Norris.

17.16). In the 1960s, one Isla Vista cliff-top property owner had at least an inkling that cliff erosion included non-marine as well as marine processes, so he covered the entire cliff face in front of his property, from top to bottom, with gunite or shotcrete. Even this was not enough to provide long-term protection because the non-marine processes attacked the concrete from the top and the waves worked away at the base. As Figure 17.17 shows,

Figure 17.17   (A) Cliff covered with gunite to protect projecting deck. Isla Vista area, March 1971. Note beginning of damage at base of concrete. (B) Erosion of gunite from bottom and sides, well above the reach of waves. Isla Vista area, February 1978.

Figure 17.17 *(Continued)*   (C) Gunite protection almost totally removed by non-marine erosion. Isla Vista area, February 1981. Note the slight seaward bulge in the cliff as a result of the reduced rate of erosion due to the gunite covering before it failed. (D) Continued cliff erosion after removal of the projecting deck. Isla Vista area, February 1990. Photos by Robert Norris.

Figure 17.18 Miramar Cove south of Santa Barbara, where a house built directly on the beach (right side of photograph) suffered major damage in the 1983 storms. Low-lying homes to the left have been armored with concrete seawalls. Photo © 2002–2004 Kenneth and Gabrielle Adelman, California Coastal Records Project, www.Californiacoastline.org.

this protection lasted only about 10 years or so until 1982, when the county required that the porch be removed, as it had become a safety hazard.

Local governments have a mixed track record when it comes to resisting owners' pressures and protecting the long-term public interest. Anyone contemplating the purchase of a cliff-top building or building site should look long and hard at the potential hazards if the property allows for less than 100 feet of setback from the cliff edge. Although much of the Santa Barbara coast is cliffed, there are several stretches of low, sandy coastline near Goleta and from Santa Barbara east to Carpinteria. Much of this area has been armored with riprap or seawalls to protect homes (Figure 17.18), industrial buildings, and the railroad. Prior to installation of these structures, storm waves cut back the beach from 150 to more than 200 feet at some places in the Carpinteria area following construction of the Santa Barbara breakwater. It is true that the armor has, for the most part, protected the houses, but it is unsightly, limits access to the beach, and occupies part of the beach itself. During severe storms, such as those that occurred in 1982–83, several armoring structures were damaged and required costly repairs. Near Sand Point in the Carpinteria area, some of the riprap was

A. Houses built on sand spit separating Carpinteria salt marsh from ocean, protected by continuous riprap.
B. Wide sandy beach backed by homes, motels, and campground.
C. Beach narrows and is backed by low rocky bluff. Rock outcrop in surf zone forms natural groin.
D. Steep cliffs with narrow beach and railroad at top of cliff. Steep cliff protected with concrete seawall.

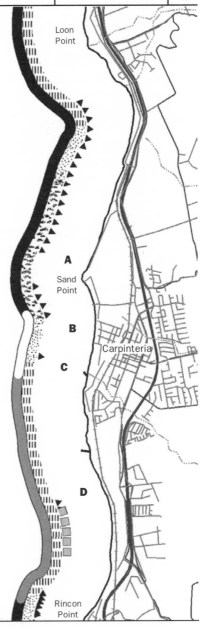

Hazard Level

☐ Stable: low risk
▨ Caution: moderate risk
■ Hazard: high risk

Erosion Rates

 Inches per year

Shoreline Environment

▨ Sandy beach
▨ Beach backed by marsh
▨ Beach backed by dunes
▨ Cliff or bluff fronted by beach
▨ Rocky coast with no beach

Armoring

▲▲▲ Riprap
■■■ Seawall
▢▢▢ Other

scattered across the beach and cost about $75,000 (in 2005 dollars) to put back in place.

The low sand spits that separate the ocean from the salt marshes at Goleta and Carpinteria are particularly unstable environments. These two areas are similar coastal environments but have very different histories. Goleta Beach is in public ownership with little in the way of construction apart from a pier and park facilities. On the other hand, the sandbars that enclose the Carpinteria salt marsh have long been used for exclusive private homes. Unfortunately, this is the very area that suffered the most severe erosion following construction of the Santa Barbara breakwater in the early 1930s. The extensive erosion dramatically reduced the margin of safety once provided by the wide, sandy beach, and in about 1960 led to the installation of the present continuous riprap barrier except at the marsh entrance (Figure 17.4). This armor has provided considerable protection during most winters, but in particularly stormy or severe El Niño winters both the riprap and the houses have been damaged, and it is a virtual certainty that this will occur in the future, particularly with a gradually rising sea level.

Although Goleta Beach, being up drift from Santa Barbara Harbor, has never suffered the severe beach sand starvation and landward shift of the high-tide line seen at Carpinteria, it is also subject to winter wave attack, and during the winter of 2002–3 erosion led the County to install a short riprap barrier to protect barbecue pits, a parking lot, and some lawn, setting off a rancorous public debate (Figure 17.2). Under natural circumstances, whenever winter storms and high tides occur together, the sea will wash over the sandbar and into the slough behind. However, the park facilities were developed on fill added to the sandbar, so each time winter storm waves occur at spring tides, the seaward edges of the park fill are eroded, threatening anything present on the sand spit. Most of the time, the riprap will protect park facilities; but inevitably, a large storm will occur at maximum spring tide and the riprap boulders will be strewn around the beach, damaging park facilities. It seems wise to acknowledge the history of this area, the inevitable wave attack that will reoccur, and to plan park facilities to be expendable and replaceable at low cost.

SUMMARY

Where railways, roads, houses, apartments, or businesses are built too close to the shoreline, the costs of protection, maintenance, and repair after storms, high tides, or even normal erosion may ultimately be so great that money would have been saved by governments and private owners alike had the oceanfront been initially acquired by some public agency and maintained as open space. Unfortunately, this was not often done in the past, so that much of this extremely valuable oceanfront property is in private ownership. Buying back this land is something very few local governments can afford today.

Along the Santa Barbara coastline, cliff retreat averages between a few inches to a foot or more per year. For this reason alone, it is unwise to permit any additional permanent construction closer than 100 feet to an eroding cliff or bluff edge. Even where it will take 40 or 50 years for cliff retreat to threaten a building, it is unlikely that building owners will then be prepared to write off their investment on the grounds that the useful life or cost of the building is fully amortized. Buildings change ownership, and few owners will be willing to move a building to a safer site or be prepared to write off the costs. Political pressure and/or legal action will inevitably be brought to bear on local governmental agencies to allow the construction of engineering structures to protect the landholders' investments.

CHAPTER EIGHTEEN

RINCON POINT TO SANTA MONICA

ANTONY R. ORME

The 72-mile-long, south-facing coast of Ventura and Los Angeles counties, between Rincon Point and Santa Monica, contains a wealth of interest for the scientist and a host of problems for coastal management. This coast extends obliquely for 35 miles across the seaward edge of the Ventura Basin, comprising 13 miles along the hilly Rincon coast and an additional 22 miles across the low-lying Oxnard Plain to Point Mugu. From there the Malibu coast extends 33 miles along the south front of the Santa Monica Mountains before turning south for 4 miles across the western Los Angeles Basin to Santa Monica (Figure 18.1).

For the most part, this is a developed coast along the margins of the greater Los Angeles area, with its population of around 15 million people. Along the Rincon and Malibu coasts, however, steep coastal mountains restrict residential growth to narrow marine terraces and crumbling hillsides whose inherent instability is aggravated by development. The intervening Oxnard coast has been progressively impacted by urban and recreational developments over recent decades. Of the 42-mile-long Ventura County coast between Rincon Point and Sequit Point, 26 miles (62 percent) are protected by riprap and over 4 miles (10 percent) by seawalls and harbor works. In its entirety, this coast presents a broad collection of management problems, but transcending these is the overwhelming pressure for development and, conversely, the urgent need for coastal conservation generated by the vast Los Angeles urban area.

PHYSICAL SETTING

The Ventura Basin and the Santa Monica Mountains are two of the many complex east–west features that form the Transverse Ranges of southern

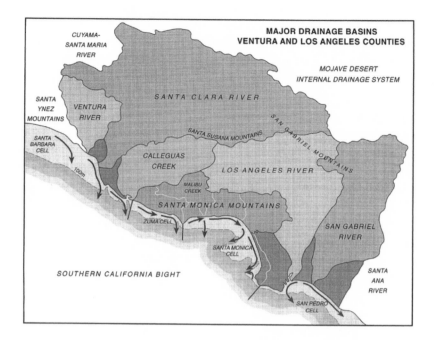

Figure 18.1   Major drainage basins and beach compartments, or littoral cells, for Ventura and Los Angeles counties. Illustration by Antony Orme.

California. Among the state's physical provinces, these ranges are unusual because their orientation departs from the normal NNW–SSE structural grain of the state. Unlike the Transverse Ranges farther east, which reveal very old metamorphic and granitic rocks, the Transverse Ranges west of the San Gabriel Mountains are composed largely of Late Cretaceous and Cenozoic sedimentary rocks, mostly of marine origin. These latter rocks are often poorly consolidated, strongly folded, and pervasively fractured, all of which favor mass wasting and erosion.

The tectonic origins of the western Transverse Ranges have occasioned much debate. In general terms, their uplift and deformation are intimately linked with subduction of the Farallon Plate and East Pacific Rise beneath the North American Plate, followed by initiation of the San Andreas transform system around 20 Ma (million years ago) and the ensuing transfer of coastal mountains to the Pacific Plate. Their east–west orientation has been attributed to about 100° of clockwise rotation during the capture of Farallon Plate fragments by the Pacific. Uplift of the western Transverse Ranges peaked in Late Cenozoic time but continues today, as shown by the magnitude (M) 6.4 San Fernando earthquake of

1971 and the M6.7 Northridge earthquake of 1994, whose epicenters lay within the eastern Ventura Basin north of the Santa Monica Mountains.

In common with much of coastal California, the western Transverse Ranges experience a modified Mediterranean-type climate characterized by warm, dry summers and cool, wet winters, with local variability attributable to elevation, aspect, and distance from the ocean. Average annual precipitation at Ventura is 12 inches, but Ojai receives 21 inches and the headwaters of the Santa Clara River over 48 inches annually. Averages are misleading, however. A few intense storms usually produce most of the precipitation and runoff; and exceptionally wet years, such as 1916, 1938, 1969, 1978, 1980, 1983, 1992, and 1998, alternate with longer dry periods. Intense storm events, such as that of 16 February 1980, when 8 inches fell in one day in the central Santa Monica Mountains amid a 10-day sequence of storms, have a predictable effect on slope stability and flooding. Major El Niño events, such as those of 1982–83 and 1997–98, are particularly hazardous because they bring inland floods to a coast beset by elevated sea levels and storm waves.

Stream flows and geomorphic processes mimic this climate regime, with winter rains, especially from January through March, generating episodic floods, mass movement, and fluvial erosion, whereas summer drought yields little activity. Indeed, much of the annual work of landscape sculpture occurs in a few stormy days, or even hours, in winter. For the rest of the year the physical landscape lies mainly dormant, recovering from storms and healing its scars.

Fractured rocks, steep hillsides, and episodic drenching rains are prerequisites for the troublesome landslides that plague the Rincon and Malibu coasts. Hillside failures vary widely in the amount of material involved, the depth of failure, and the rate of movement. Where weak materials lie unsupported above seacliffs or highway cuts, massive landslides and rotational slumps are common. Where bedrock is more consolidated, the overlying soils may fail during high-intensity rainstorms, causing soil slips that transform downslope into debris flows. The potential hazards to life and property posed by such flows have too often been ignored in the development of this coast. Fire, caused by lightning or human activity, has a damaging effect on slope stability by removing the vegetation that protects hillsides from raindrop impact and runoff. Fires are fairly frequent in the tinder-dry months of late summer and autumn, when hot, dry, desiccating Santa Ana winds from the interior promote low humidity and severe vegetal stress. Furthermore, landscape watering, septic-tank drainage, and increased surface runoff from roads and residential tracts have created slope environments that are wetter than in predevelopment years, causing old landslides to be reactivated and new mass movement to occur. Such problems represent a perennial hazard along the California coast but are nowhere more problematic than along the Malibu coast.

Stream sediment yields, so important for nourishing local beaches, also reflect the precipitation and runoff regime, as well as the availability of sediment and the extent to which stream transport is blocked by dams. With most annual precipitation falling in a few intense winter storms, large, infrequent floods account for most sediment transport. For example, most of the 10 million tons of sand delivered to the coast by the Santa Clara River between 1933 and 1938 arrived in six days of flooding in 1938. Moreover, this river discharged 52.4 million tons of sediment during the 1969 floods, compared with only 1 or 2 million tons in relatively dry years. The coastal sediment budget thus sees a few years of plenty (most recently 1969, 1978, 1980, 1983, 1992, and 1998) and many years of famine (such as the 30 years preceding the 1969 floods).

Over the past century, the delivery of river sand to beaches has been curtailed by dams built in the contributing drainage basins and by the mining of floodplain sands and gravels. About 37 percent of the Ventura River basin lies behind dams (Matilija, built in 1948, and Casitas, 1959), which has reduced sand and gravel delivery to the coast by 53 percent. About 37 percent of the Santa Clara River basin is also blocked by dams (Bouquet 1934, Piru 1955, Pyramid 1971, and Castaic 1972), which have reduced coarse sediment delivery by 27 percent. Malibu Creek, now the most important supplier of sediment to the beaches of Santa Monica Bay, has had 55 percent of its sand delivery trapped by Rindge and other, smaller dams. Because of these dams, the reduced amount of sediment reaching the shore is incapable of restoring eroding beaches, and during a succession of dry years serious beach erosion can occur. The removal of the Matilija and Rindge dams is now being seriously considered, in part because their structural integrity and function have been compromised by age, but also important, because the sand they now trap as well as future sand supplies could again contribute to the beaches. There are many environmental issues associated with dam removal that need resolution, however, such as how the sediment now stored behind the dams would be delivered to the coast and how the stream channel below the dam would be affected by the dam's removal.

The delivery of sedimentary material to the coast through seacliff erosion and rockfalls, slumps, and landslides is now much reduced as a result of the extensive protection afforded the Rincon and Malibu coasts by riprap, revetments, and seawalls. Although these processes still occur, their debris usually falls first onto the coastal highways and is usually removed by trucks and thus lost to the beaches.

The marine environment reflects climatic forcing common to California, but modified to reflect the east–west orientation of this coast. Thus, whereas most swells approach the California coast north of Point Conception from between northwest and west, most swells enter the Southern California Bight south of Point Conception from between west and southwest. Thus, 70 percent of swells pass through the Santa Barbara Channel toward the Ventura coast from due west, whereas 80 percent of swells approach the Malibu coast from WSW, through the window between the Northern Channel Islands and San Nicolas Island, or from SSW, between San Nicolas and San Clemente islands. This coast is also exposed to southerly swells from late summer tropical storms off western Mexico and winter storms in the Southern Hemisphere, and to storm waves generated by winter cyclones, all of which erode south-facing beaches. Wave heights at the shore average 3 feet, but storm waves may reach over 20 feet.

With most waves approaching shore from between west and southwest, the predominant longshore current and littoral drift move eastward. Along the Oxnard coast, however, this pattern may reverse under the influence of southerly swells during summer. The result is a net southeastward move-ment of about 1 million cubic yards of sand annually, most of which is siphoned off into Hueneme and Mugu submarine canyons. The modest amount of sediment moving east from Point Mugu is in turn deflected into submarine canyons off Point Dume, and little passes eastward toward Santa Monica. Thus, this coast may be divided into three littoral cells: the Santa Barbara cell extending from north of Point Conception to Point Mugu, the small but distinct Mugu cell between Point Mugu and Point Dume; and the Santa Monica cell east of Point Dume.

The mean tidal range along this coast is around 4 feet. Since 1933, the tide gauge at Santa Monica Pier has recorded a mean range between mean lower low water (MLLW) and mean higher high water (MHHW) of 5.4 feet and a maximum range of 10.6 feet (from 2.6 feet below MLLW in 1933 to 8 feet above MLLW in 1983).

During El Niño forcing in 1997–98, the maximum observed tidal range was 8.9 feet. Tidal flushing becomes more important when flood tides are trapped in lagoons behind partly closed barrier beaches.

DEVELOPMENT BACKGROUND AND MANAGEMENT ISSUES

Development between Rincon Point and Santa Monica has been strongly influenced by the physical setting. Mountains have restricted growth to a narrow ribbon along the Pacific Coast Highway close to the shore along the

Rincon and Malibu coasts, while the Oxnard Plain has favored more expansive growth. Most development of significance to contemporary coastal management occurred during the mid-twentieth century, and by 1970 the attendant problems had begun to attract serious attention from both citizen groups and government agencies. Since then, increased vigilance and environmental legislation have slowed the pace of coastal change, but a legacy of irrational or unwise development and proximity to the greater Los Angeles urban area continue to pose major management challenges.

The first permanent European settlements came after Gaspar de Portolá's expedition of 1769 led to the establishment of four presidios (forts), three pueblos (towns), and a chain of 21 Franciscan missions from Baja California northward to San Francisco Bay over the next 50 years. A few ranchos were established inland during the Spanish period, notably Rancho Simi (120,000 acres) in 1795 and Rancho El Conejo (50,000 acres) in 1802; but because of its rugged, inaccessible terrain, the Malibu coast saw little ranching at this time.

Spain's precarious hold over the area ended with Mexican independence in 1821, but the nominal transfer of power from one distant authority to another was similarly ineffective. When secularization of the missions began in 1834, local authority passed mostly into the hands of rancho owners. Vast areas of coastal and valley land were granted by the Mexican governor to soldiers and fortune seekers, among whom eight families formed a colony south of the Santa Clara River in the vicinity of modern Oxnard. The division of mission lands into ranchos favored expanded livestock ranching, with cattle on the coastal plains and terraces below sheep pastures on nearby hillsides. During this time much of the native chaparral was gradually converted to grassland, especially dense, shallow-rooted European grasses, with significant implications for erosion and sediment yields.

Cattle ranching continued after Alta California was transferred to the United States in 1848 but declined as the Gold Rush waned. The first California legislature divided the state into 27 counties in February 1850, three of which—San Diego, Los Angeles, and Santa Barbara—encompassed the entire coast from the new Mexican border to the Santa Maria River. Many existing Mexican land titles were confirmed, and many were not, but depressed cattle prices following the severe droughts and great floods of the 1860s wrought havoc with the cattle business. Ranches were divided again and again, and, as new settlers arrived, the coastal land began to lose its pastoral character. The small cow town of San Buenaventura, a collection of some 70 to 80 houses and a hotel in 1856, acquired a post office in 1861 and was incorporated as a city in 1866. A regular stagecoach line was established in 1868 as coaches made their way from Los Angeles to San Buenaventura before continuing along the beach toward Santa Barbara. A wharf was built in 1872. In 1873 Ventura County was carved from the southeast corner of Santa Barbara County. As the new county seat, San Buenaventura acquired a courthouse, library, bank, hotels, newspaper, churches, fire department, and other trappings of developing townhood. Farther down coast, Hueneme (the Chumash word *huenemu* means halfway place) was settled by squatters.

The arrival of the Southern Pacific Railroad from Los Angeles in 1887 promoted a speculative real estate boom along its route and inflicted the first major disturbance of the narrow coastal terrace west of San Buenaventura, now abbreviated for timetable purposes to Ventura. The railroad also encouraged a change from ranching to commercial crop farming, especially in the fertile Oxnard Plain. Initially, dry farming focused on wheat, hay, sugar beets, almonds, apricots, and later lima beans. During the twentieth century, however, with abundant groundwater resources beneath the Oxnard Plain recharged in winter by Santa Clara River floods, irrigation agriculture expanded. Citrus and avocado orchards, dairy farms, and fields of tomatoes, peppers, lettuce, celery, squash, cauliflower, broccoli, and strawberries became widespread and food-processing industries expanded. By 1980, two-thirds of farmland in Ventura County was under irrigation. Cultivation and irrigation in turn led to widespread erosion around the margins of the Oxnard Plain, deepening barrancas and increasing sediment yields to local streams by 8 to 10 times their estimated values under pre-European conditions.

Meanwhile, the tar seeps of the Ventura area, long used by Chumash peoples to caulk their boats and baskets, had prompted the search for petroleum resources as early as the 1850s. In 1857 a small refinery was constructed near Ventura to produce marketable oil from these seeps. In 1865 the first deep well was sunk near Sulphur Mountain on the Ojai road, and others followed, including some wells along the coast west of Ventura. The Union Oil Company of California had its beginnings in the recovery of oil from the Ventura Avenue anticline north of Ventura. In the early twentieth century, however, improved deep-drilling methods and the increased demand for petroleum products led to a major boom in the local oil industry. Onshore oil production has long since peaked, but the Ventura Avenue oil field and its extension to the coast between Pitas Point and Punta Gorda, remain important, as reflected in the untidy landscape of well heads and chemical plants in the Ventura River valley. Similar oil-bearing structures beneath the Santa Barbara Channel have been exploited since the 1960s. Onshore oil production, storage, and pipeline facilities add a further component to the already congested Rincon coastal corridor, while offshore oil production poses a pollution hazard to local beaches.

Today the city of San Buenaventura (104,000 people) remains a residential focus and tourist venue, but its growth is limited by the physical constraints of its site. Development on the Oxnard Plain is not so constrained. The city of Oxnard originated in 1897 and was incorporated in 1903. Its not entirely rational growth to a population of 182,000 over the past 50 years has consumed much former farmland, and the town has developed an important light industrial and service base to its economy, all of which have increased population pressure on the coast. Major growth occurred along the old *El Camino Real* following remodeling of U.S. 101 as a freeway from Los Angeles to Ventura in the 1960s and subsequent freeway construction from Ventura to Santa Paula. To protect these urban developments from occasional severe floods, levees were constructed along the Santa Clara and Ventura rivers and nearby tributaries, and Calleguas Creek was channelized. This interference with natural systems has further affected the coast by its impact on flood routing and sediment yields.

Immediately to the south of Oxnard, the old Port Hueneme, with its picturesque early homes, lighthouse, and small wharf, has since 1942 been dominated by the U.S. Naval Construction Battalion Center ("Seabees"), operating an expanded naval harbor, warehouses, and training facilities. A short distance farther southeast, Point Mugu Naval Air Weapons Station has operated a missile range since 1946. This facility, although it excludes the general public from 7 miles of shoreline, does protect the largest surviving semi-natural wetland in southern California, Mugu Lagoon, from the depredations of ill-considered development. The coast and mountains to the east of this base now form part of Point Mugu State Park. Farther east, Santa Monica was incorporated in 1886, and although an attempt by railroad interests to develop a major ocean port failed, the town witnessed considerable residential and recreational development thereafter and today has a population of about 90,000 people.

While most land between Port Hueneme and Santa Monica long remained unincorporated, mostly in private ranches, completion of what is now the Pacific Coast Highway between 1906 and 1929 paved the way for modest corridor development. The city of Malibu (13,300 people) was incorporated as recently as 1991 by citizens concerned with preserving their distinctive community.

These developments have created, belatedly, an awareness of the need for better environmental management and for conservation of remaining open space and coastal access. These issues are now being addressed by various federal, state, and county agencies and nongovernmental organizations.

A. Rincon Point to Ventura River Mouth: Low erodible bluffs fronting Holocene marine terraces are now largely protected by riprap, revetments, and short seawalls. Developments within this shore zone are constained by the relatively narrow terrace terrain such that houses, roads, and the railroad all require protection from marine erosion.

B. At Rincon Point, homes built at the seaward edge of the Rincon Creek fan-delta were extensively damaged during storms in 1983.

C. The coast between Rincon Point and Punta Gorda comprises 10-to 20-foot high bluffs that were retreating rapidly prior to the emplacement of riprap.

D. The community of La Conchita nestles below 300- to 500-foot-high early Holocene seacliffs, which are prone to frequent mass movement, most notably in a currently active landslide debris flow that has engulfed the houses nearest to the cliff base. Mudflows in January 2005 killed 10 people.

E. Although many of the oil wells immediately offshore have now been capped and their infrastructure removed, petroleum production associated with anticlinal structures continues from onshore platforms and deep-water platforms in the Santa Barbara Channel.

F. From Sea Cliff southward to Pitas Point, the road and rail corridor is protected by a poorly constructed rubble revetment and short seawall.

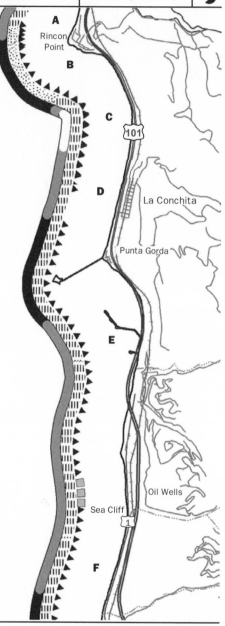

Hazard Level

☐ Stable: low risk
▨ Caution: moderate risk
■ Hazard: high risk

Erosion Rates

① Inches per year

Shoreline Environment

▨ Sandy beach
🌿 Beach backed by marsh
▨ Beach backed by dunes
⑊ Cliff or bluff fronted by beach
▨ Rocky coast with no beach

Armoring

▲▲▲ Riprap
■■■ Seawall
⦿⦿⦿ Other

Figure 18.2   The Rincon coast, 1933. Spence Collection. Reproduced by permission of the Department of Geography, University of California, Los Angeles.

Increasingly, parcels of undeveloped land within the mountains are being protected in parklands, including those of the federal Santa Monica Mountains National Recreation Area and the state-sponsored Santa Monica Mountains Conservancy. Fragmented riparian habitats and former farmlands along the lower Santa Clara River are now being considered for incorporation as a recreational parkway.

THE VENTURA BASIN, RINCON COAST, AND OXNARD COAST

THE RINCON COAST

The Rincon coast stretches in three shallow curves for 13 miles along the northern margin of the Ventura Basin from Rincon Point southeastward to the Ventura River (Figure 18.2). Continuing tectonic uplift of Rincon Mountain has raised a series of Holocene marine terraces into a narrow bench 200 to 1,000 yards wide. This bench forms a congested corridor for human activity between the unstable former cliffs that rise steeply inland to 300 to 650 feet above sea level, and a line of low, active seacliffs fronted by riprap and seawalls. The corridor contains six small development nodes at

Figure 18.3   Development on the alluvial fan–delta at Rincon Point, 1983. Photo by Linda O'Hirok.

Rincon Point, La Conchita, Punta Gorda, Seacliff, Pitas Point, and Dulah, together with a railroad completed in 1887. The old coast highway is now reduced to parkways and access roads, a major freeway (U.S. 101), and various petroleum facilities. Small parks have been developed at Rincon Beach County Park, Hobson County Park at Seacliff, Faria County Park at Pitas Point, Emma Wood State Beach, and the Ventura River mouth.

Two major problems characterize this coast: unstable cliffs and the proximity of development to the shoreline. Regarding the first problem, the coastal corridor is backed by abandoned or relict seacliffs formed in poorly consolidated claystone, mudstone, and sandstone of Miocene and Pliocene age, topped by Pleistocene beach conglomerates. These cliffs are prone to frequent mass movement and gully erosion, most notably at La Conchita, where recurrent landslides have buried or partly destroyed homes built at the cliff base, most recently in January 2005. Here and elsewhere, the instability of these old cliffs poses a continuing hazard to activities within the corridor.

The second problem is directly attributable to human intervention. The local coastline faces southwest, and wave energy associated with westerly swells in the Santa Barbara Channel is muted by refraction around sheltering headlands. Nevertheless, before development, this coast was characterized by erodible cliffs, 5 to 30 feet high, that shed debris to beaches at their base. However, homes have been placed on or against these low cliffs and on the alluvial fan–delta at Rincon Point (Figure 18.3), and railroad and high-

Figure 18.4   Seawall collapse along the Rincon Parkway, 1983. Photo by Amalie Jo Orme.

way construction has occurred above the cliff top. The infrastructure of oil production, although now declining, exacerbates the problem. All these developments have required protective riprap, revetments, or seawalls, which in turn have inhibited cliff erosion and therefore beach replenishment. As a result, wave action may cause serious damage, such as that which occurred during the El Niño storms of 1982—83, when several properties were lost, many more were damaged, and sections of highway and railroad were undermined (Figure 18.4).

THE OXNARD COAST

The Oxnard coast extends 22 miles in a broad convex arc across the seaward front of the Oxnard Plain between the Ventura River and Point Mugu. Historically, this coast comprised a complex of narrow, sandy barrier beaches and low dunes interspersed with linear lagoons, marshes, and alluvial flats. Winter floodwaters of the shifting Santa Clara River and Calleguas Creek often reshaped this complex and brought fresh sediment to the shoreline. Human intervention has changed much of this, creating a variety of management problems.

The impact of human interference has taken two main forms: restricted or reduced sediment delivery to the coast as a result of upstream dams and downstream levees, and construction directly on the shoreline. As a result of the dams mentioned earlier, the sand and gravel yield of the Ventura River

A. Backshore properties at Pitas Point are protected by an uncoordinated series of seawalls and riprap revetments, many of which fail from time to time.

B. Reinforced concrete seawall breached along a 300-foot front in winter 1983. The wall has since been replaced.

C. Emma Woods State Beach: access road to beach destroyed in winter 1983. Revetment later installed.

D. Road and rail corridor exposed to episodic marine erosion, partly protected by seawalls.

E. Although the Ventura River is partly regulated by dams upstream, the river mouth is prone to seasonal flooding and erosion during winters of heavy rainfall, as in 1983, 1992, and 1998.

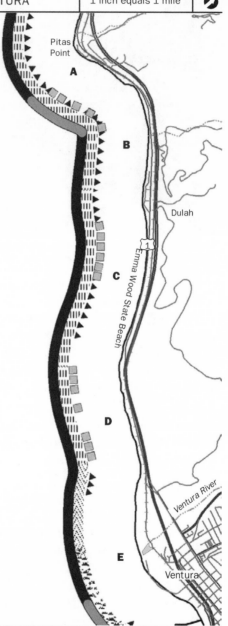

Pitas Point

A

B

Dulah

Emma Wood State Beach

1

C

D

Ventura River

E

Ventura

Hazard Level
- [ ] Stable: low risk
- Caution: moderate risk
- Hazard: high risk

Erosion Rates
- 1 Inches per year

Shoreline Environment
- Sandy beach
- Beach backed by marsh
- Beach backed by dunes
- Cliff or bluff fronted by beach
- Rocky coast with no beach

Armoring
- ▲▲▲ Riprap
- Seawall
- Other

A. San Buenaventura State Beach: severe erosion of natural beach following regulation of Ventura River has been countered by artificial nourishment within a constraining groin field. Periodic re-nourishment has been necessary since the project was first authorized in 1954.

B. Two marinas at Ventura were developed in back-barrier lagoons and marshes in the 1960s, subsequently impacted by the Santa Clara River floods in 1969, and since redesigned. Entrance problems continue, despite a succession of poorly conceived engineering works.

C. Santa Clara River mouth, usually opened by winter floods and closed by wave action during summer drought. This river is the principal source of sediment to the Oxnard coast.

D. McGrath State Beach is developed in low vegetated dunes immediately downdrift from the Santa Clara River.

E. McGrath Lake is one of several lagoons trapped behind the low but continuous barrier beach-dune system along the coast.

F. Oxnard Shores: Development on the low barrier has occasioned many problems and controversies, notably significant house damage during the 1970s followed, after lengthy debate, by reconstruction and expansion.

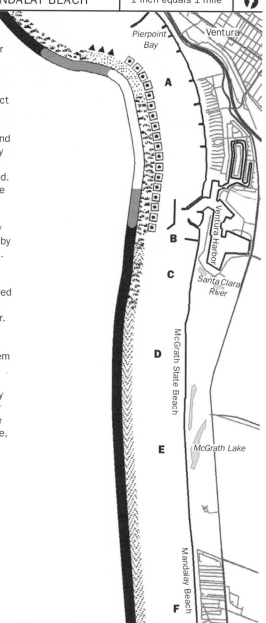

Hazard Level

- ☐ Stable: low risk
- ▨ Caution: moderate risk
- ■ Hazard: high risk

Erosion Rates

 Inches per year

Shoreline Environment

- Sandy beach
- Beach backed by marsh
- Beach backed by dunes
- Cliff or bluff fronted by beach
- Rocky coast with no beach

Armoring

- ▲▲▲ Riprap
- ■■■ Seawall
- ⊡⊡⊡ Other

Figure 18.5   Beach erosion during the winter of 1978 at the Oxnard Shores development. Photo by Antony Orme.

now averages about 100,000 cubic yards annually, or about 47 percent of its natural production. The Santa Clara River yields about 1.2 million cubic yards annually, or about 72 percent of its potential. Because the sand fraction is similar to the volume of littoral drift estimated to move along this coast annually, there is little surplus for replenishment during periods of erosion.

Reduced sediment yields from the Ventura River during the dry interlude between 1948 and 1959, which also happened to follow construction of the Matilija Dam, were largely responsible for causing the shoreline between the Ventura and Santa Clara rivers to erode 300 feet landward over that period. To offset this, an erosion control project was completed between 1962 and 1967 at San Buenaventura State Beach, comprising seven rock groins (400 to 540 feet long) and placement of 880,000 cubic yards of sand in a beach 130 to 260 feet wide and 2 miles long. This beach has since been maintained in part with sand dredged from the Ventura Marina.

Farther down coast, beach erosion between 1969 and 1973 caused significant shoreline retreat and property losses at Oxnard Shores, where many homes built close to the berm crest during the 1960s toppled into the surf (Figure 18.5). This erosion phase may have been a delayed response to diminished littoral drift during the relatively dry years between the 1938 and 1969 floods, aggravated by dam construction inland. It may also have reflected changing near-shore bathymetry and coastal circulation following the 1969 floods. In the 1990s, however, sand surpluses led to widespread coastal progradation or outbuilding, encouraging renewed construction of beach homes, now bigger and more deeply founded than before, but nevertheless in harm's way should the erosion cycle return and sea level continue to rise. Construction of beachfront homes here has been a contentious issue since the 1960s, pitting backshore homeowners against beachfront developers, local agencies seeking to increase their tax base against state agencies empowered to enforce restrictive coastal legislation, and long-term scientific wisdom against short-term commercial gain.

The other form of interference with this coast has been the construction of major harbor works for military and recreational purposes. Although a wharf was built at Ventura as early as 1872, the first real harbor here was at Port Hueneme, built by the Oxnard Harbor District as a civil project in 1938–40 and converted for use by the U.S. Navy in 1942. Located at the head of Hueneme Submarine Canyon, this harbor comprises an entrance channel, 36 feet deep and 330 feet wide, protected by two jetties, 1,000 and 1,100 feet long. After jetty construction, accretion occurred up-coast of the north jetty, and by 1948 erosion extended 7 miles down coast of the south jetty. Fill dredged from up-coast beaches and a 3,000-foot rock revetment were then placed down coast, but erosion continued beyond the revetment, and former dunes were eroded. Shoreline surveys between 1856 and 1938 indicated that this down-coast area was relatively stable over that period. Clearly, a bolder plan was needed to stabilize these beaches, and this was linked to the construction of Channel Islands Harbor in 1961 (discussed later).

Meanwhile, Ventura Marina was constructed in a poor location, without adequate research, and opened in 1963. It originally comprised an entrance channel 20 feet deep and 300 feet wide, protected by two converging rock rubble jetties (north 1,250 feet, south 1,070 feet long), a small middle groin, a turning basin, and three boat basins (10 to 16 feet deep with 520 berths), plus a nearby private residential marina. The rock for this and other nearby projects came mainly from the Conejo volcanics along the east side of the Oxnard Plain. After the marina's opening in 1963, rapid shoaling caused by sand moving into the entrance channel generated breaking waves up to 8 feet high at low water, which closed the marina for about 66 days every year. The need to bypass sand down coast every two to three years had been anticipated, but severe shoaling had not, nor did the annual dredging of about 190,000 cubic yards of sediment from the entrance channel alleviate it. Before remedial measures could be taken, the marina's problems were compounded by the 1969 winter floods of the Santa Clara River, which breached its north bank about 1,000 yards upstream of the mouth and discharged much of its flow and sediment into the marina. All but two docks were destroyed, 490 berths and 88 boats were demolished, and five 20,000-gallon gasoline tanks were washed into the harbor; in addition, with main trunk lines to a nearby sewage treatment plant destroyed, raw sewage flowed into the marina at a rate of 4 million gallons a day for over two weeks. The marina filled with sediment to MLLW. After the sediment was removed, a 1,500-foot-long detached rock rubble breakwater was completed in 1971, offset toward the north to trap up to 785,000 cubic yards of sand up coast of the north jetty. Despite maintenance dredging more or less annually ever since, shoaling remains a problem, leading in the 1990s to the construction of extensions to the north jetty and offshore breakwater, and of a further groin to allow for safer recreation and surfing down drift of the south jetty. Nevertheless, it is difficult to escape the conclusion that the Ventura Islands Harbor, with its seemingly unending series of adjustments, will continue to require expensive maintenance dredging.

In contrast, Channel Islands Harbor, begun in 1961 some 8 miles to the southeast, was carefully designed for both marina development and to address the problems caused by Port Hueneme, a mile down coast beyond the artificially constrained Silver Strand. Channel Islands Harbor was provided with two 1,300-foot-long entrance jetties and a 2,300-foot-long detached breakwater, offset up coast to provide a sand trap from which over 1 million cubic yards of sand are dredged more or less every two years to nourish Silver Strand and beaches beyond Port Hueneme. This bypassing operation has had a mostly beneficial effect on stability at Ormond Beach down drift of Port Hueneme. However, erosion problems have continued to beset Point Mugu Naval Air Weapons Station, requiring riprap protection in a more complex wave and current environment. Since 1961, Channel Islands Harbor has been developed into a large residential marina by excavation in former back-barrier lagoons.

Traces of the former barrier-lagoon system along the Oxnard Plain survive amid the artificially maintained beaches, dunes, and lagoons of McGrath State Beach south of the Santa Clara River mouth, and more extensively at Mugu Lagoon within the Point Mugu Naval Air Weapons Station (Figure 18.6). At one time or another throughout the year, this wetland is home to 310 bird species, 39 fish species, 41 mammals, 11 amphibians and reptiles, and numerous invertebrates. Mugu Lagoon is by no means a perfect wetland, subject as it is to channelized inputs from the urbanized upper and agricultural lower basin of Calleguas Creek and, historically, to interference by military activities. However, the U.S. Navy is now sensitized

Figure 18.6   Point Mugu and Mugu Lagoon, 1940. Spence Collection. Reproduced by permission of the Department of Geography, University of California, Los Angeles.

to its role as a coastal manager and maintains an environmental division concerned with maintenance of the lagoon and adjacent habitats.

Many of the problems of the Oxnard coast could have been anticipated with better understanding of the natural system, and avoided by more rational development and conservation measures. In essence, under natural conditions, frequent shoreline changes were to be anticipated along this coast because its narrow complex of low-lying barriers was exposed to westerly swells and storm waves moving up the Santa Barbara Channel, and because sediment supplies varied with the magnitude and frequency of river floods. Problems have been exacerbated because people have sought to freeze an inherently unstable shoreline with residential, recreational, and military projects that pay little attention to the fluctuations in natural processes and shoreline position in the coastal zone.

THE SANTA MONICA MOUNTAINS AND MALIBU COAST

The 33-mile-long Malibu coast extends from Point Mugu eastward to Santa Ynez Canyon and then curves southward for 4 miles to Santa Monica Pier. Of these 37 miles, nearly half have now been absorbed into state and county beaches and parks, with much of the remainder exposed to

A. Channel Islands Harbor: completed in 1961 with two entrance jetties and an offset, offshore breakwater designed to trap littoral drift north of the entrance. The accumulated sand trap is then dredged to nourish beaches downdrift of Port Hueneme.

B. Port Hueneme: completed as a civil project in 1940 and transferred to the U.S. Navy in 1942. The entrance jetties deflect littoral drift into Hueneme submarine canyon. This necessitates the periodic dredging and sand by-passing program that transfers sand from updrift of Channel Islands harbor to downdrift of Port Hueneme.

C. The low barrier coast between Port Hueneme and Calleguas Creek fronts a series of salt marshes, salt flats, and lagoons that have been variously modified.

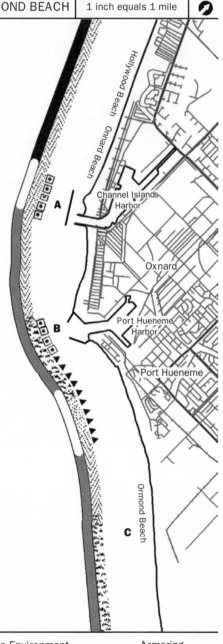

Hollywood Beach
Onnard Beach
Channel Islands Harbor
Oxnard
Port Hueneme Harbor
Port Hueneme
Ormond Beach

Hazard Level
- Stable: low risk
- Caution: moderate risk
- Hazard: high risk

Erosion Rates
1 Inches per year

Shoreline Environment
- Sandy beach
- Beach backed by marsh
- Beach backed by dunes
- Cliff or bluff fronted by beach
- Rocky coast with no beach

Armoring
- ▲ ▲ ▲ Riprap
- Seawall
- Other

A. The coast from Ormond Beach to near Point Mugu lies within Point Mugu Naval Missile Test Center. Owing to its exposure to both westerly and southerly swells, much of this coast is actively eroding, notably around Laguna Point where riprap has been used to counter erosion.
B. Mugu Lagoon, within the confines of the U.S. Navy Missile Test Center, is the largest surviving, relatively natural wetland along the southern California coast. Its western extensions have been largely obliterated by military activities, but its central and eastern portions are the focus of continuing restoration projects. Calleguas Creek provides freshwater inputs seasonally, somewhat contaminated by urban and agricultural runoff from upstream sources.
C. Point Mugu comprises an actively eroding mass of Miocene volcanic, and sedimentary rock.

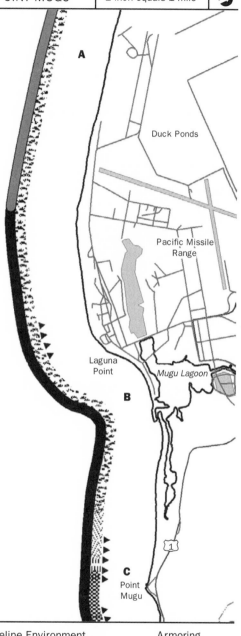

Duck Ponds

Pacific Missile Range

Laguna Point

Mugu Lagoon

B

C

Point Mugu

Hazard Level

Stable: low risk
Caution: moderate risk
Hazard: high risk

Erosion Rates

 Inches per year

Shoreline Environment

Sandy beach
Beach backed by marsh
Beach backed by dunes
Cliff or bluff fronted by beach
Rocky coast with no beach

Armoring

▲ ▲ ▲ Riprap
■ ■ ■ Seawall
● ● ● Other

A. Between Point Mugu and Sequit Point, the Santa Monica Mountains approach the shore leaving little room for development. The coastal highway is prone to landslides and rock falls from the mountains above, and to marine erosion from below, necessitating extensive use of riprap protection.

B. Climbing dunes rise against the mountainside.

C. Big Sycamore Canyon: the outlet of this major stream in the western Santa Monica Mountains introduces much coarse clastic sediment, mostly cobbles, to the shore.

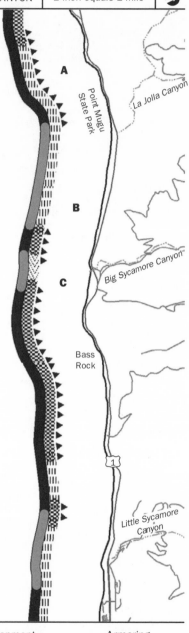

Point Mugu State Park

La Jolla Canyon

A

B

Big Sycamore Canyon

C

Bass Rock

1

Little Sycamore Canyon

Hazard Level

☐ Stable: low risk
▨ Caution: moderate risk
■ Hazard: high risk

Erosion Rates

1 Inches per year

Shoreline Environment

Sandy beach
Beach backed by marsh
Beach backed by dunes
Cliff or bluff fronted by beach
Rocky coast with no beach

Armoring

▲▲▲ Riprap
■■■ Seawall
◉◉◉ Other

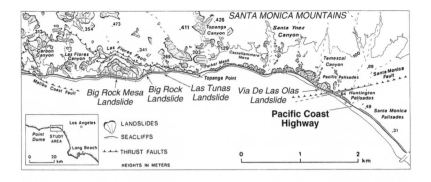

Figure 18.7   Faults and landslides along the Malibu coast. Illustration by Antony Orme.

unplanned or poorly conceived development from the 1920s to the 1970s. Management problems here are akin to those along the Rincon coast, namely unstable cliffs and close proximity of development to the shoreline, but are of far greater intensity because of the pressures exerted by the Los Angeles urban area.

The Santa Monica Mountains are a broadly anticlinal structure, 3 to 9 miles wide, extending 50 miles westward from the Los Angeles River narrows to Point Mugu. These mountains are dominated by sedimentary rocks, mostly mudstones, sandstones, and conglomerates. These mountains are still being uplifted. Pleistocene marine terraces between Little Sycamore Creek and Malibu, and between Topanga Canyon and Pacific Palisades, offer a measure of this uplift.

Much of this uplift is associated with the Malibu Coast fault system (Figure 18.7), which passes seaward at Potrero Canyon just north of Santa Monica, remains close to shore, and comes on land again between Carbon Canyon and Sequit Point. The fault system is seismically active, as demonstrated during the February 1971 San Fernando earthquake, when hot water bubbled to the surface off Malibu Pier, and by later earthquakes off Point Mugu. Furthermore, splays of this fault reveal Holocene displacement beneath Winter Mesa and Malibu Lagoon.

Most drainage within the Santa Monica Mountains developed consequent to uplift, as shown in Topanga Creek and other short, south-flowing streams. However, Malibu Creek and Big Sycamore Creek are remnants of an earlier south-flowing drainage system and have maintained sufficient erosive capacity to cut deep canyons across the still-rising mountains. Malibu Creek drains a 110-square-mile drainage basin, encompassing the Simi Hills and Conejo Valley north of the mountains, and reaches the sea through a picturesque 4-mile-long canyon. Because its upper basin has

Figure 18.8   Big Rock Mesa landslide before development, 1933. Spence Collection. Reproduced by permission of the Department of Geography, University of California, Los Angeles.

experienced extensive urban development over the past 40 years, discharge from Malibu Creek is a particular challenge for coastal management.

The Santa Monica Mountains dominate this coast, bringing steep, fire-prone, chaparral-clad hillsides directly to the coast, or at least to the Pacific Coast Highway, which has been notched into their base (Figure 2.1). Rocks exposed in coastal bluffs and canyon walls within the Malibu Coast fault zone are pervasively fractured and deformed. Hill slope failure is thus a perennial problem (Figure 18.8), but prior to human intervention, rocks tended to collapse harmlessly to the shore, where they broke down to replenish the beaches. Since the early twentieth century, however, mass movement processes have been aggravated by railroad and highway construction, which in turn has denied beaches replenishment.

The Hueneme, Malibu, and Port Los Angeles Railroad was opened along 15 miles of coastal bluffs westward to near Little Sycamore Canyon in 1908, partly to help ranch operations but mostly to deter other railroad speculators. Much of this route was later absorbed into the Roosevelt Highway (now the Pacific Coast Highway), which, after six years of construction, was opened in June 1929 between Malibu and Oxnard. Construction of this highway, and widening farther east, involved the

A. From near Sequit Point to Lechuza Point, dissected but narrow marine terraces separate the mountains from the shore. The terraces and the underlying bedrock reach the coast in moderately high cliffs that are prone to rock falls and marine erosion.

B. The mouths of small mountain streams are characterized by abundant cobble-sized materials, for example off Arroyo Sequit and San Nicholas Canyon. Beaches between canyon mouths are sandy and narrow.

C. Properties in the back shore west of Lechuza Point are subject to episodic damage from storm waves associated with storms passing along southwesterly tracks into the Southern California Bight.

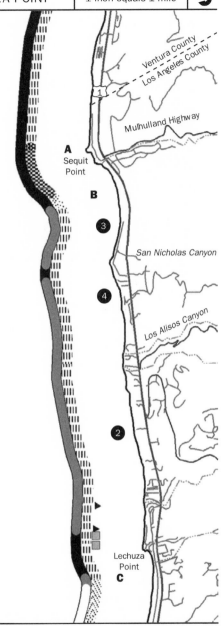

Ventura County
Los Angeles County

Mulhulland Highway

A
Sequit
Point

B

San Nicholas Canyon

Los Alisos Canyon

Lechuza
Point
C

Hazard Level

☐ Stable: low risk
▨ Caution: moderate risk
■ Hazard: high risk

Erosion Rates

 Inches per year

Shoreline Environment

▨ Sandy beach
▨ Beach backed by marsh
▨ Beach backed by dunes
▥ Cliff or bluff fronted by beach
▨ Rocky coast with no beach

Armoring

▲▲▲ Riprap
■■■ Seawall
⊡⊡⊡ Other

removal of 3 million cubic yards of rock debris from the cliff base between Santa Monica Canyon and Point Mugu for use as roadbed and coastal protection. Added protection was afforded by 24 groins, 40,000 cubic yards of riprap, and several long seawalls. Understandably, the coastal bluffs were thrown into disequilibrium, from which they are still recovering. Rockfalls and temporary closures occur with regularity along this section of the Pacific Coast Highway. For example, a bedrock bluff east of Big Rock Mesa (Figure 18.8) developed arcuate tension cracks above the highway during winter 1979, causing dangerous rockfalls and threatening to shift 2.5 million cubic yards of rock seaward. These cracks were opening at a rate of more than 1 inch per day in April 1979. The California Department of Transportation subsequently removed 150,000 cubic yards of rock, terraced the cliff face, and installed a protective barrier at the base, but complete stabilization is unlikely.

Debris flows also occur during most winters, especially when drenching rains mobilize surficial sediments on hillsides whose vegetation cover has been ravaged by fire (for example, following the major fires of autumn 1993). These flows often enter stream channels and ultimately spill to the shoreline.

More massive slope failures also occur fairly frequently, often after winter rains have had sufficient time to infiltrate coastal slopes and weaken the underlying materials. Cliff-top developments involving vegetation change, grading, road drainage, septic tanks, and landscape watering have greatly changed hill slope hydrology, frequently pushing terrain of marginal stability over the threshold into renewed landslide activity. This is well illustrated at Big Rock Mesa, between Piedra Gorda and Las Flores canyons, where a 250-acre prehistoric landslide complex went largely unrecognized until renewed movement began in the 1970s, at which time nearly 300 homes occupied the slide area. Although some homes were built as early as 1928, most development occurred during the relatively dry 1950s and 1960s in conjunction with individual sewage disposal systems. Septic tank effluent, landscape watering, and heavy winter rains in 1978 and 1980 subsequently raised the groundwater table, reduced shear strength, and triggered downslope movements. Rotation of this landslide complex along a series of intersecting slide planes caused major deformation within the slide and along the Pacific Coast Highway. Stabilization efforts, begun in 1983, involve the use of vertical wells and horizontal drains to dewater the landslide. These methods have achieved some success, but whether dewatering will be as effective during a sequence of wet years is uncertain. Furthermore, a much larger area than that originally involved is capable of shifting seaward south of the Las Flores Fault.

Today unstable cliffs protected by a variety of slope stabilization structures flank the inland side of the Pacific Coast Highway (Figure 18.9).

A. Zuma Beach County Park is a broad, largely natural beach trapped against low, now inactive, seacliffs by the natural trapping effect of Point Dume. Although characterized by strong longshore currents and often powerful rip currents, this beach is an important recreational asset.

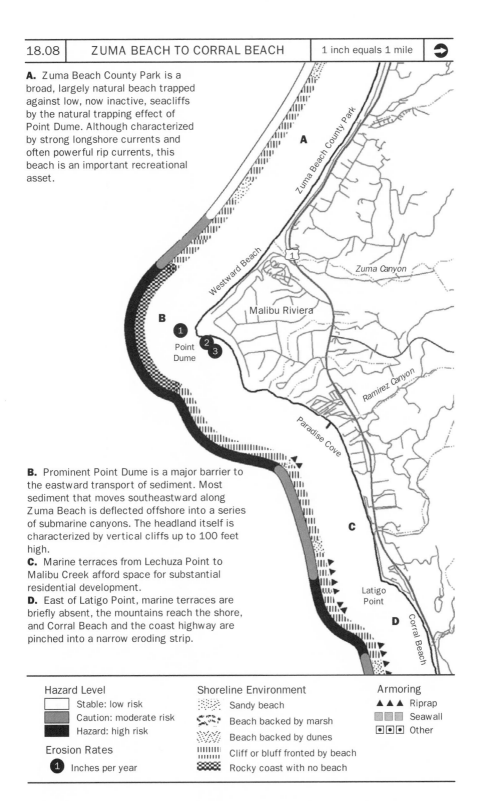

B. Prominent Point Dume is a major barrier to the eastward transport of sediment. Most sediment that moves southeastward along Zuma Beach is deflected offshore into a series of submarine canyons. The headland itself is characterized by vertical cliffs up to 100 feet high.

C. Marine terraces from Lechuza Point to Malibu Creek afford space for substantial residential development.

D. East of Latigo Point, marine terraces are briefly absent, the mountains reach the shore, and Corral Beach and the coast highway are pinched into a narrow eroding strip.

Hazard Level
- Stable: low risk
- Caution: moderate risk
- Hazard: high risk

Erosion Rates
- ① Inches per year

Shoreline Environment
- Sandy beach
- Beach backed by marsh
- Beach backed by dunes
- Cliff or bluff fronted by beach
- Rocky coast with no beach

Armoring
- ▲▲▲ Riprap
- Seawall
- Other

A. Marine terraces between Corral Beach and Malibu are partly conserved in Malibu Bluff State Park. Below the landslide-prone cliffs fronting these terraces, significant housing development has occurred along Puerco and Amarillo Beaches. These houses, often built on pilings seaward of the high-water mark, are particularly vulnerable to storm-wave damage.

B. Malibu Beach is the site of a "colony" developed since the 1920s by celebrities in the Hollywood entertainment industry. The "colony" is built on a relatively stable barrier beach fronting a largely reclaimed lagoon, but is vulnerable to direct storm-wave impacts.

C. Malibu Lagoon State Beach is the focus of continuing debate regarding conservation theory and practice (see text). In essence, this is a dynamic physical and ecological system that has been constrained by past developments such that it no longer functions well as a natural system.

D. From Malibu to Topanga, the Santa Monica Mountains reach directly to the shore, marine terraces are absent, and highway and housing developments lie mostly below crumbling, landslide-prone former seacliffs.

Developments over the past century have largely obliterated the former beach and aggravated mass movements from the cliffs behind.

E. The landslide hazard along this stretch of coast is compounded by: poorly lithified, structurally incompetent bedrock; severely fractured and faulted rocks within the Malibu Coast and Las Flores fault zones; episodic drenching from winter rains; human impact related to highway construction, housing developments, and landscape irrigation; and vibration from waves and vehicular traffic.

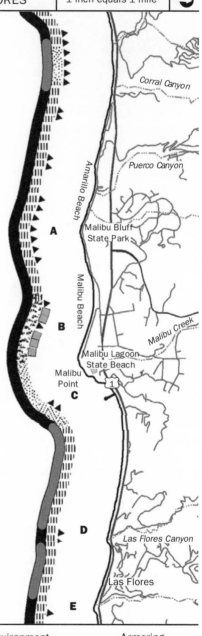

Hazard Level

☐	Stable: low risk
▨	Caution: moderate risk
■	Hazard: high risk

Erosion Rates

① Inches per year

Shoreline Environment

▨	Sandy beach
⌇	Beach backed by marsh
⩗	Beach backed by dunes
⫚	Cliff or bluff fronted by beach
▨	Rocky coast with no beach

Armoring

▲▲▲	Riprap
■■■	Seawall
⊡⊡⊡	Other

Figure 18.9   Regrading of the hillside at the Big Rock landslide along the Malibu
Coast, 1982. Photo by Kevin Mulligan.

Perpetually troublesome areas have been regraded, often several times.
Along the shoreline, homes built on fill and fractured rock often perch precariously on pilings above beaches starved of sediment, backed by a mix of
riprap, bulkheads, and seawalls.

   Further eastward, opening of public access across the Rindge Ranch in
Malibu Canyon in the 1920s, and the subsequent opening of the Roosevelt
Highway from Malibu to Oxnard in 1929, paved the way for development.
Among the first projects was the Malibu Colony on the barrier beach at the
mouth of Malibu Creek—initially a collection of modest cottages linked to
the burgeoning Hollywood film industry, now reshaped into a gated community of expensive beachfront homes protected by various engineering
structures. For many years the Malibu coast retained a detached rural
image, but after partial relocation and widening of the Pacific Coast
Highway in 1949, development along this corridor introduced the full
impact of Los Angeles to the region. Nevertheless, flat land was at a pre-
mium, and only around Malibu Lagoon and on the marine terraces behind
Point Dume did development proceed any distance inland. Recent decades
have seen many of the remaining undeveloped portions of this coast
acquired for public use by county and state agencies, but despite the constraints of the California Coastal Act, housing construction and replace-
ment, usually with larger homes on smaller lots, have continued.

A. Big Rock Mesa landslide: This 250-acre prehistoric landslide complex went largely unrecognized until renewed movement began in the 1970s, by which time nearly 300 homes occupied the area. Septic tank effluent, landscape watering, and heavy winter rains in 1978 and 1980 triggered renewed movement, which peaked at 5 inches per month in the early 1980s. Subsequent stabilization efforts have reduced this rate but the area, together with the highway and nearshore zone toward its base, remain inherently unstable.

B. Big Rock landslide: Winter 1979 saw a series of rock falls and slides, involving 2.5 million cubic yards of rock, descend onto the coastal highway. The highway department subsequently removed 150,000 cubic yards of material, terraced the cliff face, and installed a protective barrier at its base, but the area remains unstable.

C. The Las Tunas landslide complex involves several discrete slope failures that not only shed rock debris onto the coastal highway, but cause the highway and property foundations to buckle. As with the other landslides to the west, these movements are related to deep-seated, arcuate slide planes that extend beyond the present shoreline.

D. The mouth of Topanga Canyon is presently the focus of attempts to restore a small historic lagoon and wetland.

E. The seaward edge of Castellamare Mesa is the site of numerous landslides and rock falls, which pose continuing problems for coastal highway maintenance.

F. Potrero Canyon, now the site of a massive landfill, has developed where the active Malibu Coast fault zone crosses the shoreline.

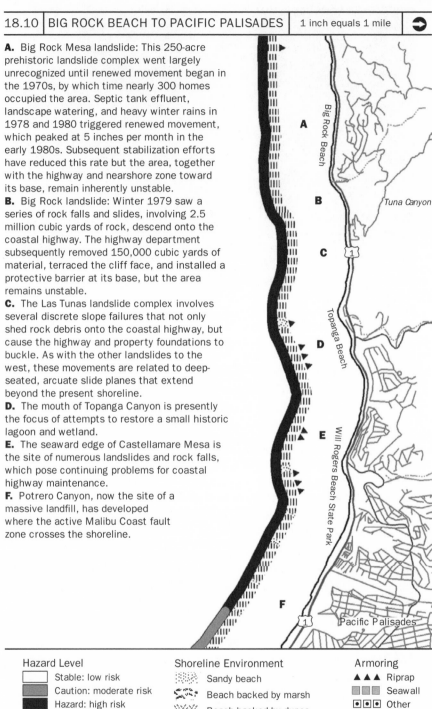

Big Rock Beach

A

B

Tuna Canyon

C

Topanga Beach

D

Will Rogers Beach State Park

E

F

Pacific Palisades

Hazard Level

☐ Stable: low risk
▒ Caution: moderate risk
■ Hazard: high risk

Erosion Rates

 Inches per year

Shoreline Environment

Sandy beach
Beach backed by marsh
Beach backed by dunes
Cliff or bluff fronted by beach
Rocky coast with no beach

Armoring

▲▲▲ Riprap
■■■ Seawall
⊡⊡⊡ Other

Figure 18.10   Malibu Creek mouth during high discharge in February 1998. Photo by Antony Orme.

Nowhere is the conflict between development and conservation more acute than around Malibu Lagoon, the subject of detailed studies in recent years. Here a prehistoric lagoon has been decimated in size by natural sedimentation and artificial reclamation, aggravated by the delivery of contaminants from the urbanizing upper Malibu Creek basin (Figure 18.10). The natural habitats formerly occupied by various invertebrates, fish, birds, and mammals have been severely impacted by increased eutrophication and invasive plant species. Many species, such as the tidewater goby and steelhead trout, have become locally endangered. Furthermore, human recreation around the lagoon and in near-shore waters is threatened by pathogens such as *Giardia* and *Cryptosporidium* derived from incomplete wastewater treatment upstream and local septic-tank effluent. Several proposals for the enhancement and restoration of this threatened resource are now being considered.

The easternmost portion of this coast is characterized by former seacliffs that descend from an elevation of 230 feet near Santa Ynez Canyon to disappear south of Santa Monica Pier (Figure 18.11). These cliffs are notably prone to rockfalls and landslides. At Pacific Palisades, the Via de las Olas prehistoric landslide complex has been a perennial problem. In March 1958 this landslide failed again after heavy rains, moving 800,000 cubic yards seaward (Figure 18.12). As opposed to removal of the toe of the landslide, the Pacific Coast Highway was relocated seaward and protected by groins. To

Figure 18.11   Pacific Palisades and Castellammare Mesa, 1931. Spence Collection.
Reproduced by permission of the Department of Geography, University of California,
Los Angeles.

Figure 18.12   Via de las Olas landslide in Pacific Palisades, 1958. Spence Collection.
Reproduced by permission of the Department of Geography, University of California,
Los Angeles.

A. A quasi-active landslide here is occupied, appropriately, by mobile homes.
B. The Via de las Olas landslide, Pacific Palisades, occupies a bowl that developed prehistorically along the front of abandoned but unstable seacliffs. It was reactivated after heavy rains in March 1958 and, rather than remove 800,000 cubic yards of collapsed material, the coastal highway was relocated seaward of the base.
C. The Huntington Palisades, south of Potrero Canyon, are 150-180-feet-high, active cliffs developed in Pleistocene fanglomerates. Cliff-top houses, developed since 1924, tumble from time to time over the cliff edge, most notably during the Northridge earthquake of 1994.
D. The Santa Monica Palisades are also developed in Pleistocene fanglomerates but decline in elevation southward from 120 feet to 60 feet high behind Santa Monica Pier. Frequent slope failures pose a continuing maintenance problem for the cliff-top park.
E. Beaches fronting Pacific and Huntington Palisades and Santa Monica have been augmented by artificial nourishment and constraining groins. A small-boat anchorage and sand trap were developed behind an offshore breakwater but this structure was largely destroyed during the El Niño storms of 1983.

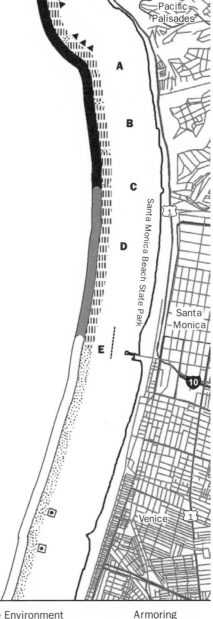

Hazard Level

  ☐ Stable: low risk
  ▨ Caution: moderate risk
  ■ Hazard: high risk

Erosion Rates

  ① Inches per year

Shoreline Environment

  ▨ Sandy beach
  〰 Beach backed by marsh
  ▨ Beach backed by dunes
  ⫴ Cliff or bluff fronted by beach
  ▨ Rocky coast with no beach

Armoring

  ▲▲▲ Riprap
  ■■■ Seawall
  ⊡⊡⊡ Other

the south, vertical cliffs formed in poorly consolidated alluvial fan and terrace deposits have long been prone to failure. The cliffs forming Huntington Palisades collapsed spectacularly during the 1994 Northridge earthquake, carrying homes with them. The shoreline here lies within Will Rogers State Beach and Santa Monica State Beach, both of which have been augmented by decades of intermittent artificial replenishment. At the latter, an offshore rubble breakwater was built in 1933 to provide a small-boat anchorage and a broader beach in its lee. This structure was later damaged and finally succumbed to the El Niño storm seas of 1983, when the outer part of Santa Monica Pier was also lost. Although the pier has since been rebuilt, no attempt has yet been made to refurbish the breakwater, in part because small boats are now accommodated farther south at Marina del Rey.

CHAPTER NINETEEN

THE COAST OF SOUTHERN CALIFORNIA

Santa Monica to Dana Point

DOUGLAS SHERMAN AND BERNARD PIPKIN

The 71 miles of mainland coast between Santa Monica and Dana Point can be divided into four main segments: (1) the Santa Monica Littoral Cell, (2) the Palos Verdes Peninsula, (3) the San Pedro Littoral Cell, and (4) the Laguna Beach Mini Cells (Figure 19.1). Large-scale geological processes control the general forms of the four segments. The Santa Monica and San Pedro littoral cells are long, sandy beach systems that formed as part of the depositional processes that filled the Los Angeles Basin following crustal downwarping. The Palos Verdes Peninsula and the Laguna Beach mini-cells are cliffed coasts caused by crustal uplift. The two beach segments total about 40 miles in length, and the uplifted reaches total about 31 miles of coast.

This part of the California coast includes the state's largest coastal plain, formed across the low-lying Los Angeles Basin. The large, relatively flat surface of the plain provided the space necessary for the explosive urban development that would make Los Angeles one of the 10 largest mega-cities on Earth, with a metropolitan population exceeding 13 million.

Juan Rodriguez Cabrillo anchored in Bahia de los Humos (the Bay of Smokes) in 1542. The natural landscape in the region comprised headlands with dramatic cliffs, sandy barrier systems protecting extensive wetlands, large dune fields, and small pocket beaches. The climate was mild and the coastal waters relatively warm, even through the winter. Abundant game and plentiful marine life supported scattered communities of Native peoples at or near the coast. It is unlikely that Cabrillo would recognize the twenty-first-century version of that anchorage, now home to one of the busiest port complexes in the world, although the atmospheric tints that once gave rise to the name Bay of Smokes might seem familiar.

Figure 19.1   Los Angeles and Orange counties, including the four littoral environments described in this chapter. Photo courtesy of Earth Sciences and Image Analysis Laboratory, NASA Johnson Space Center.

Despite the many salubrious aspects of this coastline, the native Gabrielino Indians dwelt there only sparsely, and their population numbered perhaps 5,000 across the entire basin. At the time of the Portolá expedition, in 1769, the only coastal settlements were adjacent to Ballona Lagoon, then the mouth of the Los Angeles River, and around some of the coastal wetlands associated with the San Gabriel and Santa Ana rivers. Few other locations had reliable, year-round sources of freshwater, and the natives visited the coast mainly to forage. Indeed, there was minimal development directly at the coast. The early Spanish missions were established at San Gabriel and San Fernando, well inland, where the San Gabriel and Los Angeles rivers, respectively, provided a secure water supply. Even the small Pueblo de Nuestra Señora la Reina de Los Angeles de Porciúncula, later contracted to Los Angeles, was founded at a location approximately 15 miles from the Pacific Ocean.

Anaheim Bay

Figure 19.2   Residential development at Surfside in 1935, looking toward the south. Note the extremely narrow beach and extensive use of wood groins. Spence Collection. Reproduced by permission of the Department of Geography, University of California, Los Angeles.

There was minimal coastal development until the late nineteenth to early twentieth century. Before this period of expansion, port facilities were rudimentary, the fishing fleet was small, and there were few other viable aspects of coastal economic activity. The overall regional population remained small, totaling about 15,000 for Los Angeles County by 1870. The population of Orange County was minimal. By 1900, however, about 200,000 residents called the two-county area home. Much of the rapid growth was a result of the arrival of the transcontinental railroad in Los Angeles and aggressive real estate promotion by developers, several of whom were pushing coastal property and resort development. Santa Monica, Venice, Manhattan Beach, Redondo Beach, Long Beach, Seal Beach, Huntington Beach, and Newport Beach, linked to Los Angeles by trains and trolley cars, all benefited (population-wise) from such promotion in this era. Aside from some of the grand resorts, a great deal of the development near the coast involved small cottages or bungalows on tiny lots crowded along the beach. Figure 19.2 illustrates this type of development on the narrow sand barrier at Surfside. It was common for construction to occur almost up to the mean high water line. One result was that property damage due to coastal erosion and flooding was frequent.

San Pedro Bay

Los Angeles River

Ports of Los Angeles and Long Beach

Figure 19.3   Development of the ports of Los Angeles and Long Beach transformed barrier island and marsh environments into one of the busiest harbor facilities in the world. Expansion of port facilities continues through the creation of artificial (landfill) islands. Photo courtesy of Earth Sciences and Image Analysis Laboratory, NASA Johnson Space Center.

It was also around this time that the competition for the location of the region's harbor was resolved. Several business interests had been competing stridently to establish the main port facilities at Santa Monica, Redondo Beach, or San Pedro. Although the latter afforded the best physical setting, the other sites had existing railheads for national systems and powerful supporters. San Pedro was the winner, and creation of the Port of Los Angeles was begun. Barrier islands, with names such as Rattlesnake Island and Deadman's Island, and the marshes, mudflats, tidal channels, and bays behind them, were transformed into the modern ports of Los Angeles and Long Beach. A 9-mile-long breakwater, the world's longest, protects the modern harbor. The combined ports are among the busiest in the world and have a completely developed waterfront about 28 miles long (Figure 19.3). The mouth of the Los Angeles River is inside the area protected by the breakwater, as are some of the beaches of Long Beach.

The population of greater Los Angeles doubled six times in the twentieth century. Once-sleepy coastal settlements such as Hermosa Beach, Surfside, and Laguna Beach grew rapidly. Some of the fastest urbanization occurred during a decades-long period of exceptionally benign coastal climate extending from the mid-1940s to 1978. Storms, and resulting erosion, were less frequent and less severe than in prior or subsequent years. The benign weather also encouraged development in locations that would prove to be too close to the shoreline to stand up to more severe wave conditions. The phenomenal population growth has resulted in the area's beaches having the heaviest recreational use in the state. "Going to the beach" remains a popular activity for residents of the high-density urban area and for tourists drawn to the image of southern California beaches. The beaches are an important factor in the quality of life in southern California, and their maintenance is of critical aesthetic, emotional, and economic value to the region and the state.

Population pressure has also contributed to a great need for expanded recreational and pleasure-boating facilities along this coast. To address the vocal demands of a resourceful constituency, politicians and developers promoted the establishment of new marinas and harbors, often at the expense of beaches and wetlands. Marina Del Rey, King Harbor, Huntington Harbor, Newport Bay, and Dana Point are examples of large development projects designed at least partly to meet such demand. The large landscape and ecosystem transformations associated with these developments have increased the importance of careful management of the few coastal systems that have survived relatively undisturbed.

HUMAN IMPACTS

Human impacts have been profound along this coast. In addition to the port and harbor developments just described, there are direct human interventions associated with substantial beach nourishment projects, armoring of the shoreline, construction of jetties to stabilize the mouths of inlets, construction of offshore breakwaters, and installation of groin fields. Where these interventions have been substantial, they are discussed in the context of the coastline where they occur. There have also been substantial indirect impacts caused by the interruption of sand delivery to the coast by flood control and water supply dams, by channelization of coastal rivers, and by development over coastal dune systems.

Groins, jetties, and breakwaters all tend to produce localized accretion up drift of or, in the case of detached breakwaters, behind the structure. A typical consequence is down-drift erosion. The use of mechanical means to move sediments past these obstructions and the prevalence of beach nourishment have reduced the deleterious effects of these structures. Jetties are long structures designed both to provide protection from waves for boats entering or leaving the harbor and to constrain the tidal flow, thus reducing or preventing shoaling in harbor entrances. Development on the up-drift side of jetties often enjoys the benefit of wide, sandy beaches and the enhanced shore protection that they provide. Such is the case at

Figure 19.4   Newport Beach and Balboa in 1923. Other than the small foredunes developed near the center of the image, there was little protection against overwash. Note that development north of the dunes was also seaward of the dune line. Spence Collection. Reproduced by permission of the Department of Geography, University of California, Los Angeles.

Marina Del Rey and Newport Beach, for example. It would be extremely difficult to enter Newport Harbor, down drift of the mouth of the Santa Ana River, if the jetties that mark its entrance did not keep it free from sandbars and shoals. A fringe benefit of these jetties has been the formation of a fierce body-surfing break known as "the Wedge." During most of the year this is a tranquil beach, but when big waves from distant storms arrive, the resulting surf is legendary. Body-surfing the wedge-shaped wall of water that forms along the northern jetty under these conditions is for experts only, and becomes a spectator sport even for strong swimmers.

In the long run, indirect human impacts will have a larger influence on the beaches of Los Angeles and Orange counties. Part of this influence will result from the urban development that has covered large tracts of formerly mobile sand surfaces. This has been the case in Santa Monica Bay, where the large dune field backing this coast has been almost entirely urbanized. The rolling hills of El Segundo, Manhattan Beach, and Redondo Beach, for example, consist of the dunes created by sand blowing landward from the beach. Under natural conditions, parts of these sand bodies would be reclaimed by the waves during periods of severe shoreline erosion. Now these sands are locked away under houses, streets, and yards. The sand spit at Newport Beach that forms the ocean side of Newport Bay is another example. The entire spit was a wave-and-current–formed feature whose shape would have changed in response to different wave and current regimes. This can be seen clearly in Figure 19.4, an image taken when residential development was incomplete. The spit is now locked in place under a grid of city streets and houses.

The coastal watersheds in this region include some of the most altered river systems in the United States. All of the rivers, most of the permanent streams, and many of the ephemeral streams are controlled by structures. The construction of dams represents the most substantial method of control. Although the purpose of the dams is to reduce the volume of flood flows downstream or to trap and retain water for drinking supply, another consequence is the impoundment of sand that might otherwise be transported to the ocean. The reservoirs and lakes behind the southern California dams are nearly 100 percent efficient in their sand trapping, except for the atypical instances when they become sand filled. It has been determined that only about 33 percent of the natural sediment loads of the Los Angeles, San Gabriel, and Santa Ana rivers still reach the coast. This is a reduction of about 530,000 cubic yards of sand per year received by beaches in the San Pedro Littoral Cell—equivalent to losing a beach roughly 2 miles long, 135 feet wide, and 10 feet thick every year! This has important implications for the long-term status of these beaches and for coastal managers concerned with their maintenance. There has been no substantial river discharge into the Santa Monica Cell since the Los Angeles River last changed its course away from Ballona Lagoon in 1825. Consequently, reductions of fluvial sediment supply have not occurred in the Santa Monica Cell.

THE SANTA MONICA LITTORAL CELL

The Santa Monica Littoral Cell reaches from Point Dume to the Palos Verdes Peninsula. Before 1825, the primary source of beach sand in the cell was the Los Angeles River. The modern supplies of beach sediments are much more modest, including small volumes from the coastal streams along the Malibu coast, some transport from the inner continental shelf, and most important, sand from beach nourishment projects. Transport in and near the surf zone is generally toward the Palos Verdes Peninsula, and most of the sand is lost offshore through the Redondo Submarine Canyon. The beaches through most of this reach are eroding only slowly and in many instances appear to be stable over decadal time scales. There are several erosion hazard areas where the threat is usually a result of human action.

The Santa Monica Municipal Pier is a long-standing icon of California beach recreation (Figure 19.5). The state beach adjacent to the pier is wide,

A. Wide, sandy beach, backed by recreation infrastructure. The beach has been nourished to enhance recreation space, and has exhibited long-term stability. Convenient access has made Santa Monica State Beach one of the most popular recreation beaches in the Los Angeles region.

B. A detached breakwater built in 1934 has caused localized accretion at Dockweiler State Beach. Some riprap has been installed to protect recreation facilities. This beach is stable, although some facilities are vulnerable to wave attack in severe storm seasons.

C. This stretch of sandy beach is wide and partially stabilized by the jetties at Marina Del Rey. Development is set well back from the water line. Small dunes are developing at several locations at Dockweiler State Beach, where beach cleaning operations do not prevent the establishment of vegetation.

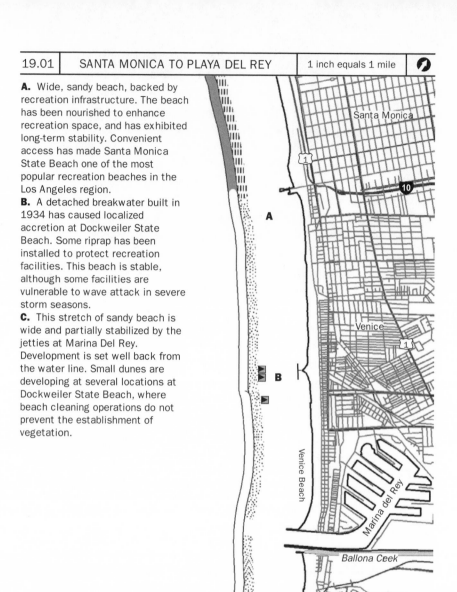

Santa Monica

A

Venice

B

Venice Beach

Marina del Rey

Ballona Ceek

C

Playa del Rey

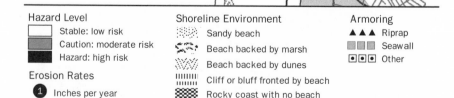

Hazard Level		Shoreline Environment		Armoring	
☐	Stable: low risk	▦	Sandy beach	▲▲▲	Riprap
▨	Caution: moderate risk	〜	Beach backed by marsh	▨▨▨	Seawall
■	Hazard: high risk	⩗	Beach backed by dunes	⊡⊡⊡	Other
Erosion Rates		‖‖‖	Cliff or bluff fronted by beach		
❶	Inches per year	▓	Rocky coast with no beach		

Figure 19.5   Amusement park and recreational facilities on the Santa Monica Municipal Pier. Note the large expanses of "on-beach" development, especially for parking. Photo © 2002–2004 Kenneth and Gabrielle Adelman, California Coastal Records Project, www.Californiacoastline.org.

high, and sandy and is an extremely popular tourist and day-use venue. But the beautiful beach is not the result of natural processes. In 1933 a detached breakwater was built just north of the pier to create an anchorage for pleasure craft—Santa Monica Harbor. The immediate result was substantial accretion in the lee of the structure extending southward past the pier, and extensive erosion farther down drift. The accretion caused by the breakwater reduced the harbor's utility substantially, and Will Rogers is said to have commented, "It looks as if it might need irrigating before long." Dredging has been used to maintain the structure's utility. In addition to the breakwater-induced accretion, large-scale beach nourishment has occurred since the late 1940s. Beach nourishment has prevented critical erosion problems in this area. However, there is some threat to the pier and ancillary facilities associated with severe storms. Such damage, including damage to the breakwater, occurred during the El Niño storms of 1983.

About 2 miles south of the Santa Monica Pier is another breakwater at Venice. A sand spit commonly connects this structure to the coast, and the complex has created one of the area's better surfing spots. This stretch of coastline owes its present condition to large-scale beach nourishment projects and, toward the south, to sediment trapped by the northern jetty at Marina Del Rey. The urban development at Venice began as a planned community with residences built along a series of canals dredged from the

Figure 19.6　Ballona Lagoon and wetlands ca. 1960. The jetties were completed in 1958, but dredging for Marina Del Rey had not yet begun. Coastal development is still sparse, and extensive wetlands are present. Ballona Creek occupies a former channel of the Los Angeles River. Photo courtesy of the Los Angeles County Department of Beaches.

marshes of Ballona Lagoon. The beach itself is the remnant of the barrier spit that enclosed the lagoon. It is the location of the modern version of Muscle Beach, and a very busy tourist destination. Severe beach erosion and damage to housing and infrastructure in the 1930s and 1940s, caused by urban development almost up to the mean high water line, were major factors in the adoption of beach nourishment as a management strategy here. Few erosion threats exist in this region because of the substantial protection afforded by the wide, high beach. However, during the winter, sand berms are often bulldozed along the beach to replicate the overwash protection of foredunes.

Marina Del Rey was dredged from the wetlands of Ballona Lagoon, where the Los Angeles River once debouched (Figure 19.6). It is one of the largest human-made recreational boating and residential marinas in the world, providing moorings for 6,000 boats and facilities for thousands more. The marina was created in the 1960s in response to the demand for recreational boating by the rapidly growing population in the region. The shelter offered by the Santa Monica breakwater was inadequate for this purpose, especially as there were no shore facilities associated with the structure. Similarly, King Harbor at Redondo Beach was too small to meet the demand.

The U.S. Army Corps of Engineers designed the marina after close consideration of local wave energy, longshore currents and littoral drift, impact on

A. Wide sandy beach from extensive beach nourishment. The El Segundo dune field backs the recreation facilities of Dockweiler State Beach. Parking lots and lifeguard towers susceptible to damage during severe storms.

B. Parking lots and wastewater tanks at the Hyperion Sewage Treatment Plant are at risk from coastal erosion, and are protected by an artificial berm.

C. Narrow, low sandy beach. Power plant and refinery facilities are protected by riprap, groin, and beach nourishment. Severe erosion during winter storms of 1983. Some mechanical bypassing of sand over the groin to replenish downdrift beaches.

D. Manhattan Beach State Park. Wide sandy beaches from nourishment projects, some backed by small dunes.

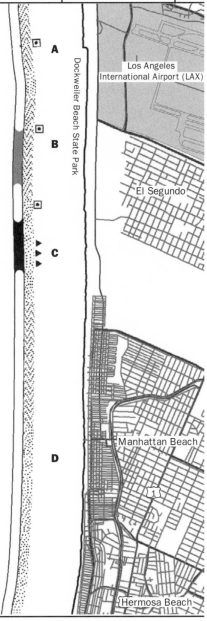

Hazard Level

 Stable: low risk

Caution: moderate risk

Hazard: high risk

Erosion Rates

 Inches per year

Shoreline Environment

 Sandy beach

 Beach backed by marsh

 Beach backed by dunes

 Cliff or bluff fronted by beach

Rocky coast with no beach

Armoring

▲ ▲ ▲ Riprap

 Seawall

 Other

Figure 19.7 Marina Del Rey in 1981. Compare this photo with Figure 19.6 and note the addition of the detached breakwater across the harbor entrance, the loss of wetlands, and the reduction in beach width in the two decades after the jetties were built. Photo courtesy of the Los Angeles County Department of Beaches.

the remnants of Ballona Creek and its lagoon, and general geological studies of site conditions. Wave climate was hindcast in order to estimate the likely magnitude of the largest storm waves that could be expected to break across the jetties that would stabilize the entrance channel. The jetties were designed so wave energy would be dissipated outside of the marina area, allowing calm conditions to prevail within the anchorage. After construction was completed and the marina occupied by happy boaters, nature complicated the scene. Waves of a particular height, period, and direction of approach set up resonance patterns within the channel and in the marina proper. At points within the harbor, waves as high as 6 feet occurred, causing extensive damage to boats and slips. Many boat owners fled, and concessionaires sued Los Angeles County for lost revenue.

Temporary baffles were installed across and within the main channel to reduce the transmission of wave energy into the anchorage. A large-scale model of the marina was built in a wave tank to better assess the problem. A 1,200-foot detached breakwater was installed just offshore of and across the ends of the jetties in an attempt to remedy the situation (Figure 19.7).

This solved the problem of seiches in the marina, and a safe harbor was created. However, that was not the end to the complications. The jetties and breakwater interfered with alongshore sand transport and caused localized deposition of sediments discharged by the now-controlled mouth of Ballona Creek immediately adjacent to the south jetty. Shoaling in the southern approach to the inlet requires maintenance dredging for safe navigation. There are similar, although lesser, problems caused by sand percolating through or blowing across the top of the northern jetty.

The Marina Del Rey today is a fine facility that supports boating and other coastal recreation activities for thousands of residents and visitors. The complex of restaurants, high-rise condominiums, and shops has transformed the wetland site of former Native American villages into an urban coastal development. Few today are aware of the design issues and the continuing maintenance challenges associated with the facility. However, the economic costs of Marina Del Rey are far offset by the revenues produced by the development.

The beaches at Playa Del Rey, south of Ballona Creek, are part of Dockweiler State Beach. For the most part, the beaches are wide and high enough that there is minimal threat from erosion or storm flooding to residences and other development behind the beach. This is also an example where the protection is entirely due to human activity—namely beach nourishment. Beginning in the 1930s, plans were initiated to widen all of the beaches between Santa Monica and the Hyperion Sewage Treatment Plant in El Segundo (just to the south of Playa Del Rey) using sand excavated from the large relict dune field at El Segundo (Figure 19.8). This beach nourishment project would bring immediate relief to the coastal communities to the north that were experiencing damaging erosion and flooding, and indirect relief to communities to the south as some of the nourished sand was transported alongshore. In the 1940s, large-scale nourishment, involving 14 million cubic yards of sand, was begun. Most of this sand is still in place along this coastline, representing one of the nation's most successful beach nourishment projects (Figure 19.9). It has also provided Los Angeles with a set of extraordinary recreational beaches. Later construction projects in the dune field, such as the enlargement of the Hyperion Sewage Treatment Plant and the construction of the Scattergood Steam Generating Station, have led to additional sand being delivered to the beaches.

Small dunes are present along the backshore of much of Dockweiler State Beach, hinting at the aeolian activity that built the large El Segundo dune field. In Figures 19.8 and 19.9, the first signs of urban development can be seen on the larger dunes—residential streets and some houses. With the expansion of the runways at Los Angeles International Airport in the 1960s, the houses were removed, but the streets remain, and the Hyperion Sewage

Figure 19.8 The El Segundo dune field in 1931, view toward the north. Note the out-fall pipelines from the Hyperion sewage treatment facility and the naturally narrow beaches. Spence Collection. Reproduced by permission of the Department of Geography, University of California, Los Angeles.

Figure 19.9 The El Segundo dune field, ca. 1940, view toward the south. The pipeline for pumping nourishment sand is recognizable along the beach in the foreground, and beach width has increased substantially since 1931. In the foreground, large waves can be seen propagating up Ballona Creek. The Hyperion outfall and a refinery pipeline can be seen in the distance. Photo courtesy of Pat Merriam, Pacific Air Industries.

A. Breakwater at King Harbor. Alongshore sand transport is interrupted by the structure, causing beach erosion to the south. Harbor and pier facilities are damaged during severe storms.

B. Redondo State Beach and Torrance County Beach, both prone to erosion. This is the southern end of the Santa Monica Littoral cell, and a groin has been built here to reduce sand loss to the Redondo Canyon. Despite beach nourishment in 1969, severe erosion occurred in 1983.

C. North shore of the Palos Verdes Peninsula. Cliff heights increase rapidly toward the south. Moderate risk to houses from bluff erosion other than high risk associated with the landslide area at Bluff Cove. Localized cliff failure, with slow rates of retreat.

D. Development at the edge of the Palos Verdes Point is threatened by cliff failure.

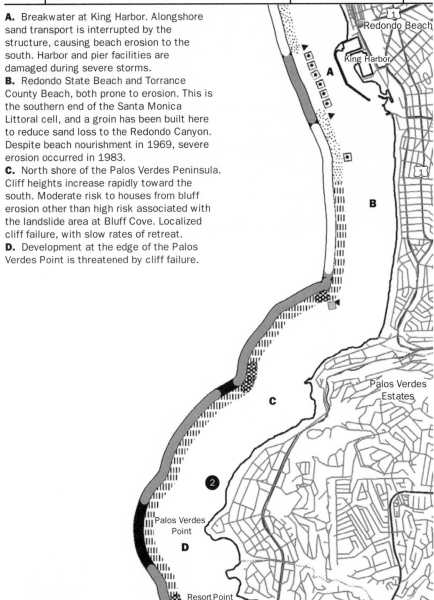

Hazard Level

☐ Stable: low risk
▨ Caution: moderate risk
■ Hazard: high risk

Erosion Rates

① Inches per year

Shoreline Environment

▨ Sandy beach
▨ Beach backed by marsh
▨ Beach backed by dunes
▥ Cliff or bluff fronted by beach
▨ Rocky coast with no beach

Armoring

▲▲▲ Riprap
■■■ Seawall
⊡⊡⊡ Other

Figure 19.10   The Hyperion sewage treatment facility in 2002. Note that the nourished beaches are still much wider than the natural state depicted in Figure 19.8. Photo
© 2002–2004 Kenneth and Gabrielle Adelman, California Coastal Records Project,
www.Californiacoastline.org.

Treatment facility has grown to occupy the entire width of the former dune
field (Figure 19.10).

In 1913 Standard Oil established a refinery and tanker-loading facility at
a site that was named El Segundo because it was the second refinery to be
built in California. The city that grew up around the site was incorporated
as El Segundo a Nada, meaning "second to none." As the Standard Oil property became threatened by shoreline erosion, a groin system was installed to
widen and stabilize the beach. During the severe 1982–83 winter storm season, extensive beach erosion again threatened the facility. The California
Coastal Commission permitted the construction of a 900-foot-long, permeable rock groin just down drift of the site, and allowed the up-drift beach
to be nourished with about 600,000 cubic yards of sand. This project
enhanced the protection of the oil facilities substantially (although some
risk remains). However, the project destroyed one of the area's premier surf
breaks. One condition of the permit was that Chevron fund the construction of an artificial surfing reef.

Almost two decades later, construction of Pratte's Reef began, the first
such project in the United States. Named after the co-founder of the
Surfrider Foundation, the reef was funded at an inadequate $300,000. The
reef was built with 110 geotextile bags that weighed approximately 14 tons

Figure 19.11 Passengers carrying luggage on a long walk to a waiting steamer at Redondo Beach, 1910. Note the narrow beach. Photo courtesy of the Redondo Beach *Daily Breeze.*

each when filled with sand. Depth across the crest of the reef was 6 feet at mean lower low water. Unfortunately, this meant that waves would rarely break over the structure. The first incarnation of Pratte's Reef was a failure. In April 2001 82 bags were added to the reef, increasing its size and reducing its depth to about 3 feet at mean lower low water. By the fall of 2002 some degradation of the bags and leakage of sand was observed. Waves do break across the reef, but the jury is still out on the success of the effort.

The southerly transport of nourishment sands has fed the shorelines of Manhattan Beach and Hermosa Beach. For the most part, these beaches are wide enough that small dunes have developed on the back beach, especially in Manhattan Beach State Park, where they have escaped the destructive cleaning activities that occur along many of the Los Angeles area beaches. Some of these beaches also have artificial dunes bulldozed up during the winter to reduce the frequency and magnitude of coastal flooding.

Redondo Beach is the only reach in this littoral cell that experiences chronic erosion, coastal flooding, and storm wave damage. Most, if not all, of these problems can be attributed to the construction of King Harbor. In 1890 the head of a submarine canyon—subsequently named Redondo Canyon—was discovered close to the shore. A pier was built out near the canyon so that ocean-going ships could dock (Figure 19.11). Indeed, there was substantial pressure to make Redondo Beach the primary port for the Los Angeles region, driven in part by the Santa Fe Railroad railhead at Redondo, poised to handle the freight traffic to and from Los Angeles proper and points east. The railroad also established passenger service to

(A)

(B)

Figure 19.12 *(Opposite and above)*   (A)–(C) Storm waves pound Redondo Beach in the 1950s. Photos courtesy of the Redondo Beach *Daily Breeze*.

Redondo and built a 225-room hotel to attract tourists. Although San Pedro became the regional port, Redondo Beach won out in the tourism department. It is claimed that surfing was introduced to California when George Feth was brought in from Hawaii to entertain the tourists.

Although the submarine canyon provided deep water near the shore, there was little in the way of protection against storm waves. In 1938 President Roosevelt approved $50,000 for the construction of a long, shore-perpendicular breakwater to provide some degree of protection for small craft. The impacts on local sediment supply were immediately apparent. According to commentary in the February 2, 1940, edition of the *Redondo Breeze*, "It was men against the sea at dawn today. . . . A mass of billowing ground swells rolled onto the sand and pounded against the underpinning of several homes. The water took huge chunks of sand from the foot of Second Place and Third Street. Damage became less as one neared the breakwater." Further storm damage to waterfront residences over the next two decades (Figure 19.12) prompted the city to acquire beachfront property and to form a redevelopment district focused on improving the harbor. King Harbor, named after a local congressman, was built by adding a dogleg to the existing breakwater to provide additional protection from waves (Figure 19.13); another segment was built from the shoreline to form the south side of the harbor entrance and to protect a turning basin and shopping mall.

Figure 19.13   Redondo Beach and King Harbor in 1947, view south toward the Palos Verdes Peninsula. The breakwater has disrupted alongshore sand transport, causing accelerated erosion north to the Municipal Pier. Spence Collection. Reproduced by permission of the Department of Geography, University of California, Los Angeles.

Although King Harbor, completed in 1959, added a much-needed anchorage for recreational boating (Figure 19.14) and provided a tourist attraction with waterfront restaurants, bars, and shops—some built over the water on the Redondo Beach Municipal Pier—it has not been an unqualified success. It has been the cause of substantial down-drift erosion. This erosion is partly a result of sediments being trapped along the northern margins of the breakwater but is also due to the deflection of much of the sand offshore and into Redondo Submarine Canyon, which acts as the dominant sediment sink for the Santa Monica Littoral Cell. The harbor itself has sustained repeated damage from storm waves that overtop and damage the breakwater, and damage boats and facilities within the harbor (Figure 19.15). Severe storms in 1962 and 1963, with waves approaching 30 feet in height, overtopped the breakwater. Projects were implemented to raise the height of sections of the breakwater. This site also suffered extensive damage during the El Niño winter of 1982–83, when the breakwater once again experienced substantial overtopping, causing millions of dollars of damage inside the harbor. The Portofino Inn, a major landmark in the harbor, was inundated by waves. Beaches to the south of the harbor experi-

King Harbor

Municipal Pier ➝

Figure 19.14   King Harbor and the Redondo Beach Municipal Pier in 1994. Photo courtesy of the U.S. Geological Survey.

enced extensive erosion. In January 1988, another storm caused extensive damage to facilities on the municipal pier as waves broke over the structure. A fire later that year almost put an end to the pier; instead it was rebuilt to continue the long tradition of tourism. South of Redondo Beach, the sandy beaches grade into the coastal cliffs of the Palos Verdes Peninsula.

PALOS VERDES PENINSULA

The Palos Verdes Peninsula is a large headland separating the Santa Monica and San Pedro littoral cells. Separated from the geologically downwarped Los Angeles Basin by the Palos Verdes Fault, the area has been uplifted to almost 1,500 feet about sea level. This uplift led to the for-

(A)

(B)

Figure 19.15   Storm waves overtop the King Harbor breakwater during severe storms in 1983. (A) The Portofino Inn is damaged by wave surge inside the breakwater. Photo courtesy of Suzanne Butler, Emergency Services Coordinator, City of Redondo Beach. (B) Damage to the breakwater. Photo by B. Pipkin.

Figure 19.15 *(Continued)* (C) Waves overtopping the breakwater, setting up surge within the harbor. Photo by B. Pipkin.

mation of the dramatic cliffs that characterize the coastline (Figure 19.16). The hills have the distinction of being part of the first Mexican land grant, the Rancho San Pedro, allotted in 1784. Although real estate development in the area began in the 1920s, growth was slow until about the middle of the twentieth century because of the distance from downtown Los Angeles (about 25 miles) and the steepness of slopes along the fault line. The terraced hillsides of the ocean sides of the peninsula made it a developer's dream. They also speak to the complex tectonic history of this site. Each of the 13 distinct terraces was formed at an ancient shoreline when sea level was high and wave erosion created an intertidal rock platform. A succession of relatively rapid uplift events raised the platforms and left the stairstep morphology (Figure 19.17).

Impressive cliffs, up to 200 feet high in places, and widespread slope instability are the result of rapid uplift. Landslides are common across the peninsula and occur along 53 percent of the shoreline between Abalone Cove and Point Fermin. Some form of downslope mass movement is apparent along 90 percent of this shoreline, mostly soil creep and rockfalls. Almost all of the slides are predictable based on geological investigation, and occur in areas where layers of sedimentary rocks tilt seaward out of the cliffs. Wave erosion at the base of the cliffs keeps them oversteepened, and thus unstable and prone to failure. Much of the movement occurs along beds of altered volcanic ash that becomes slippery when wet. Large landslides of this nature are found at Portuguese Bend, South Shores, and Point Fermin.

A. Moderate risk to bluff top development from localized landslides and rock falls. Numerous pocket beaches separated by rocky headlands. Slow cliff retreat rates.

B. Cliff heights decrease from 140 feet at Point Vincente to 40 feet at Long Point. Minimal cliff top development, and localized landslides. Boulder and cobble beaches common.

C. Cliff heights increase to about 160 feet in Abalone Cove. Extensive ancient and modern landslides from north of Portuguese Point to south of Inspiration Point. No development at the shoreline, inland structures threatened by the slides.

D. Sandy pocket beaches, localized landslides, minimal bluff top development south to Royal Palms Beach Park, where the Palos Verdes Shores Mobile Home Park sits on the South Shores landslide.

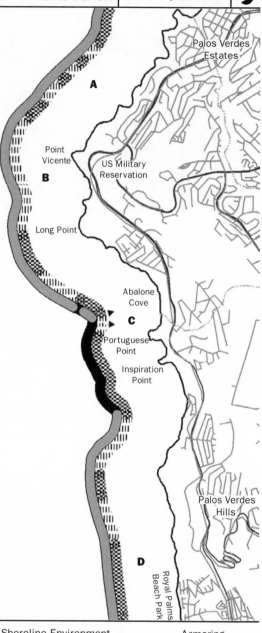

Palos Verdes Estates

Point Vicente

US Military Reservation

Long Point

Abalone Cove

Portuguese Point

Inspiration Point

Palos Verdes Hills

Royal Palms Beach Park

Hazard Level

☐ Stable: low risk
▨ Caution: moderate risk
■ Hazard: high risk

Erosion Rates

1 Inches per year

Shoreline Environment

▒ Sandy beach
🌊 Beach backed by marsh
░ Beach backed by dunes
‖‖‖ Cliff or bluff fronted by beach
▨ Rocky coast with no beach

Armoring

▲▲▲ Riprap
■■■ Seawall
⊡⊡⊡ Other

Figure 19.16   Cliff-top dwelling on Palos Verdes Point. Note the debris from many small landslides along the base of the cliff, and the pocket beach in Lunada Bay. Photo © 2002–2004 Kenneth and Gabrielle Adelman, California Coastal Records Project, www.Californiacoastline.org.

The largest and most destructive of these slides occurred at Portuguese Bend. The U.S. Geological Survey mapped an ancient landslide, covering about 1,000 acres, at that location. Developers and home buyers alike ignored the well-publicized report, with tragic consequences. Reactivation of downslope movement occurred in 1956 (Figure 19.18). What began as a series of cracks in driveways and streets ended three years later with the total destruction of 150 homes on 300 acres. Although a few of the houses survived the unusual journey, about $68 million in property damage (in 2005 dollars) occurred. In the adjacent Abalone Cove, another part of the ancient landslide was reactivated in 1976. Again, cracks in a roadway signaled the beginning of movement. In this case, the failure involved about 1,200 feet of shoreline and a total of about 80 acres. Moving with the slide were 25 houses, a county beach facility, and the famous Wayfarer's Chapel. Initiation of the movements has been attributed to destabilization caused by a road cut across the slope, and to the effects of water draining from septic tanks, cesspools, and landscape irrigation. Wells have been drilled to lower the groundwater table in an attempt to stabilize the slide. The reactivation of the ancient slide is seen as an inevitable consequence of urbanization of the

Figure 19.17   The Lighthouse at Point Vicente is built on a marine terrace. Another terrace is in the upper left-hand corner of the image. Cobble and boulder beaches are common along this shore. Photo © 2002–2004 Kenneth and Gabrielle Adelman, California Coastal Records Project, www.Californiacoastline.org.

hillsides. Parts of the slide continue to move, requiring frequent repairs to roads and utility lines.

The landslide at South Shores is at least 16,000 years old, involves about 3,000 feet of the shoreline, and extends inland 4,000 feet. The slide has similar geological characteristics to those at Portuguese Bend, but real estate development here has been better planned. After extensive geological analysis, it was decided that the site was appropriate for development as a mobile home park, with the assumption that mobile homes could be relocated in the event the slide were to become active again. No permanent dwellings were to be built on the site, although a substantial recreation center was constructed. However, if the slide were reactivated with rapid movement, major property loss might still result because the park has a high density of high-end mobile homes. There is also urbanization, including multi-story condominiums, right up to the southern edge of the ancient slide, where some degree of risk still remains.

Slide movement has occurred at Point Fermin since 1929, destroying houses and sections of roadway. Total downslope movement has exceeded 200 feet and continues, especially during years with heavy rain. The geology

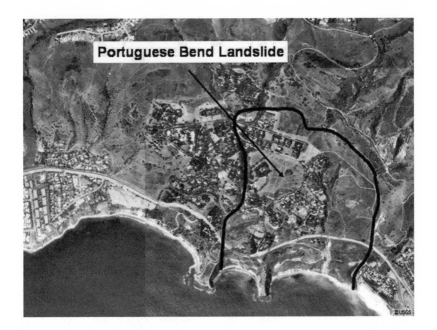

Figure 19.18   The landslide at Portuguese Bend was reactivated in 1956, with disastrous consequences for homeowners. Photo courtesy of the U.S. Geological Survey.

of the area, and the forces driving the slide, are similar to those at other failure zones. This is a relatively small slide, but it has caused the area to be known locally as Sunken City because of the remnants of streets, foundations, and trees marking the locations of destroyed development (Figure 19.19).

Because marine erosion around the base of the cliffs of the Palos Verdes Peninsula continues, and because material delivered to the shoreline by the landslides is removed about as quickly as it arrives, the cliffs will continue to be oversteepened and thus unstable. The litany of failure-prone cliffs and slopes should be warning enough against building too close to the cliff edge. Although average cliff retreat rates for the Palos Verdes Peninsula are about 2 inches per year—relatively slow—local rates associated with landslides can be much faster. Most recent development has taken the dangers of slope failure into account. Parks and golf courses are now common along the perimeter of the southern edge of the peninsula, in particular.

THE SAN PEDRO LITTORAL CELL

The San Pedro Littoral Cell contains the longest barrier beach and estuary system in southern California. The original barrier islands, where the Los Angeles and Long Beach ports are now, have been modified extensively or

A. Riprap protects the beach front parking lot at White's Point Nature Preserve.

B. Several houses built along the base of, or into, the cliff, at risk from either erosion or landsliding. Development along the edge of the cliff top is also threatened by cliff failure. Some property loss in this reach.

C. Localized landslides and some Point Fermin buildings jeopardized by cliff retreat.

D. Point Fermin landslide, locally referred to as "Sunken City." Cliffs are about 100 feet high here.

E. Cabrillo Beach Park, a sandy recreation beach stabilized by a groin and the San Pedro Breakwater.

F. San Pedro breakwater, built between 1899 and 1912, has withstood numerous storms. In February and March 1983, waves overtopped the structure, displacing 10 to 20 ton granite blocks, and causing $7.2 million damage (in 2004 dollars).

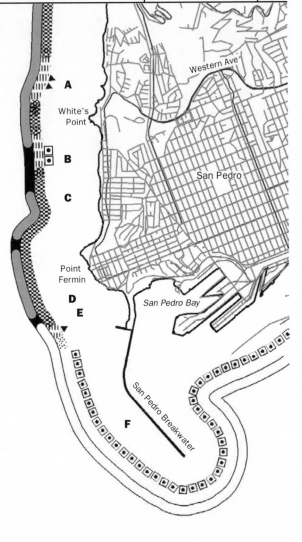

Hazard Level
☐ Stable: low risk
▨ Caution: moderate risk
■ Hazard: high risk

Erosion Rates
 Inches per year

Shoreline Environment
Sandy beach
 Beach backed by marsh
Beach backed by dunes
 Cliff or bluff fronted by beach
Rocky coast with no beach

Armoring
▲▲▲ Riprap
 Seawall
 Other

Figure 19.19   The 1929 slope failure at "Sunken City" caused the destruction of a residential neighborhood. Photo © 2002–2004 Kenneth and Gabrielle Adelman, California Coastal Records Project, www.Californiacoastline.org.

eliminated by development. All of the modern barriers are spits, and all of the estuaries have undergone substantial development for marinas or harbors. The once-extensive marsh systems have been largely eliminated by commercial, industrial, and residential construction. Blue waters, white sands, fresh air, and separation from urban congestion have made the real estate along this coastline some of the most valuable in the state. However, hand in hand with development has come a suite of coastal hazards associated with beach erosion and storm flooding. In 1962 Congress authorized the U.S. Army Corps of Engineers to initiate the San Gabriel River to Newport Bay (Surfside-Sunset and Newport Beach) project to study the 17 miles of coast between these two points and to develop and implement strategies to reduce or eliminate the threats from erosion and flooding.

Under natural conditions, the San Pedro Littoral Cell began at the southern edge of the Palos Verdes Peninsula. Because of the major environmental changes associated with port development, the present northern end of the cell has been displaced about 3 miles to the east, at the mouth of the Los Angeles River. The cell's southern terminus is at the Newport Submarine Canyon. The three largest coastal rivers in southern California, the Los Angeles, San Gabriel, and Santa Ana, empty into the San Pedro Cell. Although their drainage systems have been altered greatly by human activities, and their sediment loads reduced substantially as a result, the rivers remain the leading sources of sediments for the area's beaches.

The Los Angeles River is one of the most heavily altered fluvial systems in the United States. Dams, debris basins, and extensive channelization have

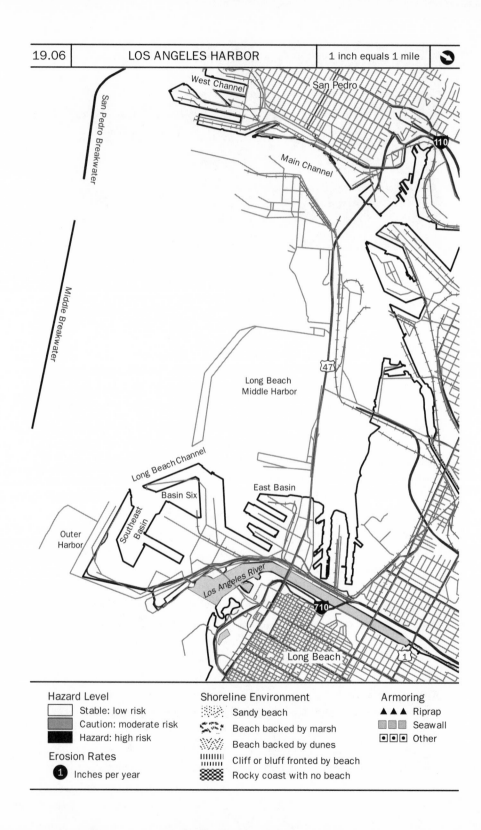

West Channel

San Pedro

San Pedro Breakwater

Main Channel

110

Middle Breakwater

47

Long Beach
Middle Harbor

Long Beach Channel

Basin Six

East Basin

Southeast Basin

Outer
Harbor

Los Angeles River

710

Long Beach

1

Hazard Level

Stable: low risk

Caution: moderate risk

Hazard: high risk

Erosion Rates

1 Inches per year

Shoreline Environment

Sandy beach

Beach backed by marsh

Beach backed by dunes

Cliff or bluff fronted by beach

Rocky coast with no beach

Armoring

▲▲▲ Riprap

Seawall

Other

A. Wide, sandy beach from extensive beach nourishment, protected by the Long Beach Breakwater.

B. Some risk of coastal flooding of the Belmont Plaza Swimming Pool, built directly on the beach.

C. Narrow sand beach along the Alamitos Peninsula (barrier spit). Houses and infrastructure threatened by severe storms. Maintenance nourishment by moving sand from the Alamitos Bay jetty.

D. San Gabriel River mouth. Primary source of beach sands along this reach. Wide sandy beaches on the south side of the jetty. Shelter from Long Beach Breakwater ends here.

E. Severe erosion and flooding hazard from the Seal Beach Pier to Surfside. Beaches are narrow, with some beach nourishment used to build high berms. Wider beaches occur on the south side of the Anaheim Bay jetties, partly a result of nourishment.

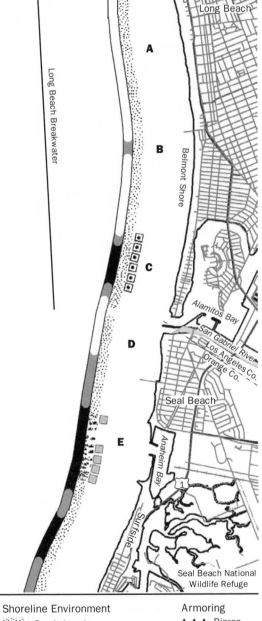

Hazard Level
- ☐ Stable: low risk
- ▨ Caution: moderate risk
- ■ Hazard: high risk

Erosion Rates
- ① Inches per year

Shoreline Environment
- Sandy beach
- Beach backed by marsh
- Beach backed by dunes
- Cliff or bluff fronted by beach
- Rocky coast with no beach

Armoring
- ▲▲▲ Riprap
- Seawall
- ⦿⦿⦿ Other

Figure 19.20 The jetties and breakwaters of the northern reaches of the San Pedro
Littoral Cell have pronounced impacts on local sediment budgets. Photo courtesy of the
U.S. Geological Survey.

reduced the transport of sand to the coast to a negligible volume. The San
Gabriel River, still carrying some sand to the coast during floods, meets the
Pacific Ocean about 4 miles to the east, at Seal Beach. The Long Beach City
Beach lies between the two river mouths. Most of this beach is protected
against coastal erosion and flooding by the breakwaters built for the ports.
However, along the Alamitos Peninsula, parts of the beach are not sheltered by
the breakwaters. In particular, there is about a half-mile of eroding beach, in
the area known as Peninsula Beach, stretching north from the jetties at the
entrance to Alamitos Bay (Figure 19.20). The natural source of beach sand at
this location is the San Gabriel River. However, the jetty reduces that sediment
supply, so that Peninsula Beach is eroding, with commensurate accretion
about a mile to the north. About 100 homes are threatened by shoreline erosion in this area. To combat this problem, the City of Long Beach operates a
sand back-passing program that moves about 100,000 cubic yards of sand per
year from the accretion zone back to Peninsula Beach. Other remedial actions
are being contemplated to eliminate or reduce the necessity of sand back-passing.

Alamitos Bay used to be the outlet for the San Gabriel River. The bay has
been altered extensively for residential and marina development, including

A. Narrow beach with threat of erosion and flooding. Some small sand dunes in front of houses.

B. Moderate risk from erosion or overwash to recreational facilities on the low barrier beach that protects the Bolsa Chica Ecological Reserve.

C. Poor quality, degrading riprap protects bluff-top parking lot. Sandy beach is narrow and low.

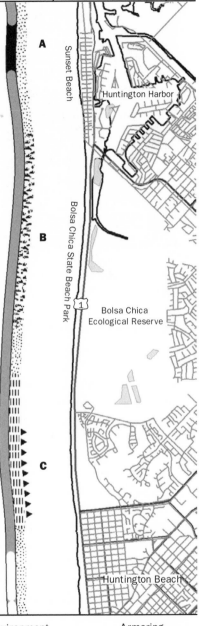

Sunset Beach

Huntington Harbor

Bolsa Chica State Beach Park

Bolsa Chica Ecological Reserve

Huntington Beach

Hazard Level

☐ Stable: low risk
▨ Caution: moderate risk
■ Hazard: high risk

Erosion Rates

 Inches per year

Shoreline Environment

Sandy beach
Beach backed by marsh
Beach backed by dunes
Cliff or bluff fronted by beach
Rocky coast with no beach

Armoring

▲▲▲ Riprap
☐☐☐ Seawall
◉◉◉ Other

Figure 19.21 The construction of Huntington Harbor transformed wetlands into marina and residential developments. The Seal Beach National Wildlife Refuge protects remnants of the wetlands. Photo © 2002–2004 Kenneth and Gabrielle Adelman, California Coastal Records Project, www.Californiacoastline.org.

the creation of the island community of Naples, built largely on sediment dredged from the surrounding wetlands. After severe flooding in 1938, the river was rerouted to the south to protect the development within the bay (Figure 19.20). There is a set of three jetties protecting the bay's entrance and the river's mouth. The jetties have profoundly altered the sediment transport patterns, and the consequent erosion has required intervention. In the city of Seal Beach, between the San Gabriel River and Anaheim Bay, about 250,000 cubic yards of sand were used to nourish the otherwise low, narrow beach. In addition, because of chronic overwash and flooding, a sand berm (artificial dune) has been built every winter since the 1960s. The berm does not prevent all flooding—overtopping events occurred in 1983 and 1997, for example—but dramatically reduces its frequency.

The entrance to Anaheim Bay is protected by a pair of jetties that also provide anchorage. The harbor is the site of the 5,256-acre Seal Beach Naval Weapons Station, established in 1944. About one-fifth of the site has been set aside as the Seal Beach National Wildlife Refuge. Further inland, up one branch of Anaheim bay, is the large marina and residential development of Huntington Harbor, constructed in 1962. The development of the Naval Weapons Station and Huntington Harbor resulted in the loss of wetland

and marsh environments (Figure 19.21). Anaheim Bay and Huntington Harbor are behind a northwest-pointing barrier spit that supports the communities of Surfside and Sunset Beach. Both locales have experienced severe beach erosion, usually attributed to the jetties and harbor works at Anaheim and Alamitos bays. Some shore protection structures have been installed along this reach, but beach nourishment has been the preferred means of mitigating beach losses. Since 1945, about 20 million cubic yards of sand has been placed on these beaches in an effort to maintain their integrity. The real culprit in this fight against erosion has been inappropriate development of the spit. Beach erosion has been a problem here since the earliest developments, and nothing has been done to address the problem. Figure 19.2 depicts the area in 1935, well before the extensive development of Anaheim Bay. However, an underlying rationale for the continuing large-scale nourishment of these beaches is that the alongshore transport regime feeds nourishment sands to the beaches to the south, and helps prevent erosion hazards in Huntington Beach and Newport Beach.

Bolsa Chica (Little Pocket) is an estuary whose natural inlet has closed, and it is now artificially linked to Huntington Harbor. Much of the lagoon has been set aside by the state of California as the Bolsa Chica Ecological Reserve. Other parts of the estuary have been developed extensively, and there are contentious plans for additional development. The physical and environmental problems connected with the project are legion and range from increasing tidal currents throughout the lagoon and erosion of its banks to blocking littoral drift at a proposed restored ocean entrance. The barrier beach at Bolsa Chica Beach State Park is narrow and low lying, and vulnerable to overtopping during storms. This was the case in 1983, when there was flooding of the park, the Pacific Coast Highway (U.S. 101), and oil production facilities, and almost 500 homes experienced some degree of inundation-related damage.

Most of the city of Huntington Beach, also known as "Surf City, USA," is built on a low Pleistocene bluff—the only stretch of cliffed coast in the littoral cell (Figure 19.22). Because of the transport of nourishment sediments from the north, most of the beaches are wide and high, and are a prime recreation resource. Where the beach is narrow, bluff retreat is a concern. The beaches are managed as Huntington City Beach and Huntington State Beach. Most development has been on top of the bluff, and the primary erosion hazard is the result of some construction too close to the ocean, mainly recreation support facilities and a residential development just north of the Huntington Beach Pier. Long-term retreat of the bluffs has caused some infrastructure damage and may eventually threaten the Pacific Coast Highway.

A. Wide sandy beaches from nourishment and Santa Ana River.

B. Santa Ana River mouth. Largest sediment source for the San Pedro Littoral Cell under present conditions. The channel is artificial.

C. Severe erosion and inundation hazard to houses built on the beach. Narrow, low beach, stabilized with rock groins and periodic beach nourishment.

D. Narrow beach with moderate risk to recreation support facilities built on the beach next to the Newport Beach Pier.

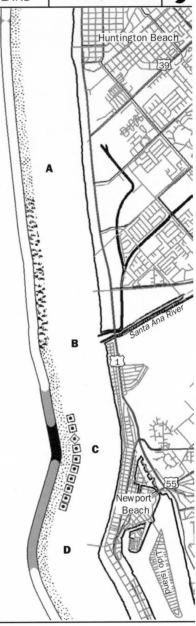

Hazard Level
- Stable: low risk
- Caution: moderate risk
- Hazard: high risk

Erosion Rates
-  Inches per year

Shoreline Environment
- Sandy beach
- Beach backed by marsh
- Beach backed by dunes
- Cliff or bluff fronted by beach
- Rocky coast with no beach

Armoring
- ▲ ▲ ▲ Riprap
- Seawall
- Other

Figure 19.22   Huntington Beach. (A) The Huntington Beach Pier and beaches in 1931. Note the narrow beach, small dunes, and low bluffs. Spence Collection. Reproduced by permission of the Department of Geography, University of California, Los Angeles. (B) The modern beach, high and wide as a result of nourishment. Dunes are gone and development is encroaching on the beach wetlands. Photo © 2002–2004 Kenneth and Gabrielle Adelman, California Coastal Records Project, www.Californiacoastline.org.

The mouth of the Santa Ana River is at the southern end of Huntington State Beach and is controlled by a pair of short jetties. There is no inlet to protect here; this course of the river is artificial, implemented in 1929 to divert the flow from its natural drainage into Newport Bay. Severe flooding in 1938 (Figure 19.23) inundated large tracts of

Figure 19.23   1938 Santa Ana River flooding. (A) Levees were breached in numerous locations, causing widespread flooding. Courtesy of the Fairchild Aerial Photography Collection, Whittier College. (B) The low-lying barrier south of Huntington Beach was overwashed by the flood. Courtesy of the Fairchild Aerial Photography Collection, Whittier College.

Figure 19.23 *(Continued)* (C) Modern development on the floodplain wetlands. Photo © 2002–2004 Kenneth and Gabrielle Adelman, California Coastal Records Project, www.Californiacoastline.org.

coastal marsh between Huntington Beach and Newport Beach, most of which were subsequently developed. Most of the coast south of the river mouth to the Newport Bay inlet is a highly developed barrier spit. Newport Beach is seriously affected by large waves generated by Antarctic storms and tropical storms off Baja California because the coast is south facing and lacks protection from offshore islands. Erosion is often most severe in the summer. In the summer of 1965, for example, the beach eroded 165 feet before being stabilized by sandbags just 5 feet from property lines. Similarly severe erosion occurred in 1968. The greatest threat has been at West Newport Beach, and this area has been subject to numerous beach protection and restoration efforts, mainly as a result of Corps of Engineers regional projects. Rubble mound and sheet pile groins were installed in the 1960s and 1970s, and several nourishment projects have added more than 1.5 million cubic yards of sand to this beach. In 1991, another 1.3 million cubic yards of sand, dredged from the Santa Ana River, were used to build an offshore mound. Some of the sediment from the mound migrated onshore to widen this beach. This area is still under consideration for further remedial actions. Again, the low barrier spit offers little natural protection against the nearshore processes

Figure 19.24   Development on Balboa Island. Compare this to conditions in 1923 (Figure 19.4), when the island was a mudflat wetland. Photo © 2002–2004 Kenneth and Gabrielle Adelman, California Coastal Records Project, www.Californiacoastline.org.

that created it. Within Newport Bay, residential development on former wetlands created some of the most valuable real estate in southern California (Figure 19.24).

THE LAGUNA BEACH MINI CELLS

From Corona Del Mar to Dana Point the coast consists of a series of head-lands, cliffs, and pocket beaches. This reach has been termed the "Laguna Beach Mini Cells" to reflect the presence of many small, confined, littoral systems (Figure 19.25). Erosion rates are minimal because of the resistance of the cliff rocks. There are, however, coastal hazards at several locations resulting from development too close to cliff tops, too close to the mean high water line, or in an area prone to landslides.

Corona Del Mar (Crown of the Sea) is across the inlet from Newport Beach. Corona Del Mar State Beach lies next to the jetty; however, south of the beach the coastal cliffs become prominent. In the 1920s the area sported the largest surf club in the United States, attracted by the quality of the surf at "Killer Corona," and among the membership were some of the best surfers in the world. Surfing took a hit when the Newport jetties were extended and the present beach formed. Occasionally, a large southwest swell will still provide an excellent break along the southern jetty.

A. Wide, sandy beach, partly a result of beach nourishment, south to the Newport Bay jetty (location of "the wedge"). Recreational facilities on the beach at the Balboa pier at moderate risk during storms.

B. 60 to 80 foot coastal cliffs. Some threat from localized landslides and cliff retreat to houses built into or at the edge of the cliffs. Sandy pocket beaches between rocky headlands.

C. Beach front development at Crystal Cove threatened by severe storms.

D. Low, narrow beach. Beach front mobile park at El Moro Beach threatened by severe storms. Some riprap and bulkheads installed to protect highway and mobile homes.

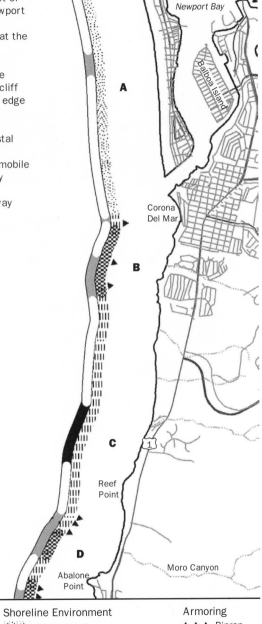

Hazard Level
- Stable: low risk
- Caution: moderate risk
- Hazard: high risk

Erosion Rates
- **1** Inches per year

Shoreline Environment
- Sandy beach
- Beach backed by marsh
- Beach backed by dunes
- Cliff or bluff fronted by beach
- Rocky coast with no beach

Armoring
- ▲▲▲ Riprap
- Seawall
- Other

Figure 19.25   The Laguna Beach Mini Cells. (A) Bluff-top development at Abalone Point, with pocket beaches such as Irvine Cove in the background. (B) Residential development in Laguna Beach spills down the coastal bluff to the water's edge. Photos © 2002–2004 Kenneth and Gabrielle Adelman, California Coastal Records Project, www.Californiacoastline.org.

This area is also where the San Joaquin Hills abut the Pacific Ocean to create the rocky shoreline that extends to Dana Point. Compared with the cliffed Palos Verdes Peninsula, this coast has lower cliffs and is more irregular, with many more pocket beaches. The irregular coastline is due in part to the composition of the San Joaquin Hills and in part to the different geological structure, as this is part of the central block of the Los Angeles Basin.

The coast in the vicinity of Arch Rock is scalloped in a series of small pocket beaches cut into cliffs about 60 feet high. Some houses are built directly into the cliff faces and are exposed to a degree of hazard from landsliding. To the south, the coast becomes relatively straight as layered shales outcrop in cliffs that are fronted by some of the least disturbed sandy beaches in the region. The beaches protect the cliffs, minimizing erosion hazard, and there is minimal urban development on the marine terrace above the cliffs. This is a remnant of the 77,000-acre Irvine Ranch, which was put together by James Irvine from Mexican land grants. Continued development is planned, but at present the coast in the vicinity of Pelican Point probably still looks much as it did when the first European explorers arrived. Part of this reach has been set aside as Crystal Cove State Park.

A small community was built as a resort at Crystal Cove in the 1920s. Although the beachfront cottages have survived decades of storms and erosion, they remain perilously close to the water and are considered to be at risk. The community is on the National Register of Historic Places and exemplifies southern California coastal development through the first half of the twentieth century.

From Abalone Point to Laguna Beach bays separate a series of rugged points, formed from resistant volcanic rocks, and coves cut into softer shales. Beautiful examples can be seen at Emerald Bay and Crescent Bay, in the northern Laguna Beach area. The erosion hazard in this area is mainly the result of extremely expensive houses that have been built right to the edges of coastal cliffs that are about 100 feet high, so that almost any degree of cliff retreat will cause substantial property damage. Secluded, sandy beaches are present in the coves, although access is extremely limited because of the bluff-top development and the ruggedness of the coast. Most of these beaches appear to be relatively stable through time, as they are "trapped" between sets of headlands.

Laguna Beach was once a relatively secluded artists colony. Although it is still an art center, it has also become an enclave for the wealthy of Orange County. The relatively large main beach at Laguna Beach, a recreational focus for the area, is also stabilized between headlands. At the southern end of the beach and continuing to Cactus Point, riprap has been installed to protect the toe of the coastal bluffs from wave attack. The hazard is exaggerated here because of the high property values and because some of the houses are built almost to the water's edge.

Between Laguna Beach and Dana Point the coastal cliffs consist mainly of sandstones of different degrees of resistance to wave erosion. The risk along this part of the coast is similar to that near Laguna Beach. Many expensive dwellings are built right to the edge of the cliff, where any retreat

A. Narrow beach in Emerald Cove. Some threat to houses, roads, and recreation facilities.

B. Development at edge of cliff top threatened by cliff failure.

C. Moderate risk to cliff development in coves. Small slope failures and localized protection structures common.

D. Low, narrow beach with houses built into the cliff and down to the beach, revetments and riprap offer localized protection against severe storms. Almost no dry beach at high tide.

E. 40 to 60 foot cliffs with some structures threatened by localized landslides. Some houses built down to small pocket beaches at risk from storm erosion and innundation.

F. Low, narrow beach with houses built into the cliff and down to the beach level. A beachfront resort, protected by riprap, extends seaward beyond the high water line. Substantial risk from severe storms.

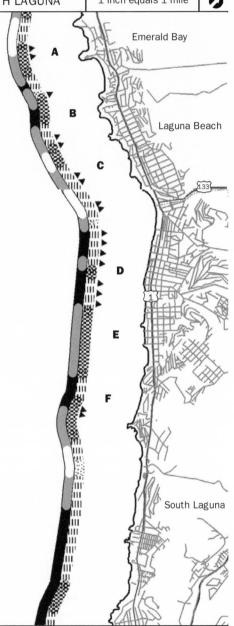

Emerald Bay

Laguna Beach

133

1

South Laguna

Hazard Level
- ☐ Stable: low risk
- ☐ Caution: moderate risk
- ■ Hazard: high risk

Erosion Rates

 Inches per year

Shoreline Environment
- Sandy beach
- Beach backed by marsh
- Beach backed by dunes
- Cliff or bluff fronted by beach
- Rocky coast with no beach

Armoring
- ▲▲▲ Riprap
- ☐☐☐ Seawall
- ◉◉◉ Other

Figure 19.26   Dana Point, with the harbor in the middle distance, and Doheny State Beach in the background. Note recent slope failures (outlined) below residences. Photo © 2002–2004 Kenneth and Gabrielle Adelman, California Coastal Records Project, www.Californiacoastline.org.

will result in property damage. In some locations where there are narrow beaches, such as the south end of Victoria Beach, houses or hotels have been built right down to the beach. The only location where there is minimal risk from erosion or coastal flooding is in the vicinity of Aliso Creek, where the stream delivers enough sand to maintain a beach between the headlands. Some minor threat to parking lots built behind the beach at the county park exists. In other locations where narrow beaches afford a degree of protection to the cliffs, development has still eliminated any margin of safety by building close to the cliff top.

Just north of Dana Point, about a mile of riprap has been emplaced at the base of the cliffs to provide stabilization. Continuing real estate development behind this stretch of coast will raise the degree of risk, although some building has taken place at what appear to be prudent distances landward of beaches and cliff edges. Nevertheless, slope failure remains a threat along this coast (Figure 19.26). There is a controversial project planned for the old

A. Development along pocket beaches and small headlands threatened by severe storms or local cliff failure. Many small structures and one large residential unit built on beaches. Many houses built into the sides of cliffs.

B. Houses at edge of 80 to 100 foot cliff top. Localized landslides. Ongoing construction into the cliff at Mussel Cove.

C. Some development set back from cliff, but many structures built into the cliff or close to the edge of the top. Toe of the cliff at Salt Creek Beach is protected by riprap. Very narrow beach at normal high tide.

D. Dana Point Headland. New development planned along the beach just north of the headland, including additional riprap. Small landslides on Dana Point. Ground water seepage threatens cliff top development along the coastline protected by the Dana Point Harbor Breakwater.

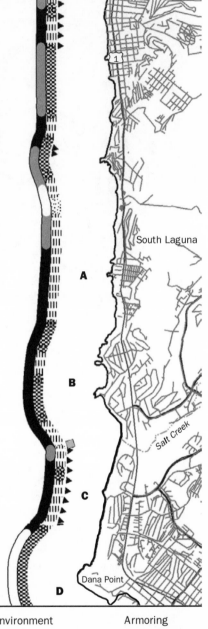

Hazard Level
- ☐ Stable: low risk
- ▨ Caution: moderate risk
- ■ Hazard: high risk

Erosion Rates
- ① Inches per year

Shoreline Environment
- ▨ Sandy beach
- ⛰ Beach backed by marsh
- ▨ Beach backed by dunes
- ⦀ Cliff or bluff fronted by beach
- ▨ Rocky coast with no beach

Armoring
- ▲▲▲ Riprap
- ■■■ Seawall
- ◉◉◉ Other

trailer park in this area. Developers want to install additional shore protection to protect the bluff, as more building is being planned.

Dana Point itself is a relatively undeveloped promontory. However, the construction of the small craft harbor (Dana Point Harbor) in the lee of the point has been one factor driving extensive urbanization in the area. Cliff erosion is active here, but represents a threat to only a few structures.

CHAPTER TWENTY

DANA POINT TO THE INTERNATIONAL BORDER

REINHARD FLICK

INTRODUCTION

The coast and beaches are the San Diego region's most important natural assets. When we think of the region's positive image, we most often think of the climate and the shoreline. Beaches are by far San Diego's largest attraction, and we depend on the sand beaches to buffer homes, businesses, and public improvements from impact and inundation by ocean waves.

Most of the San Diego shoreline consists of narrow beaches backed by steep seacliffs. The beaches and cliffs have for thousands of years been subject to erosion from waves abetted by a rising sea level. During the stormy period of 1884 to 1893, the U.S. Coast and Geodetic Survey noted, "New erosion during each winter storm is the characteristic feature of this coast."

Beach and cliff erosion have become a vexing problem in the San Diego region, particularly in places that have been densely developed since the end of World War II. In light of better understanding, much coastal development now appears to have been less than prudent, especially in the long term. This is true of both public and private improvements that were built too low, too close to the beach, or with insufficient setback from the cliff edge.

Awareness of local erosion problems increased after severe storms and flooding occurred during the winters of 1978, 1980, 1982–83, 1988, and 1997–98. The El Niño winter of 1982–83 was notable for the consecutive or "cluster" storms that produced heavy rainfall, elevated sea levels, and high, long-period waves. These events produced hundreds of millions of dollars in damage in California, with a proportional share in San Diego.

The most important factor that characterizes the San Diego coast is the regional geology, especially the tectonic history. The next most important factors are the relevant coastal processes, those physical mechanisms that alter

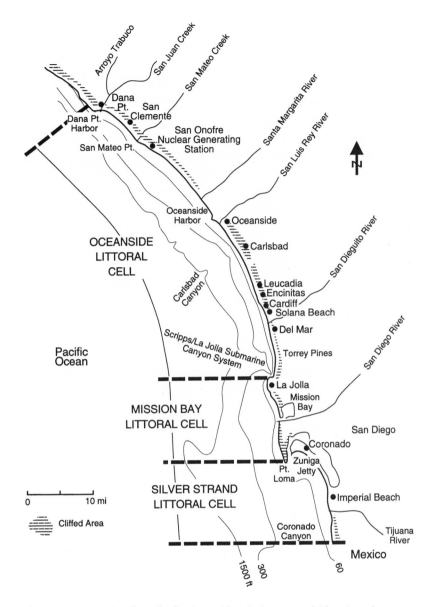

Figure 20.1   San Diego littoral cells: Oceanside, Mission Bay, and Silver Strand.

the shoreline and add or remove sand. The most important of these are waves. Tides, sea-level changes, weather, and climate also play a role, making waves more or less severe and modulating the amount of sand reaching the beaches.

The San Diego region can be divided into three littoral cells, or beach compartments. The littoral cell concept brings together the important elements of geography, nearshore processes, sand supply, and wave forces. From north to south, these compartments are the Oceanside Cell, the Mission Beach group of subcells, and the Silver Strand Cell (Figure 20.1). The Oceanside Cell extends about 50 miles from Dana Point to La Jolla Shores, where it terminates at the Scripps–La Jolla Submarine Canyon. The rocky headlands of La Jolla form the boundary between the Oceanside Cell and the Mission Beach group of subcells, which include the Mission Bay entrance and the adjacent San Diego River mouth. The Silver Strand Cell extends south from Zuniga Jetty, at the entrance to San Diego Bay, to a headland about 3 miles south of the international border.

Human actions have altered the San Diego coast in many ways, as they have the rest of southern California. Whereas some projects have slowed erosion and widened beaches, others have increased erosion and narrowed beaches. Several human activities stand out, such as extensive public and private construction, including roads, bridges, the railroad, power plants, harbors, jetties, houses, and seawalls. Beach sand supply has been reduced due to the construction of flood control and water storage dams, and beach sand supply has increased both from by-products of some coastal developments and from dedicated beach nourishment projects.

REGIONAL GEOLOGICAL SETTING

San Diego, and the rest of southern California, is geologically young and seismically active. The San Andreas Fault zone forms a conspicuous fracture at the boundary between the North American Plate and the Pacific Plate. Until about 20 million years ago, the two plates were colliding, with the oceanic plate dipping under the continent, causing tectonic uplift. Although most of the plate movement occurs along the San Andreas Fault far to the east of San Diego, a considerable portion also occurs on the San Jacinto, Elsinore, Rose Canyon, Coronado Bank, San Diego Trough, and San Clemente fault systems. These faults form a series of earthquake zones that stretch across the width of southern California and into the ocean areas offshore. The complicated horizontal and vertical motions of the blocks of land surrounded by these faults, together with long-term sea-level changes, define the outline of the region that we see today.

The same tectonic movement and erosion are responsible for many of the features that make the San Diego coast so geographically attractive. The up-and-down topography of the San Diego urban shoreline was formed as a result of bends in the Rose Canyon Fault zone and the alternate raising and lowering of blocks as they moved past each other. From north to south, these lows and highs include the La Jolla Submarine Canyon, Point La Jolla and Mount Soledad, Mission Bay, Point Loma, San Diego Bay and Silver Strand, and the hills of Tijuana.

Figure 20.2   Rocky low-tide terrace and cliff structure in Solana Beach, ca. 1997.
Photo courtesy of Terra Costa Consulting Group.

MARINE TERRACES

Marine terraces and cliffs comprise the most prominent features of the region's coast. Raised terraces form a series of broad, easily developed flat areas in San Diego, locally known as mesas. These terraces were formed by wave erosion during successive prolonged periods of high sea level. The platforms were subsequently uplifted by tectonic activity, resulting in the series of raised mesas observed today.

The marine terraces near the shoreline include the submerged platform near low-tide level, which is being cut by wave action at the present time (Figure 20.2). This low-tide terrace started forming about 6,000 years ago, during the most recent relative stillstand of sea level. It comprises the flat, rocky, shallow part of the shore face common along much of San Diego and often visible during low tide. These areas are often hazardous for beachgoers, but they offer excellent hard-bottom habitat for a diverse variety of tide pool organisms. A veneer of sand over the low-tide terrace forms most of the beaches in the area.

COASTAL CLIFFS

The steep coastal cliffs that back most of the San Diego shoreline are simply the seaward edges of the eroding marine terraces. The cliffs are composed of a wide variety of sedimentary rocks. The compositions vary from mudstones and shale to sandstones that contain boulders, cobbles, and

Figure 20.3   Point La Jolla, showing resistant bedrock and low marine terrace. The high-rise building in the background was one of the developments that led to voter approval of the California Coastal Act and eventually to the establishment of the California Coastal Commission to regulate coastal development. Photo © 2002–2004 Kenneth and Gabrielle Adelman, California Coastal Records Project, www.Californiacoastline.org.

shells. Depending on the local tilt and uplift patterns, sedimentary rocks of different hardness and grain size are exposed and subject to erosion. At most locations, the lowest portions of the cliffs are up to 60 million years old (Eocene) and are more resistant to erosion than the overlying, younger (up to 120,000 years old) Pleistocene terrace materials. From La Jolla to Point Loma there is relatively hard, 90-million-year-old Cretaceous bedrock exposed at sea level and offshore. This material is more resistant to erosion than the Eocene rocks found up and down coast, and accounts for the presence of the headlands at these two locations (Figure 20.3).

The cliffs are an aesthetically pleasing element of the San Diego landscape, setting off the coastline from the inland areas. They offer commanding ocean views while providing isolation and solitude for the beach user. Most San Diego area cliffs are not made of especially resistant rock, and this makes them vulnerable to episodic marine and subaerial erosion. The steep coastal cliffs are geologically unstable because most of them consist of weak sedimentary rocks. The cliffs are also heavily faulted and cracked. These breaks and joints are weak and easily eroded by wave action, forming caves and arches that periodically collapse, causing the upper cliff to fail. It has

Figure 20.4   Encinitas, showing cliff-top residential development and localized land-sliding and erosion. Photo © 2002–2004 Kenneth and Gabrielle Adelman, California Coastal Records Project, www.Californiacoastline.org.

become increasingly clear that there is a price associated with the development of these scenic areas.

Cliff erosion occurs over a large time scale and is usually associated with severe winter storms, high rainfall, high tides, and elevated sea level. Erosion of the cliffs is highly site specific and episodic, and areas have been documented where several feet of retreat occurred in a few days at one part of a property, with no retreat at all 50 or 100 feet away (Figure 20.4).

BEACHES AND COASTAL PROCESSES

Coastal or nearshore processes are those physical mechanisms that alter the shoreline and supply, remove, or shift sand. Coastal processes are responsible for the details of the coastline that we see, and for both the rapid changes and the slower, imperceptible evolution that modifies the shoreline over time. Wave action provides almost all the energy that drives local shoreline processes. Waves generated in the Pacific Ocean erode the coast of the San Diego region and also cause most of the cyclical changes, such as seasonal beach width fluctuations, for example. Tides and other sea-level changes also play a role by making wave damage episodically more or less severe. Weather and climate, in particular wind and rainfall, also are factors that influence the region's coastal processes, especially the sand supply.

SAND SUPPLY AND TRANSPORT

Sand supply is the crucial element that determines the health of the beaches and indirectly influences cliff and bluff erosion rates. The inflow and outflow of sand to and from a stretch of shoreline constitute the local sediment budget. The difference between the rate of sand inflow and outflow determines whether a beach will narrow or widen over time. Understanding the regional sand supply and its history, especially where the sand comes from and how much comes from each source, is important for understanding the San Diego region's beach erosion problems, and for formulating sensible strategies to cope with them.

In the San Diego region, significant amounts of sand can enter or leave a segment of shoreline only from up or down coast, or to and from the offshore area. Furthermore, in southern California, there are only three sources of beach sand: discharge from rivers, cliff or bluff erosion, and nourishment by humans. Quantifying the natural inputs and losses is difficult, and many studies have been conducted that yield very different results. Calculating the on/offshore and longshore sand transport is also much more difficult and uncertain than it may seem. This is mainly because the relationships between the wave action and the sand motion are not well understood, and because sufficiently detailed site-specific wave measurements and statistics do not exist.

Nevertheless, the overall conclusion is that the rivers between Dana Point and the international border deliver approximately 40 percent as much sand currently as they did under more natural conditions, before dams were built and urbanization of the watersheds occurred. Further, both the rivers and cliff erosion north of Oceanside Harbor supply much more sand to the shoreline than do the rivers and cliffs to the south. The difference is substantial, with a total of 360,000 to 400,000 cubic yards per year of sand produced north of the harbor, and only 65,000 to 105,000 cubic yards per year produced to the south.

The difference between production in the northern and southern part of the Oceanside Cell is important since the jetties at Oceanside Harbor interrupt the southward transport of sand. It has also been suggested that the jetties might deflect large quantities of sand offshore. Either way, this has been mitigated to a variable degree by the intermittent dredging of the harbor and by concerted efforts to bypass sand around it. However, much of the massive amount of sand produced by cliff erosion at Camp Pendleton likely has not reached the beaches south of Oceanside Harbor for one reason or another.

The rates of sand supply from beach nourishment, on the other hand, are greater in the southern half of the Oceanside Cell than north of the harbor. Without this nourishment, the North County beaches would be in even worse shape than they are at the present time. In the northern part of the cell, just over 2 million cubic yards of sand was placed on Doheny Beach in the late 1960s. Another 2.6 million cubic yards resulted from construction of the San Onofre Nuclear Generating Station from 1964 to 1985. South of Oceanside Harbor, 6.5 million cubic yards of sand came from dredging the harbor from 1942 to 1968. Another 1 million cubic yards was trucked from the San Luis Rey River to nourish the beaches in 1982. An additional 4 million cubic yards resulted from dredging Agua Hedionda Lagoon in 1954. The Batiquitos Lagoon Enhancement Project added 1.8 million cubic yards to the south Carlsbad shoreline beginning in 1994, and the San Diego Regional Beach Restoration project distributed about 1.8 million cubic yards between Oceanside and Torrey Pines in 2001. Thus, nearly 20 million cubic yards of sediment has been placed on the beaches of northern San Diego County over the past 40 years, or about 500,000 cubic yards per year on average.

The effect of beach nourishment on the San Diego shoreline is most dramatic in the Silver Strand Littoral Cell. About 26 million cubic yards of sand was placed on the Silver Strand in 1946 as a by-product of naval base dredging in San Diego Bay. Other, smaller projects have brought the annualized rate of beach nourishment to an astounding 810,000 cubic yards since 1946. This nourishment rate dwarfs the sand yield from the Tijuana River, both before and after dams were built. The large amount of sand changed the Silver Strand from a narrow, marginal barrier spit to a substantial peninsula. The shoreline south of the Hotel del Coronado widened by about 1000 feet, and the stretch from the hotel groin to Zuniga Jetty at North Island became one of the widest and most stable beaches in southern California.

However, there are serious problems in the Oceanside Littoral Cell since the large natural sources of sand are distant and isolated from the places where the sand is most needed for property protection and recreation. This has been compounded by a subtle change in wave climate. Beginning in the late 1970s, the relative amount of wave energy coming from the southwest increased compared with the period from 1945 to about 1978. As a result, the balance of northward and southward sand transport in the Oceanside Cell has been much closer, and waves don't seem to spread out the sand supply along the coast as effectively. This may explain why there are areas such as Solana Beach or parts of Encinitas that often have virtually no sand, while up- and down-coast beaches still have some.

WAVES

Waves provide nearly all of the energy input that drives shoreline processes along the California coast. Understanding the effects of waves is key to understanding and dealing with coastal processes in southern California. The Pacific Ocean is the largest ocean in the world and therefore has plenty of space, or fetch, for storms to generate high and long waves. Incoming waves along the southern California coast fall into three main categories: Northern Hemisphere swell, Southern Hemisphere swell, and seas generated locally. Waves in the swell categories originate either north or south of the equator and arrive in southern California after traveling great distances.

Seas in the region are produced by storms sweeping through the area, strong atmospheric pressure differences that induce strong winds, or daytime sea breezes. Santa Ana winds blowing from east to west are a fairly common event in southern California. They result from high atmospheric pressure over inland areas and can produce winds with speeds up to 50 mph reaching 100 miles offshore. These conditions can produce damaging sea waves at Avalon Harbor, located on the east side of Catalina Island, but have little consequence in the San Diego region.

The Southern California Bight is noted for its offshore islands, shallow banks, submarine canyons, and generally complicated bathymetry. Coastal orientation and the islands and banks greatly influence the swell propagating toward shore by partially sheltering southern California, including the San Diego region. The San Diego coastline generally faces west, with the notable exception of the southern tips of Dana Point, Point Loma, and Coronado, which are south facing. Because of the complicated effects of bathymetry and island shadowing, the wave height at the shoreline is sensitive to relatively small changes in the incoming direction of the deep ocean waves. Likewise, coastal wave heights can vary drastically over short distances along the shore.

DESCRIPTIONS OF INDIVIDUAL COASTAL AREAS

The following is a descriptive discussion of the characteristics and history of the major subregions and coastal features of the San Diego region. It is meant to complement the accompanying set of maps that present the coastal hazard level, shoreline environment characteristics, and degree and type of armoring present on the coast.

DANA POINT TO SAN MATEO POINT

Dana Point forms the northern end of the Oceanside Littoral Cell. It is a rocky headland formed by several different rock types of different ages that have been brought together by fault-controlled uplifting. The point is undercut in places by wave activity, and landslides have been noted. Dana Point Small Craft Harbor was constructed between 1963 and 1970 immediately down coast

Figure 20.5   Dana Point Small Craft Harbor, completed in 1970. Photo by Gary Griggs and Kiki Patsch.

of and in the lee of Dana Point (Figure 20.5). This is an ideal location for a harbor since it is both well sheltered from waves and at the up-coast end of the littoral cell, where it does not interfere with coastal sand transport. As a result, Dana Point Harbor has no sand accumulation problem and has never been dredged. This is in striking contrast to many of California's other small craft harbors, which were built in the middle or at the down-coast end of littoral cells and therefore require expensive annual dredging. On the negative side, the harbor eliminated a popular surfing spot at Doheny Beach.

San Juan Creek discharges immediately down coast of the harbor and remains an important source of beach sand to this coastal reach, even though it is increasingly leveed and altered.

Doheny State Beach has excellent beach access and facilities, including parking. Capistrano Beach is largely developed with a mixture of mobile homes, older small beach houses, and newer luxury homes, in some cases built directly on the back beach (Figure 20.6). Risks from cliff erosion, landsliding, and coastal flooding are high, although the area is moderately sheltered from northwesterly waves by Dana Point. A rise in groundwater levels in the cliffs from extensive landscape watering has contributed to cliff failure.

The coastline of San Clemente is characterized by a relatively narrow sandy beach and moderate to high risks of beach erosion, cliff landsliding, and coastal flooding. The cliff toe is largely armored with riprap to protect the railroad, which runs along a low bench just above the beach. Serious

A. Dana Point Harbor was constructed between 1963 and 1970 as a small boat harbor and harbor of refuge by the U.S. Army Corps of Engineers and the Orange County Harbor District. The breakwater was overtopped but undamaged during the intense storm of January 1988.

B. San Juan Creek occupies a half-mile wide channel that filled with sediment as sea level rose over the past 18,000 years. The creek is now channelized with levees.

C. Capistrano: The developments seaward of the cliff are sited on what was historically the active backbeach area of the profile and what was the natural shore protection for the cliffs. The cliffs are, in effect, sheltered from marine erosion by the development on the backbeach.

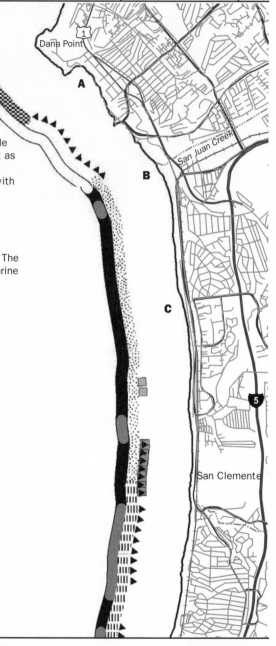

Hazard Level

☐ Stable: low risk
◼ Caution: moderate risk
◼ Hazard: high risk

Erosion Rates

① Inches per year

Shoreline Environment

▨ Sandy beach
▨ Beach backed by marsh
▨ Beach backed by dunes
▥ Cliff or bluff fronted by beach
▨ Rocky coast with no beach

Armoring

▲▲▲ Riprap
◼◼◼ Seawall
◼◼◼ Other

Figure 20.6 Capistrano Shores area, showing seacliff isolated from shoreline by highway, railroad, and residential housing built directly on the back beach. Photo © 2002–2004 Kenneth and Gabrielle Adelman, California Coastal Records Project, www.Californiacoastline.org.

damage to beach facilities and structures occurred during the high tides and storm waves of the 1982–83 El Niño winter storms.

Sediment discharged from San Mateo Creek forms a wide, low delta and floodplain at San Mateo Point, home of the famous "Trestles" surf spot.

SAN ONOFRE TO OCEANSIDE

Northern San Onofre Beach is a low-hazard, sandy beach backed by recreational facilities reserved for military personnel. The southern part comprises San Onofre State Beach, where the beach itself, as well as the limited facilities and back-beach parking, are at higher risk of erosion and flooding. This is a very popular beach access and surfing mecca. The San Onofre Nuclear Generating Station (SONGS) was built in a large space excavated out of the bluffs between 1964 and 1985. Over 1 million cubic yards of sand was produced by construction activities. This widened the beaches to the north and south, and two laydown pads used for construction stabilized the fill on the north side, greatly benefiting the state park (Figure 20.7A). SONGS facilities are not at risk from erosion or flooding owing to massive double-seawall protection (the design preserves lateral access between the walls). The adjacent beaches, however, have narrowed since the last pad was removed in 1985 (Figure 20.7B).

Figure 20.7 (A) San Onofre Nuclear Generating Station (SONGS) in 1979, showing lay down pad for units 2 and 3 and the resulting beach width increase. (B) SONGS in 1989 showing narrowing of beach following removal of lay down pad in 1985. Photos © 2002–2004 Kenneth and Gabrielle Adelman, California Coastal Records Project, www.Californiacoastline.org.

A. San Clemente: Narrow to moderately wide sandy beaches are backed by park facilities, railroad tracks, and high coastal cliffs. The cliff rim and elevated coastal plain are developed with residential buildings. The cliff toe is now protected by the elevated railroad line. The railroad bed is founded on the backbeach, and almost continuously fronted by rock revetments offering varying levels of protection. The railroad was overtopped in many places during the winter storms of 1982-1983.

B. San Mateo Creek is marked by a sand spit and beach backed by wetlands and a wide flood plain. The mouth of the creek is frequently closed by alongshore transport during dry years.

C. This coastal segment is relatively undeveloped and contains no coastal protection. It is part of the Camp Pendleton Marine Corps Reservation which extends roughly from Basilone Road south to Oceanside. The Camp Pendleton recreational beach facilities are located on a sloped bluff that is 20-40 feet high. The beach width and condition of this section has benefited from sand supply and construction activity at San Onofre Nuclear Generating Station (SONGS).

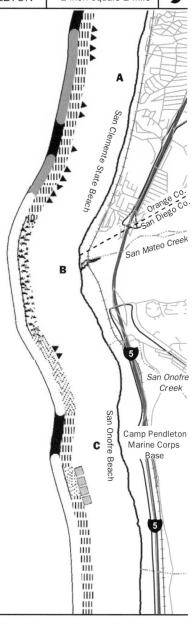

Hazard Level

☐ Stable: low risk

▨ Caution: moderate risk

■ Hazard: high risk

Erosion Rates

 Inches per year

Shoreline Environment

▒ Sandy beach

▚ Beach backed by marsh

▨ Beach backed by dunes

⫼ Cliff or bluff fronted by beach

▩ Rocky coast with no beach

Armoring

▲▲▲ Riprap

 Seawall

⊡⊡⊡ Other

A. Agra: In this area a narrow, sandy beach is backed by 60- to 120-foot-high cliffs topped by a flat, gently southwest sloping marine terrace. This coastal segment is relatively undeveloped and contains no coastal protection. Interstate 5 and the railroad are built on the terrace and are the only transportation corridors through this segment. Old Highway 101 parallels the freeway but is restricted to bicycles.

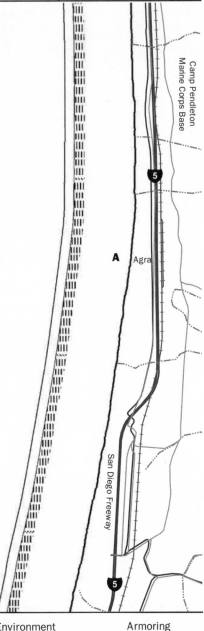

Camp Pendleton Marine Corps Base

A Agra

San Diego Freeway

Hazard Level

 Stable: low risk
Caution: moderate risk
Hazard: high risk

Erosion Rates

 Inches per year

Shoreline Environment

 Sandy beach
 Beach backed by marsh
Beach backed by dunes
 Cliff or bluff fronted by beach
 Rocky coast with no beach

Armoring

 Riprap
 Seawall
Other

Figure 20.8 Oceanside Harbor, ca. 1987, showing Camp Pendleton Del Mar Boat Basin in the background, entrance jetties and Harbor Beach on the left, and the lower reach of San Luis Rey River at the bottom. Photo courtesy of the California Department of Boating and Waterways.

The Camp Pendleton reach has a narrow, sandy beach backed by steep 60- to 100-foot-high bluffs and stretches nearly 15 miles from SONGS to Oceanside Harbor (the camp boundaries also include the 1-mile segment north of SONGS, nearly to San Mateo Point). This section of coast is mostly undeveloped, except for several military support installations, and represents the last large, more or less pristine expanse of coastal land in southern California. It is considered entirely low risk for coastal erosion and flooding since essentially no development is threatened. In most places, the railroad and Interstate Highway 5 (I-5) are set back at least 1,000 feet from the bluff edge.

Large gullies (arroyos) and bluff-face erosion provide essentially all of the local beach sand supply. The concentration of runoff in culverts underlying I-5 and the railroad erode through the loosely consolidated sediments during heavy rains. This process can quickly produce large volumes of sediment. In 1980 a canyon was lengthened over 200 feet in 24 hours, contributing about 50,000 cubic yards of sand to the adjacent beach.

A relatively wide, sandy beach backed by dunes and marsh extends across the Santa Margarita River mouth. Sand discharge from the river has been known to build large deltas that persisted for years after major floods.

The Oceanside Harbor complex was constructed in stages beginning with excavation of the Del Mar Boat Basin in 1942 (Figure 20.8). The boat

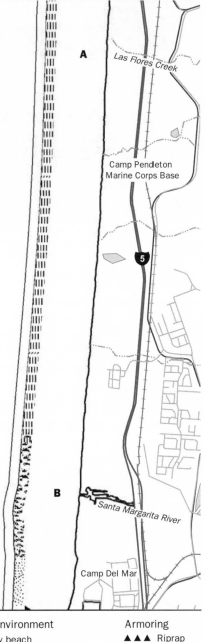

A. A sandy beach reaches across the mouth of Las Flores Creek, which occupies Las Pulgas Canyon. The creek is incised into the flat, gently southwest-sloping marine terrace. The beach is subject to erosion during high wave and flood conditions. This area is undeveloped and there are no shoreline protection structures. It is reserved for Camp Pendleton military purposes, and except for a freeway off-ramp leading to a viewpoint, it is not accessible to the public.

B. A relatively wide and sandy beach with active dunes lies on the spit across the wetland that forms the mouth of the Santa Margarita River. A delta forms at the river mouth during peak river discharge and may persist for years. The area is subject to beach retreat during high wave conditions and to flooding after heavy rainfall. This beach is stable and has accreted because of the influence of the breakwater at Oceanside Harbor just to the south.

Hazard Level

 Stable: low risk

Caution: moderate risk

Hazard: high risk

Erosion Rates

 Inches per year

Shoreline Environment

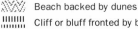 Sandy beach

Beach backed by marsh

Beach backed by dunes

 Cliff or bluff fronted by beach

Rocky coast with no beach

Armoring

 Riprap

 Seawall

Other

basin was constructed to accommodate wartime amphibious training exercises and was built with two small jetties to help keep the entrance open. At the time, little attention was devoted to understanding the environmental consequences, including rapid infilling with sand and interruption of the sand supply to down-coast beaches. The larger north breakwater was started in 1942 to prevent rapid channel clogging.

Construction of the civilian small craft harbor south of the boat basin was begun in 1958 with completion of several extensions of the north breakwater. By 1963, when the harbor was completed, about 6.8 million cubic yards of dredged sand had been placed on the adjacent beaches. The sand considerably widened the beaches immediately adjacent to the harbor, but had little or no beneficial effect for any appreciable distance to the south. The breakwaters stabilized the harbor beaches, as well as the stretch up to 5 miles to the north, and the harbor has been a major economic benefit to Oceanside. However, the breakwaters also caused a severe redistribution of sand that has contributed to beach erosion toward the south, and large volumes of sand have also been diverted offshore.

Oceanside Beach is a moderately wide, sandy beach near the harbor in the north that gradually narrows toward the south, disappearing about a mile from the south harbor jetty. It is backed by park facilities, an access road called The Strand, businesses, motels, and numerous beach bungalows. The erosion and flooding risk varies, depending on the development in danger, and on the amount and quality of shore protection. Maintenance dredging of Oceanside Harbor provides an average of several hundred thousand cubic yards of sand to various parts of the beach on an annual or two-year cycle.

Large volumes of cobbles underlie the beach sand veneer and become exposed when the sand is stripped away. Large cobbles were picked up by the great storm waves of the early 1980s and thrown against the low-lying oceanfront structures. This battering produced considerable erosion and damage in areas where construction has occurred directly on the former beach.

South Oceanside has a narrow, fine-grained sand and cobble beach that is generally at moderate to high risk of erosion and flooding during storms. Flooding, wave overtopping, and projectile damage from cobbles occurred during the storms of 1941, 1978, 1980, and 1983, prompting construction of a long, high riprap revetment of heavy stone.

CARLSBAD TO DEL MAR

The northern part of Carlsbad is fronted by a narrow, fine-grained sand beach backed by 30- to 60-foot-high cliffs that are susceptible to wear from runoff, groundwater, and animal burrowing, in addition to wave erosion. The cliff tops are intensively developed, and a wide variety of shoreline protection exists. The 4,000-foot-long Carlsbad Seawall was built in 1988 to protect the

A. Oceanside Harbor is a small-craft harbor with about 900 slips and harbor services. Its facilities include restaurants, fuel dock, haulout, and boat charter. The north breakwater was originally constructed starting in 1942 to protect the Del Mar Boat Basin. The civilian small craft harbor was built starting in 1958 when the 4,350-foot-long north breakwater was completed. Between 1942 and 1963, over 6.8 million cubic yards of sand and cobbles were dredged from the basins and mostly placed on local beaches.

B. A moderately wide sandy beach is backed by city park facilities, "The Strand," and development on the beach and cliff tops. This section has been extensively modified by a combination of factors, including Oceanside Harbor and other developments that interrupt the natural distribution of littoral sand, massive amounts of beach nourishment, and a decrease in the long-term sand supply from the Santa Margarita and San Luis Rey rivers due to dam construction. The Strand and adjacent structures were built on the 1916 and 1927 flood deltas of the San Luis Rey River.

C. A narrow, low sand and cobble spit beach is flanked north and south by the inlets of Agua Hedionda Lagoon and backed by Carlsbad Boulevard. The beach and road are subject to flooding damage, debris accumulation, and closure during high wave conditions. Jetties serve to stabilize the lagoon inlets, which are used to draw in and discharge cooling water for the adjacent Encina power plant. The north inlet and lagoon have to be dredged every 2 to 3 years.

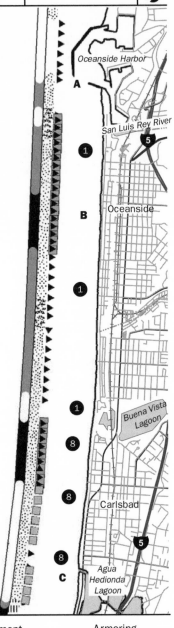

Oceanside Harbor

San Luis Rey River

Oceanside

Buena Vista Lagoon

Carlsbad

Agua Hedionda Lagoon

Hazard Level

☐ Stable: low risk
▨ Caution: moderate risk
■ Hazard: high risk

Erosion Rates

① Inches per year

Shoreline Environment

Sandy beach
Beach backed by marsh
Beach backed by dunes
Cliff or bluff fronted by beach
Rocky coast with no beach

Armoring

▲▲▲ Riprap
■■■ Seawall
◼◼◼ Other

coast highway and underlying utilities. Erosion risk varies but is generally moderate or low. The beach widens at the southern end of the seawall due to the sand retention effects of the north jetty of Agua Hedionda Lagoon.

Agua Hedionda Lagoon is a highly modified wetland separated into three basins by a railroad trestle and I-5. The outer basin has two openings to the ocean, both stabilized by rock jetties. The openings serve to circulate seawater for cooling a large power plant. Sand is pulled into the lagoon and must be regularly removed by dredging, and this has caused some controversy over the optimal distribution of the material. Beach sand losses and flooding of the coast highway have instigated construction of a seawall that protects the sand spit fronting the lagoon between the two entrances. Flooding risk is low to moderate.

South Carlsbad stretches between Agua Hedionda and Batiquitos lagoons. It has a narrow, sandy beach backed by 10- to 70-foot-high coastal cliffs that make up South Carlsbad State Beach. South Carlsbad has one of the longest cobble beaches in southern California, which was extensively mined for abrasives between 1912 and 1945.

To the south, Batiquitos Lagoon is one of the series of lagoons lying in the coastal valleys cutting across the uplifted terraces that form the San Diego region. Until 1994, when a massive restoration effort was begun, it represented another example of a seriously degraded wetland that had been negatively impacted by filling and dikes built for the railroad, roads, and I-5. Tidal flushing was limited in the 610-acre lagoon, and water quality and habitat value were poor for that reason. Between 1994 and 1997, the Batiquitos Lagoon Enhancement Project dredged and stabilized the lagoon inlet as part of an off-site mitigation restoration by the Port of Los Angeles. About 3 million cubic yards of sand was placed on the beach adjacent to the lagoon.

The city of Encinitas was incorporated in 1986 and included the San Diego County towns of Leucadia, old Encinitas, Cardiff-by-the-Sea, and (inland) Olivenhain. The beach is narrow to nonexistent, and of medium to high risk for erosion. Various stretches of cliff are unstable due to landslides, damage from excess groundwater accumulation due to cliff-top landscape watering, poor drainage, animal burrowing, and undercutting and cave formation by wave action.

Very little cliff erosion apparently took place along this section of coast between 1950 and 1970. Under San Diego County permits, essentially the entire cliff edge was developed with private homes or public access points, street ends, or parking between the 1930s and about 1980, but mostly during the latter decades, which coincided with a period of benign weather. Between 1978 and 1983, however, there was extensive localized retreat, especially during the El Niño storms of early 1983.

Figure 20.9   A variety of different types of seawalls have been constructed in an attempt to halt erosion of the cliffs in the Leucadia section of Encinitas. Photo © 2002–2004 Kenneth and Gabrielle Adelman, California Coastal Records Project, www.Californiacoastline.org.

In the subsequent years, numerous seawalls were constructed to protect the cliffs from failing (Figure 20.9). Besides causing property damage, cliff failures in Encinitas have proven fatal. In January 2000 a young woman died in a landslide near Stone Steps, south of Beacons, while sitting on the beach watching her husband surf. Her death prompted the city to redouble efforts to limit cliff-top landscape watering and to promote the increase of sand supply to the beach to lessen the potential for cliff failure.

Moonlight Beach is a deep pocket beach formed in a fault-controlled drainage. Coastal indentations such as this, as well as intertidal rock outcrops, are very important for retaining sand along the coastline. Enhancement and replication of these natural obstructions could be used to improve sand retention in the future. Sand retention may become critical as sand replenishment grows in importance, both to provide needed recreational space and access, and for mitigation of the negative effects of the growing number of seawalls.

The Self Realization Fellowship Temple, commonly known as "Swamis," was built in 1937 by Paramahansa Yogananda, widely revered as one of the preeminent spiritual figures of that time. Swamis is susceptible to erosion and cliff failure. In winter 1941 the temple toppled off the cliff as the result of a massive landslide.

A. A narrow beach with a high cobble berm is backed by Pacific Coast Highway (PCH) and Batiquitos Lagoon. A low-lying road is subject to flooding and closure during storms. During the severe winter of 1982–1983, large quantities of cobbles were moved onto the road, closing PCH.

The 610-acre lagoon had been expected to fill completely within 50 years until a large restoration was begun in 1994 to open the lagoon to tidal flow. A pair of short riprap jetties were built across the beach to help stabilize the opening. Lagoon dredging produced approximately 3 million cubic yards of beach sand that widened the adjacent beaches. The project was completed in 1997.

B. A medium-width, sandy pocket beach has formed in the floodplain of a small creek and is backed by a cliff, road, and park facilities. Sand level and beach width are subject to large changes, particularly in response to large storm waves and occasional sand nourishment. The backbeach is subject to flooding, and structures are susceptible to damage during extreme events such as the winters of 1982–1983 and 1988. Access to this heavily used beach and park is excellent, but parking is limited.

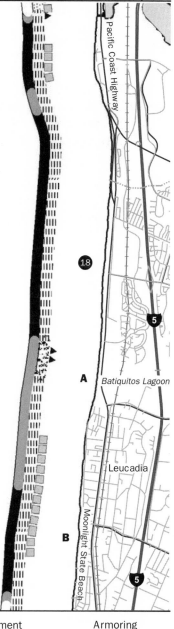

Hazard Level	Shoreline Environment	Armoring
☐ Stable: low risk	▒ Sandy beach	▲▲▲ Riprap
▨ Caution: moderate risk	⌇ Beach backed by marsh	☐☐☐ Seawall
■ Hazard: high risk	▨ Beach backed by dunes	⦿⦿⦿ Other
Erosion Rates	⦀ Cliff or bluff fronted by beach	
❶ Inches per year	▨ Rocky coast with no beach	

Figure 20.10   Restaurant Row and Pacific Coast Highway fronting San Elijo Lagoon in the Cardiff section of Encinitas. Photo © 2002–2004 Kenneth and Gabrielle Adelman, California Coastal Records Project, www.Californiacoastline.org.

San Elijo Lagoon is fronted by a spit comprising Cardiff State Beach, which has excellent access and is very popular, especially in the summer. At 900 acres, the lagoon is one of San Diego County's largest coastal wetlands, fed by a 77-square-mile watershed. San Elijo is primarily a shallow-water estuary and is divided into basins by Highway 101, the railroad, and I-5. These constrictions cause the inlet to clog with sand, reducing tidal exchange with the ocean. Periodic efforts to open the inlet by dredging have improved water quality.

Three oceanfront restaurants exist on the barrier spit, and several more stand on the east side of the coast highway (Figure 20.10). The seaward restaurants were damaged during the El Niño storms of 1982–83, after which existing riprap protection was expanded. Flooding and wave overtopping occurred again in 1988 and during the El Niño winter of 1997–98. Beach cobbles, sand, kelp, and other debris were thrown through the buildings and onto the low-lying highway, closing it to traffic for a time. Extensive beach erosion and damage to the state park access and parking facilities occurred in February 1998.

Solana Beach has a narrow, fine-grained sand beach backed by steep coastal cliffs about 75 feet high. Beach width and back-beach elevation in Solana Beach have decreased since about 1980. Beach sand has been removed by storm-driven wave action and has not been replaced, either naturally or by nourishment, at nearly the rate necessary to maintain a reasonable beach width. Without beach sand cover, the foreshore at many places

A. A wide pocket beach is backed by San Dieguito Lagoon and Pacific Coast highway and bounded on the north by cliffs and on the south by a riprap training wall and barrier spit across the lagoon. This beach is subject to large variations in width, depending on wave conditions and flood outflows from the San Dieguito River through the lagoon. The lagoon mouth is often closed for extended periods, especially during drought. The beach is heavily used for recreation, and access is excellent, but limited by parking.

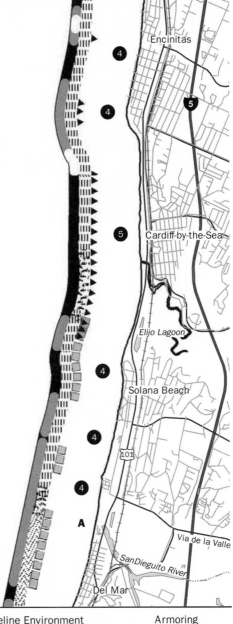

Encinitas

Cardiff-by-the-Sea

Elijo Lagoon

Solana Beach

Via de la Valle

San Dieguito River

Del Mar

Hazard Level

☐	Stable: low risk
▨	Caution: moderate risk
■	Hazard: high risk

Erosion Rates

❶ Inches per year

Shoreline Environment

▒	Sandy beach
≋	Beach backed by marsh
░	Beach backed by dunes
▥	Cliff or bluff fronted by beach
▓	Rocky coast with no beach

Armoring

▲▲▲	Riprap
▥▥▥	Seawall
⊡⊡⊡	Other

Figure 20.11   San Dieguito River mouth, located between Solana Beach and Del Mar. In the background are the railway, Del Mar Race Track, and fairgrounds, all built on fill. Note the large sand storage capacity of this embayment. Photo © 2002–2004 Kenneth and Gabrielle Adelman, California Coastal Records Project, www.Californiacoastline.org.

consists of an exposed, hard bedrock shore platform. Lateral beach access is restricted during periods of high tides, since the water at the cliff base can be several feet deep. The narrow beach combined with breaking waves and storm surge can make beach access difficult and hazardous.

When storms coincide with periods of high tides, waves breaking at the cliff face undermine it, causing failure of the lower bluff and eventual collapse. The upper bluff is subsequently destabilized and ultimately fails. These conditions have led to construction of numerous seawalls to protect cliff-top homes that were built too close to the edge before the inherent risks were fully realized. Ongoing demand for property protection is certain. However, several local groups oppose seawall construction, citing additional loss of beach width through passive erosion, and advocate eventual retreat instead.

San Dieguito is a typical southern California estuarine lagoon and forms the lower part of the San Dieguito River Valley, which drains a watershed area of 325 square miles. Only 45 square miles lie below Lake Hodges, which limits peak flood flows and traps sediment. The present-day lagoon is a 140-acre wetland located on the northern edge of the city of Del Mar. The average water depth in the lagoon is about 3 feet below mean sea level (MSL). The lagoon has been extensively filled, and dikes were constructed for roads, I-5, and the railroad (Figure 20.11). This reduced the tidal prism well below values that existed under more natural conditions. Beach sand mobilized by waves tends to clog the inlet, whereas tidal and occasional river flood flows tend to flush the sand out and keep the inlet open. To enhance tidal exchange and water quality, and to prevent flooding, San Dieguito was opened artificially 17 times in the 24-year period from January 1978 to December 2001.

The 1-mile portion of Del Mar Beach north of 15th Street is a moderately wide, sandy beach occupying the spit and former dune ridge fronting San Dieguito Lagoon, and represents the range historically swept by the San Dieguito River. This beach has excellent access, both laterally from the north and south, and vertically from street ends. The entire reach has essentially been completely armored with seawalls before and especially since flooding and wave damage occurred during the El Niño storms of 1982–83. Street end access points are still subject to wave flooding, as are the lower-lying residential developments landward of the first dune crest.

The southern portion of Del Mar is characterized by upper and lower marine terraces fronted by a receding seacliff and a narrow, sandy beach. Residential and commercial development in Del Mar is built on the upper terrace; the lower terrace was graded in 1910 for the railroad. Numerous slope failures have occurred, and on several occasions trains have fallen off the cliff, resulting in loss of life and severe damage to the railroad. These slope failures have required armor protection in the form of riprap. Access to this section is poor and limited to either end.

DEL MAR TO LA JOLLA

Peñasquitos Lagoon is a remnant coastal lagoon formed in the sediment-filled river channel of Sorrento (or Soledad) Valley. Like the other San Diego coastal lagoons, Peñasquitos has been altered by filling and by dikes constructed for the coast highway and railroad. These alterations and sedimentation reduced the tidal prism or volume, negatively affecting tidal flushing and water quality, and occasionally making artificial opening necessary.

During most summers, the sand spit fronting the lagoon has a moderately wide sand beach that is heavily used for recreation. Access is excellent, both laterally from up- and down-coast as well as from the Torrey Pines State Beach facilities located on the low-lying sand spit just south of the lagoon mouth. The park facilities are protected by inadequate riprap that is subject to overtopping and damage during large wave conditions. Substantial damages occurred to the parking lot, rest rooms, and access stairs during the El Niño storms of 1997–98.

Torrey Pines is a narrow to moderately wide sand beach backed by low-lying dunes and 300-foot-high, steep cliffs. Unlike the other cliffs in the San

A. Torrey Pines: A narrow- to medium-width, fine-grained sand beach is backed by low, active dunes and high, steep, eroding cliffs. The near shore is comprised of a thin veneer of fine to very fine sand. The cliff top along the northern part of this section is occupied by the Torrey Pines State Park and Reserve and the Torrey Pines City Park, including a golf course, glider port, and scenic overlooks. During severe winter storm periods the dunes and the cliff base are eroded by waves.

The cliffs reach heights of over 300 feet—the highest of any in the region, other than Point Loma. These over-steepened cliffs are inherently unstable in the long term. Numerous landslides, including several very large slides, have occurred along this section.

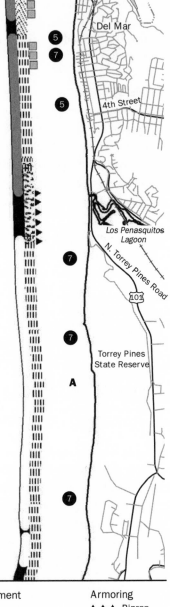

Del Mar

4th Street

Los Penasquitos Lagoon

N. Torrey Pines Road

101

Torrey Pines State Reserve

A

Hazard Level
 Stable: low risk
 Caution: moderate risk
 Hazard: high risk

Erosion Rates
1 Inches per year

Shoreline Environment
 Sandy beach
 Beach backed by marsh
 Beach backed by dunes
 Cliff or bluff fronted by beach
Rocky coast with no beach

Armoring
▲ ▲ ▲ Riprap
 Seawall
 Other

Diego region (besides those at Camp Pendleton), cliff erosion at Torrey Pines supplies a substantial amount of sand to the local sand budget. This comes in the form of large and small landslides and block falls. A massive 585-foot-long slide in 1982 deposited an estimated 1.8 million cubic yards of rock, sand, and soil onto the beach. An earlier landslide in 1949 was even larger and extended alongshore for approximately 1,700 feet.

Much of the cliff-top area in the northern portion is public, including Torrey Pines State Park and Reserve, Torrey Pines Golf Course, and the glider port, which all offer spectacular views. Vertical access is limited, and often dangerous, with numerous cliff rescues necessary each year. No shore protection structures exist along this reach.

Proceeding down coast, a narrow bedrock and cobble beach forms a transition between the sandy beach at Torrey Pines and the one at La Jolla Shores to the south. Dike Rock, an outcrop of volcanic rock, acts as a natural breakwater protecting a small headland that dominates this area. The only shore protection is a short section of riprap high on the back beach protecting the "Mushroom House," a novel, private guesthouse near Dike Rock.

The National Marine Fisheries Service Laboratory was built in the early 1960s on the cliff edge at the north end of the Scripps Institution of Oceanography (Figure 20.12). Soon afterward, it was determined that the building straddled an active landslide whose movement was correlated with distant earthquakes. Shifts of several inches in the seaward fisheries building have become evident. In addition, the top of the bluff continues to erode. Because of increasing concerns with bluff erosion and failure, a planning process began in 2003 to examine alternatives to this large cliff-top building.

The La Jolla Shores reach is located between the Scripps Submarine Canyon on the north, and the La Jolla Submarine Canyon and Point La Jolla on the south, which mark the southern boundary of the Oceanside Littoral Cell. The canyon and inter-canyon bathymetry greatly influence the local wave height distribution. The Rose Canyon fault intersects the coast at the southern end of the reach and controls the local geomorphology.

The sandy beach is moderately wide, and backed by the Scripps Institution of Oceanography and residential development in its northern half; the low-lying Kellogg Park, a filled lagoon, in the south-center; and a hotel-resort at the southern end. The developed northern and southern sections are completely armored with various types of concrete seawalls, yet are still vulnerable to wave overtopping. The Marine Room restaurant in particular has a history of being occasionally damaged by wave inundation, such as occurred in 1941 and 1983. Beach access is excellent, especially at

A. Point La Jolla: The shoreline of Point La Jolla is characterized by rocky, wave-cut low-tide terraces, vertical cliffs, rocky headlands, and small, narrow pocket beaches. This is a scenic and heavily used section of the coast.

B. Bird Rock: Rocky headlands and rocky, wave-cut low tide terraces are fronted by perched, narrow sandy beaches or narrow cobble and boulder beaches. This segment is backed by eroding cliffs, 20 to 60 feet high, topped with single-family residences, except for occasional street ends and two small, cliff-top parks.

C. Pacific Beach: A narrow to moderately wide sandy beach is backed by steep, eroding cliffs with extensive residential development. Structures, park facilities, and the pier are subject to flooding and damage during high wave conditions. The cliffs are about 60 feet high in the north and decrease in height to beach level at the southern end of this section.

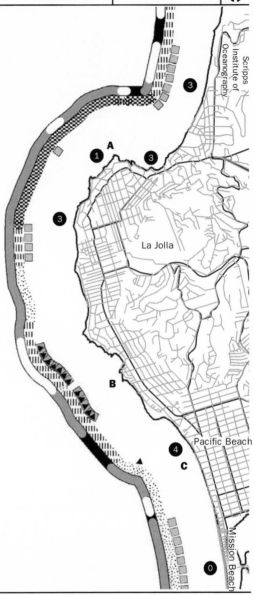

Scripps Institute of Oceanography

La Jolla

Pacific Beach

Mission Beach

Hazard Level

 Stable: low risk

Caution: moderate risk

Hazard: high risk

Erosion Rates

 Inches per year

Shoreline Environment

Sandy beach

Beach backed by marsh

Beach backed by dunes

Cliff or bluff fronted by beach

Rocky coast with no beach

Armoring

▲▲▲ Riprap

Seawall

Other

Figure 20.12   The National Marine Fisheries Service Building perched on the cliff edge on the Scripps Institution of Oceanography campus. Note the "Mushroom House" built on the back beach on the lower left in the photo. Photo © 2002–2004 Kenneth and Gabrielle Adelman, California Coastal Records Project, www.Californiacoastline.org.

Scripps, where parking is restricted, and at Kellogg Park, with its large parking lot. Along with Pacific Beach and Mission Beach to the south, La Jolla Shores provides a major portion of the available recreational beach area in the city of San Diego.

The shoreline of Point La Jolla is rocky and characterized by vertical cliffs with extensive caves, narrow pocket beaches, and uplifted wave-cut terraces at several elevations. The point has relatively hard, Cretaceous-age rock exposed at sea level. The caves on the north side of Point La Jolla are remarkable since the lack of a wave-cut terrace at the base of the sandstone cliffs suggests that the caves are the result of a process other than wave erosion alone. It does not seem probable that wave erosion by itself could have extended so deeply into this sandstone, and it is likely that some type of groundwater solution of the sandstone cement is also involved.

Examination of the topography and outcrops at the west end of the La Jolla cliffs and comparison with photographs taken around 1900 show a conspicuous lack of erosion. A single arch at Goldfish Point has shown very little change since 1909. The same is true of some massive boulders a little to the east, near La Jolla Cove, which are exposed to large waves. West of the cliffs, however, old photographs of the cove show changes in the well-known arches that existed in the early part of the century and were partially reinforced with concrete in later years. They finally collapsed, the last one falling in 1978.

Increasing numbers of coastal protection structures have been emplaced to protect private development along this reach in recent years. Coastal access is moderately good, especially at several parks, including La Jolla Cove, Scripps Park, Windansea, La Jolla Hermosa, and Bird Rock; however, parking is limited, especially in summer.

PACIFIC BEACH TO POINT LOMA

Located south of La Jolla, Pacific Beach is a 1.5-mile-long moderately wide, sandy beach backed by cliffs, bluffs, and dunes that gradually decrease in height from about 60 feet to beach level from north to south. Pacific Beach, and Mission Beach to the south, rest on the former delta of the San Diego River. The slightly elevated bench at Crown Point is an extension of the La Jolla marine terrace. The back beach is heavily urbanized, with residential development in the north and with commercial development toward the south. The cliffs and dunes are largely unprotected, except at Crystal Pier. The entire reach presents a moderate risk for beach erosion, while the lower-lying sections are subject to wave overtopping and flooding, both directly from the ocean and from seawater backing up the storm drains into the streets. Coastal access is excellent, but parking is difficult.

Contiguous to Pacific Beach, Mission Beach has a moderate to wide sandy beach on the west (ocean) side, and a narrow to moderately wide sand beach on the sheltered east (bay) side. Both beaches are low lying and prone to flooding, with the ocean-side beach also subject to wave attack and erosion. The oceanfront beach becomes increasingly wide from north to south because of the sand retention effects of the northernmost of three jetties located at the entrance to Mission Bay. This entire beach section is backed by a concrete seawall and promenade that are subject to overtopping and wave damage during major storm events such as those of 1982–83, 1988, and 1997–98 (Figure 6.5).

Mission Beach is completely urbanized, primarily with residential development. It has a long history of heavy public use as a gateway to the beach because of its accessibility and its location near the population centers of San Diego. Beach access is excellent, both laterally from north and south, and vertically from nearly every street end.

Mission Bay is an extensively modified wetland that was once called False Bay, where development for boating and other water-related recreational and tourist activity began as early as 1921. Work was substantially completed by the 1970s and resulted in the largest aquatic park in the world, which generates

A. Mission Beach: A variable-width, sandy beach is backed by an esplanade, parks, and residential and commercial developments, including hotels and an amusement park, all located on the barrier spit fronting Mission Bay. The area is densely developed and heavily utilized because of its history, accessibility, and proximity to the population centers of San Diego. This entire section is protected by a concrete seawall at an average height of about 13 feet above mean sea level. The seawall was overtopped during the winter of 1982–1983 and during the unusual storm of January 1988.

B. San Diego River: Estimates of average yearly sand yield from the San Diego River vary widely, ranging from 6,000 to 110,000 cubic yards. Three parallel jetties form the entrance channel to Mission Bay and stabilize the mouth and lower reach of the San Diego River. The north jetty was constructed in 1950 to stabilize the Mission Bay entrance. The middle and south jetties were completed in 1970. Minor sand accumulation in the entrance channel requires periodic maintenance dredging.

C. Sunset Cliffs: High, rocky, nearly vertical cliffs with many rocky coves and narrow, sandy pocket beaches are backed by a road and residential area. Cliff erosion in this area is critical. Buildings have been condemned and others are poised on the rim of the cliff.

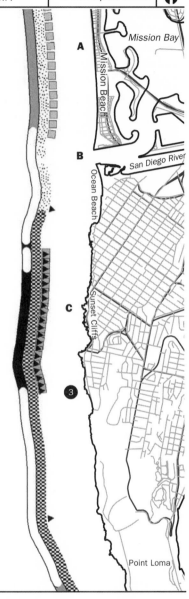

Mission Bay

Mission Beach

Ocean Beach

San Diego River

Sunset Cliffs

Point Loma

Hazard Level

☐ Stable: low risk
▨ Caution: moderate risk
■ Hazard: high risk

Erosion Rates

① Inches per year

Shoreline Environment

▨ Sandy beach
▨ Beach backed by marsh
▨ Beach backed by dunes
▨ Cliff or bluff fronted by beach
▨ Rocky coast with no beach

Armoring

▲▲▲ Riprap
■■■ Seawall
▣▣▣ Other

Figure 20.13   Mission Bay Aquatic Park in 1999, including Point La Jolla in the background, Pacific Beach and Mission Beach along the left center, and the lower reaches of the San Diego River and Ocean Beach in the foreground. Photo courtesy of the San Diego Historical Society.

significant income for the city of San Diego (Figure 20.13). Between 1948 and 1984, approximately 1.5 million cubic yards of sand was dredged from the bay and deposited on the adjacent areas of Mission Beach, greatly widening this previously marginal barrier spit. Public access to the shoreline and waters of Mission Bay is excellent, and San Diegans as well as tourists use the area heavily.

Mission Bay was originally one of the two natural outlets of the San Diego River, the other being San Diego Bay to the south. The river outlet was stabilized adjacent to and south of the Mission Bay entrance channel through a series of three rock jetties several thousand feet long. The river mouth is often closed by littoral drift, and this has led to concern about the possibility of flooding in Mission Valley, the natural course of the San Diego River, during extreme rainfall and runoff events. Large, devastating, historic floods occurred in 1862, 1884, 1889, 1891, and 1916. A lowered weir or overflow section was built into the middle jetty to allow potential floodwaters to exit through both the river channel and the bay entrance.

POINT LOMA

The residential and light commercial communities of Ocean Beach and Sunset Cliffs, the extensive naval facilities and Fort Rosecrans National Cemetery, the San Diego regional wastewater (sewage) treatment plant, and the Cabrillo National Monument occupy the ocean edge of Point Loma. The point is similar to Point La Jolla, in that it has harder Cretaceous-age bedrock exposed at sea level due to uplift associated with the Rose Canyon Fault. The steep cliffs along much of Point Loma are 300 feet high and susceptible to wave attack and erosion.

The densely built-up northern sections pose a moderate to high risk for coastal development, depending on the location. Various types of seawalls and other slope protection structures differing in design and effectiveness were built over the years. Army Corps of Engineers investigation of coastal erosion at Sunset Cliffs has determined that about 40 feet of landward retreat occurred between 1962 and 1976 at the toe of Del Mar Avenue. Erosion was the result of surface runoff, overwatering, pedestrian traffic, and animal burrowing. Erosion has occurred in several places back to the coastal road, and city authorities have dumped concrete debris and asphalt at these locations in an attempt to slow erosion. In March 1968, as a result of an earthquake, a section of Sunset Cliffs separated from the cliff and slumped, endangering the coast road. During the storms of January to March 1983 many sections of the cliffs were undercut and began to collapse.

Except at Ocean Beach, there are no long, sandy beaches on Point Loma, only sparse, short pocket beaches with essentially no practical access from above. In contrast, coastal access is excellent, but controlled, at the south end through Cabrillo National Monument.

SAN DIEGO BAY TO THE INTERNATIONAL BORDER

San Diego Bay represents one of the finest natural harbors in the world since it requires little dredging to maintain the 41-foot design depth. The bay and its shore have been heavily modified and developed. It is home to the largest naval facilities and fleet on the West Coast. San Diego Airport was built on its filled tidelands, and extensive commercial, shipping, residential, and tourist development exists along the shoreline. Construction of the 7,500-foot-long Zuniga Jetty was completed in 1904 on the south side of the entrance channel. This rock structure channelizes the tidal currents between itself and Point Loma, and also largely prevents littoral sand from entering the channel from the south.

Prior to 1821, the San Diego River entered San Diego Bay most of the time. In the fall of 1821 a flood changed the river channel in one night, and the greater volume of flow was diverted into Mission (False) Bay, leaving only a small stream still flowing into the harbor. This flood was remarkable in that no rain was reported along the coast at the time. The river was later observed to flow into San Diego Harbor in 1849, and again in a 1859 survey.

In 1853 the federal government diverted the flow of the river into Mission Bay and built an earthen levee extending from near Old Town to Point

A. Point Loma: Narrow, sandy pocket beaches are backed by wave-cut cliffs with many sea caves and sparse development. Block falls and surficial slope failures have occurred along the base, face, and top of the cliff. Most of the area is a U.S. Naval Reservation and includes Fort Rosecrans National Cemetery. The southern end is Cabrillo National Monument and includes the historic lighthouse, book shop, and other amenities. Access and parking at the monument are excellent.

B. San Diego Bay Entrance: One of the finest harbors in the world, San Diego Bay entrance requires very little dredging to maintain the 41-foot channel depth. Tidal flow asymmetry in the constricted channel and the lack of adjacent sand sources are the main reasons for this. Zuniga Jetty, the rock breakwater on the east side of the entrance, is 7,500 feet long and was completed in 1904.

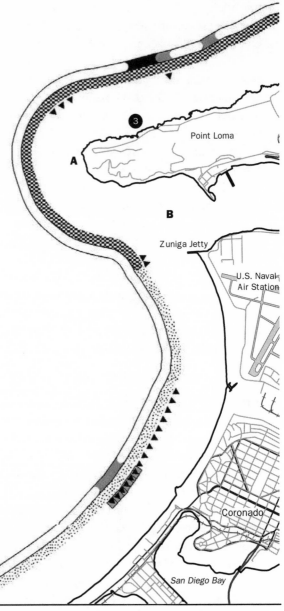

Point Loma

A

B

Zuniga Jetty

U.S. Naval Air Station

Coronado

San Diego Bay

Hazard Level

- ☐ Stable: low risk
- ▨ Caution: moderate risk
- ■ Hazard: high risk

Erosion Rates

- ① Inches per year

Shoreline Environment

- ▦ Sandy beach
- ⌇ Beach backed by marsh
- ▨ Beach backed by dunes
- ‖‖‖ Cliff or bluff fronted by beach
- ▨ Rocky coast with no beach

Armoring

- ▲▲▲ Riprap
- ■■■ Seawall
- ▣▣▣ Other

A. Silver Strand: Moderately wide to wide, sandy beaches form the barrier spit separating San Diego Bay from the ocean. This is a highly modified shoreline greatly affected by sand nourishment. Between 1948 and 1990, about 37 million cubic yards of sand from dredging projects in San Diego Bay were deposited on this section and on the Coronado Shores section to the north. Naval Amphibious Base facilities occupy the backbeach, as well as the large fill area in San Diego Bay, east of the highway. During autumn and winter, a large sand berm is often pushed up for protection against high wave conditions. South of the amphibious base, a wide sandy beach backed by dunes extends to the end of the section.

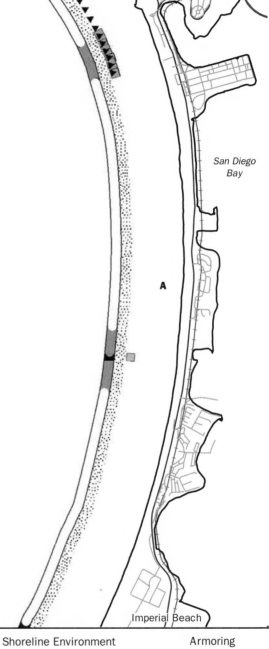

San Diego
Bay

A

Imperial Beach

Hazard Level	Shoreline Environment	Armoring
☐ Stable: low risk	⠿ Sandy beach	▲▲▲ Riprap
▨ Caution: moderate risk	⌇ Beach backed by marsh	▨▨▨ Seawall
■ Hazard: high risk	▨ Beach backed by dunes	⊡⊡⊡ Other
	⫼ Cliff or bluff fronted by beach	
Erosion Rates	▨ Rocky coast with no beach	
① Inches per year		

Loma because of the sediment deposited by the river during floods. Later that year, heavy rains caused the river to change course once again, washing out part of the levee and resuming its old course into San Diego Bay. The great flood in 1862, appropriately called the "Noachian Deluge," inundated San Diego; houses in the lower Old Town section of Mission Valley were flooded when severe coastal winds from the south backed up the water from the bay into the river. The levee was reconstructed in 1876, and no further diversions into San Diego Bay have occurred.

North Island has a moderately wide, sandy beach with no public access and is backed by low, active dunes adjacent to North Island Naval Air Station. Zuniga Jetty stabilizes this section of coast. Over 2.2 million cubic yards of sand dredged from San Diego Bay was deposited between the jetty and Coronado City Beach in 1941, widening this reach by nearly 900 feet. Prior to about 1945, North Island was separated from Coronado to the south by a swampy area that was subsequently filled.

Coronado City Beach to the west and Coronado Shores to the southwest are highly modified, wide, sandy beaches. The western portion is backed by a 1-mile-long, continuous riprap revetment built in 1905 after a severe southerly storm caused about 100 feet of beach retreat, washed out the coastal road, and threatened to destroy the Hotel del Coronado. A curved groin was built adjacent to the hotel in 1900 in an unsuccessful attempt to create a small craft anchorage. This structure and Zuniga Jetty essentially create a long pocket beach that is very stable. This stability is aided by the generally northward alongshore sand transport that prevails along this section of coast owing to the sheltering effect of Point Loma on waves approaching from the northwest and west.

Beginning in 1946, with drastic increases in the ship capacity of San Diego Bay made necessary by Navy requirements, about 37 million cubic yards of sand were dredged from the bay and deposited on Coronado and Silver Strand to the south. As a result, Coronado City Beach and Coronado Shores, the area just south of the hotel, are some of the widest beaches in southern California.

Coronado Shores is a wide beach that does not appear to be so since a large, high-rise condominium development was built directly on the accreted beach beginning in about 1967 (Figure 20.14). This was one of the developments that galvanized public reaction, leading to passage of the California Coastal Act and creation of the California Coastal Commission. The Coronado Shores towers are fronted by a large revetment that is subject to wave overtopping during intense storm wave events.

Silver Strand is a moderately wide to wide barrier sand spit separating the southern portion of San Diego Bay from the Pacific Ocean, and stretches from Coronado to Imperial Beach. Like Coronado, this is a highly modified

Figure 20.14   Coronado Shores, 1979, including the high-rise condominiums built on the back beach, and the Hotel del Coronado and its groin. Photo courtesy California Department of Boating and Waterways.

Figure 20.15   Silver Strand as seen from the cupola of the Hotel del Coronado in 1898. Photo by Ulysses S. Grant IV.

shoreline. The ocean side has benefited from the 37 million cubic yards of sand dredged from the bay beginning in 1946 and deposited between Silver Strand and Coronado. Historical photos taken about 1898 looking south from the Hotel del Coronado suggest that Silver Strand was a narrow, marginal barrier subject to frequent wave overtopping and sand overwash (Figure 20.15). The bay side has been altered by filling, including the area

Figure 20.16   Imperial Beach, showing rock revetment built to protect dense residential development. Tijuana Estuary channels appear in the background. Photo © 2002–2004 Kenneth and Gabrielle Adelman, California Coastal Records Project, www.Californiacoastline.org.

occupied by the Naval Amphibious Base and the Coronado Cays residential development.

Imperial Beach is a narrow, sandy beach backed by dense residential and commercial development (Figure 20.16). This small city has been subject to beach erosion problems for many decades, and represents a high risk for wave-induced coastal flooding and erosion damage during major storm events. Imperial Beach is the subject of numerous studies by local, state and federal agencies, notably the Army Corps of Engineers. Solutions to beach erosion range from offshore breakwaters, to groins to beach nourishment (Figure 20.16). More or less routine, albeit infrequent, beach sand replenishment occurs whenever the San Diego Bay entrance is dredged and suitable sand spoils are deposited, either just offshore or on the beach. The oceanfront residences, businesses, and public access at the municipal pier are protected by a wide variety of seawalls and revetments. This includes the nearly 1-mile-long massive rock revetment at the southern edge of the city.

TIJUANA ESTUARY

The Tijuana Estuary is a large wetland formed in the drowned river valley of the Tijuana River, a major southern California river that has been highly modified by channelization and dams on both sides of the U.S.-Mexico border. The

A. Imperial Beach: A narrow, sandy beach with good to excellent access is backed by recreation facilities and a densely developed residential and commercial area. This beach has been vulnerable to erosion for many years and the subject of numerous studies and proposed alternative solutions, including groins, offshore breakwaters, and beach sand nourishment. In 1977 about 1 million cubic yards of sand dredged from San Diego Bay was placed on the beach in the vicinity of the pier. The area remains vulnerable to wave-induced flooding and damage during high wave conditions, such as those during January 1988.

B. A narrow, sandy beach with good to excellent access is backed by dunes, park facilities, the Tijuana Estuary, and the naval reservation. The area is subject to flooding and damage during high wave conditions.

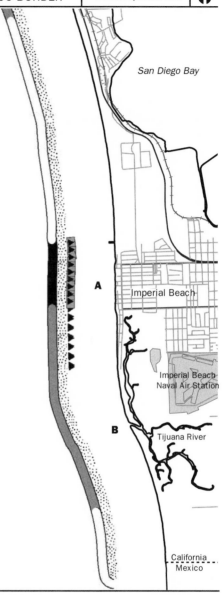

San Diego Bay

A

Imperial Beach

Imperial Beach
Naval Air Station

B

Tijuana River

California
Mexico

Hazard Level
☐ Stable: low risk
▨ Caution: moderate risk
■ Hazard: high risk

Erosion Rates
1 Inches per year

Shoreline Environment
Sandy beach
Beach backed by marsh
Beach backed by dunes
Cliff or bluff fronted by beach
Rocky coast with no beach

Armoring
▲▲▲ Riprap
☐☐☐ Seawall
◉◉◉ Other

natural yield of the river has been reduced by about 50 percent. This reduction in natural supply has caused a sand shortage in this reach that was not compensated by the massive beach nourishment that occurred farther north in the Silver Strand Littoral Cell. During large wave storm events, such as the El Niño winter of 1982–83 and January 1988, wave-induced sand overwash clogged portions of the wetland channels.

The littoral sand drift is closely balanced at this location between northward and southward transport. For this reason, and because of the sand shortage, the remnant delta of the Tijuana River has been reduced, leading to a tendency for general retreat of the entire shoreline in this reach.

APPENDIX A: USEFUL REFERENCES

BEACHES AND BEACH PROCESSES

Bascom, W. 1982. *Waves and Beaches.* Garden City, N.Y.: Doubleday.

California Department of Boating and Waterways and State Coastal Conservancy. 2002. *California Beach Restoration Study.*

Carter, R. W. G. 1988. *Coastal Environments: An Introduction to the Physical, Ecological, and Cultural Systems of Coastlines.* New York: Academic Press.

Carter, R. W. G., and C. D. Woodroofe (Eds.). 1994. *Coastal Evolution-: Late Quaternary Shoreline Morphodynamics.* Cambridge, UK: Cambridge University Press.

Davis, R. A., Jr. 1996. *Coasts.* Englewood Cliffs, N.J.: Prentice-Hall.

Davis, R. A., Jr., and Duncan M. Fitzgerald. 2004. *Beaches and Coasts.* Oxford, UK: Blackwell.

Dean, R. G. 1988. Managing sand and preserving shorelines. *Oceanus* 31:49–55.

Flick, R. E. 1993. The myth and reality of Southern California beaches. *Shore and Beach* 61:3:3–13.

Griggs, G. B. 1990. Littoral drift impoundment and beach nourishment in northern Monterey Bay California. *Journal of Coastal Research* (special issue on beach nourishment) 115–126.

King, Philip G. 1999. *The Fiscal Impact of California's Beaches.* San Francisco State University, Public Research Institute.

Komar, P. D. 1998. *Beach Processes and Sedimentation,* 2nd ed. Englewood Cliffs, N.J.: Prentice-Hall.

Nordstrom, K. F. 2000. *Beaches and Dunes of Developed Coasts.* Cambridge, UK: Cambridge University Press.

Runyan, K. B., and G. B. Griggs. 2003. The effects of armoring sea cliffs on the natural sand supply to the beaches of California. *Journal of Coastal Research* 19:2:336–347.

Wiegel, R. L. 1994. Ocean beach nourishment on the USA Pacific Coast. *Shore and Beach* 62:1: 11–35.

Willis, C. M., and G. B. Griggs. 2003. Reductions in fluvial sediment discharge by California dams and implications for beach sustainability. *Journal of Geology* 111: 167–182.

COASTAL HAZARDS: EROSION, SEA-LEVEL RISE, AND LAND USE

Benumof, B., L. Moore, and G. B. Griggs. 1998. Coastal erosion: The state of the problem and the problem of the state. In *Proceedings of California and the World Ocean '97*, pp. 505–514.

Ewing, L., and Douglas Sherman. 1998. *California's Coastal Natural Hazards.* University of Southern California, Sea Grant Program.

Federal Emergency Management Agency. 2000. *Evaluation of Erosion Hazards.* U.S. government publication in association with the Heinz Center for Science, Economics, and the Environment.

Fulton, Kim. 1981. *A Manual for Researching Historical Coastal Erosion.* California Sea Grant College Program Report No. T-CSGCP-003.

Griggs, G. B. 1994. California's coastal hazards. *Journal of Coastal Research* (special issue no. 12) 1–15.

Griggs, G. B., and D. Scholar. 1997. Coastal erosion caused by earthquake-induced slope failure. *Shore and Beach* 65:4: 2–7.

Hable, J. S., and G. A. Armstrong. 1977. *Assessment and Atlas of Shoreline Erosion along the California Coast.* California Department of Navigation and Ocean Development.

Inman, Douglas L. 1976. *A Summary of Man's Impact on the California Coastal Zone.* California Department of Navigation and Ocean Development (now Boating and Waterways).

Kaufman, W., and O. Pilkey, Jr. 1984. *The Beaches Are Moving.* Durham, N.C.: Duke University Press.

Klee, Gary A. 1999. *The Coastal Environment: Toward Integrated Coastal and Marine Sanctuary Management.* Englewood Cliffs, N.J.: Prentice-Hall.

National Committee on Property Insurance. 1988. *America's Vanishing Coastlines.*

Robinson, G. D., and A. Speiker. 1978. *Nature to Be Commanded.* U.S. Geological Survey Professional Paper 950.

Shepard, F. P., and H. R. Wanless. 1971. *Our Changing Coastlines.* New York: McGraw-Hill.

U.S. Army Corps of Engineers. 1971. *National Shoreline Study: California Regional Inventory.* House Document 93–121.

Williams, S. J., K. Dodd, and K. F. Gohn. 1990. *Coasts in Crisis.* U.S. Geological Survey Circular 1075.

STORMS AND STORM DAMAGE

Domurat, G. W. 1978. Selected coastal storm damage in California, winter of 1977–78. *Shore and Beach* 46:15–20.

Griggs, G. B. 1999. California's coastline: El Niño, erosion, and protection. In *California's Coastal Natural Hazards*, CSBPA Conference, ed. by Lesley Ewing and Douglas Sherman. USC Sea Grant Program, pp. 36–55.

Griggs, G. B., and R. E. Johnson. 1983. The impact of the 1983 storms on the coastline of northern Monterey Bay. *California Geology* 36:163–174.

Griggs, G. B., and UCSC/USGS Coop. 1998. A collaborative program to investigate the impacts of the 1997–98 El Niño along the California coast. *Shore and Beach* 66:3:24–32.

Howe, Steve. 1978. *Wave Damage along the California Coast, 1977–1978*. San Francisco: California Coastal Commission.

Seymour, R. J. 1983. *Extreme Waves in California During Winter 1983*. Report to California Department of Boating and Waterways.

Storlazzi, C. D., and G. B. Griggs. 1998. The 1997–98 El Niño and erosion processes along the central California coast. *Shore and Beach* 66:3:12–17.

Storlazzi, C. D., and G. B. Griggs. 2000. The influence of El Niño–Southern Oscillation (ENSO) on the evolution of central California's shoreline. *Geological Society of America Bulletin* 112:2:236–249.

Storlazzi, C. D., C. M. Willis, and G. B. Griggs. 2000. Comparative impacts of the 1982–83 and 1997–98 El Niño winters on the central California coast. *Journal of Coastal Research* 16:4:1022–1036.

Swisher, Mary Lou. 1983. *Preliminary Report on January 1983 Coastal Storm Damage*. San Francisco: California Coastal Commission.

SHORELINE ENGINEERING AND PROTECTION

Dean, R. G., and R. A. Dalymple. 2002. *Coastal Processes with Engineering Applications*. Cambridge, UK: Cambridge University Press.

Fulton-Bennett, K. W., and G. B. Griggs. 1986. *Coastal Protection Structures and Their Effectiveness*. Joint Publication of California State Department of Boating and Waterways and Institute of Marine Sciences, University of California, Santa Cruz.

Griggs, G. B. 1986. Littoral cells and harbor dredging along the California coast. *Environmental Geology* 10:7–20.

Griggs, G. B. 1998. The armoring of California's coast. In *Proceedings of California and the World Ocean '97*. Reston, VA: American Society of Civil Engineers, pp. 515–526.

Griggs, G. B. 1999. The protection of California's coast: Past, present, and future. *Shore and Beach* 67:1:18–28.

Griggs, G. B. 2003. Headlands and groins: Replicating natural systems. *Journal of Coastal Research* (special issue) 33:280–293.

Griggs, G. B., and K. W. Fulton-Bennett. 1988. Riprap revetments and seawalls and their effectiveness along the central coast of California. *Shore and Beach* 56:3–11.

Griggs, G. B., and J. F. Tait. 1988. The effects of coastal protection structures on beaches along Monterey Bay, California. *Journal of Coastal Research* (special issue no. 4) 93–111.

Griggs, G. B., J. F. Tait, and W. Corona. 1994. The interaction of seawalls and beaches: Seven years of monitoring, Monterey Bay, California. *Shore and Beach* 62:3:21–28.

Kraus, N. C. 1988. The effects of seawalls on beaches: An extended literature review. *Journal of Coastal Research* (special issue no. 4) 691–701.

Kraus, N. C. 1996. The effects of seawalls on the beach: Part 1. An updated Literature Review. *Journal of Coastal Research* 12:691–701.

Moffatt and Nichol, Engineeers. 1983. *Construction Materials for Coastal Structures.* Fort Belvoir, Va.: Coastal Engineering Research Center, U.S. Army Corps of Engineers.

Pilkey, O. H., and K. L. Dixon. 1966. *The Corps and the Shore.* Washington, D.C.: Island Press.

Pilkey, O. H., Jr., Orrin H. Pilkey, Sr., Walter D. Pilkey, and William J. Neal. 1983. *Coastal Design: A Guide for Planners, Developers, and Homeowners.* New York: Van Nostrand Reinhold.

Pilkey, O. H., and H. L. Wright. 1988. Seawalls versus beaches. *Journal of Coastal Research* (special issue no. 4) 41–67.

Tait, J. F., and G. B. Griggs. 1990. Beach response to the presence of a seawall: A comparison of observations. *Shore and Beach* 58:11–28.

CONSERVATION AND PLANNING

Cicin-Sain, Biliana, R. W. Knecht, and N. Foster, 1999. *Trends and Future Challenges for U.S. National Ocean and Coastal Policy.* Proc. of a workshop. Washington, D.C.: NOAA.

Griggs, G. B. 1986. Reconstruction or relocation: Viable approaches for structures in areas of high coastal erosion. *Shore and Beach* 54:8–16.

Griggs, G. B. 1995. Relocation or reconstruction of threatened coastal structures: A second look. *Shore and Beach* 63:2:31–36.

Klee, Gary A. 1999. *The Coastal Environment: Toward Integrated Coastal and Marine Sanctuary Management.* Englewood Cliffs, N.J.: Prentice-Hall.

McHarg, Ian. 1971. *Design with Nature.* Garden City, N.Y.: Doubleday.

National Ocean Service, NOAA. 1990. *Coastal Management Solutions to Natural Hazards.* NOAA Technical Assistance Bulletin No. 103.

Tyler, M. B. 1994. *Look before You Build: Geologic Studies for Safer Land Development in the San Francisco Bay Area.* U.S. Geological Survey Circular 1130.

White, G. F., et al. 1976. *Natural Hazard Management in Coastal Areas.* Washington D.C.: U.S. Department of Commerce/National Oceanic and Atmospheric Administration, Office of Coastal Zone Management.

LEGISLATION AND REGULATIONS

Bowden, G. 1975. Legal battles on the California coast: A review of the rules. *Coastal Zone Management Journal* 2:3:273–296.

California Coastal Commission. 1975. *California Coastal Plan.* San Francisco: California Coastal Commission.

California State Lands Commission. 2001. *Shoreline Protective Structures.* Staff Report to the California State Lands Commission.

Graber, P. H. F. 1981. The law of the coast in a clamshell: Part III. The California approach. *Shore and Beach* 49:20–25.

Griggs, G. B. 1998. California needs a coastal hazards policy. *California Coast and Ocean* 14:3:30–33.

Griggs, G. B., J. Pepper, and M. E. Jordan. 1991. California's coastal hazards: A critical look at existing policies and practices. In *Proc. Coastal Zone '91.* Special volume on the California coastal zone experience, pp. 89–107.

Griggs, G. B., J. E. Pepper, and M. E. Jordan. 1992. *California's Coastal Hazards: A Critical Assessment of Existing Land-Use Policies and Practices.* Special publication of California Policy Seminar Program.

RECREATION

California Coastal Commission. 2003. *California Coastal Access Guide,* 6th ed. Berkeley: University of California Press.

Dixon, Sarah, and Peter Dixon. 1979. *West Coast Beaches: A Complete Guide.* New York: E. P. Dutton.

Kelley, D. G. 1971. *Edge of a Continent.* Palo Alto, Calif: American West.

GEOGRAPHICAL REFERENCES

NORTHERN CALIFORNIA COAST: OREGON TO SAN FRANCISCO

Domurat, W., et al. 1979. Beach erosion control study: Ocean Beach, San Francisco, California. *Shore and Beach* 47:20–32.

Hayden, M. 1976. *Exploring the North Coast from the Golden Gate to the Oregon Border.* San Francisco: Chronicle Books.

Huffman, M. E. 1972. *Geology for Planning on the Sonoma County Coast between the Russian and Gualala Rivers.* California Division of Mines and Geology Report.

Huffman, M. E., and C. F. Armstrong. 1980. *Geology for Planning in Sonoma County.* California Division of Mines and Geology Special Report 120.

Kilbourne, R. T., et al. 1980. *Geology for Planning Eureka and Fields Landing 7.5 Quadrangles, Humboldt County, California.* California Division of Mines and Geology Open File Report 80-9 SF.

Muller, Barbara D. 1981. *The Mendocino Coast.* Mendocino, Calif.: Community Land Trust.

Rust, Derek J. 1984. Coastal faulting and erosion hazards in Humboldt County, Northern California. In *Ocean Studies Symposium*. Asilomar, Calif.: California Coastal Commission, pp. 821–842.

Tuttle, D. C. 1981. *Investigation of Methods for Determining Coastal Bluff Erosion: Historical Section, Gold Bluffs to the Little River, Humboldt Co.* Humboldt County Public Works Sea Grant Report.

Tuttle, D. C. 1982. *Investigation of Coastline Retreat at Shelter Cove, California.* Humboldt County Public Works Sea Grant Report.

Wagner, D. L. 1977. *Geology for Planning in Western Marin County, California.* California Division of Mines and Geology Report.

Williams, John W., and Trinda L. Bedrossian. 1976. Coastal zone geology near Mendocino, California. *California Geology* 232–237.

Williams, John W., and Trinda L. Bedrossian. 1977. Coastal zone geology near Gualala, California. *California Geology* 27–34.

CENTRAL CALIFORNIA COAST: SAN FRANCISCO TO POINT CONCEPTION

Asquith, D. O. 1982. *Rates of Coastal Bluff Retreat, Pismo Beach, California.* In *Coastal Zone '83*. New York: American Society of Civil Engineers, pp. 1195–1207.

Best, T. C., and G. B. Griggs. 1991. A sediment budget for the Santa Cruz littoral cell, California. In *From Shoreline to the Abyss*. Society for Sedimentary Geology Special Publication No. 46, pp. 35–50.

Bowen, A. J., and D. L. Inman. 1966. *Budget of Littoral Sands in the Vicinity of Point Arguello, California.* U.S. Army Coastal Engineering Research Center Technical Memo 19.

California Coastal Commission. 1995. *ReCAP Pilot Findings and Recommendations: Monterey Bay Region.* San Francisco: California Coastal Commission.

Gallagher, J. 1996. *The Late Holocene Evolution of the Chorro Delta, Morro Bay, California.* Ph.D. dissertation, University of California, Los Angeles.

Griggs, G. B., and Brown, Kristin. 1998. Erosion and shoreline damage along the central California coast: A comparison between the 1997–98 and 1982–83 winters. *Shore and Beach* 66:3:18–23.

Griggs, G. B., and Fulton-Bennett, K. W. 1987. Failure of coastal protection at Seacliff State Beach, Santa Cruz County, California, *Environmental. Management* 11:175–182.

Griggs, G. B., and R. E. Johnson. 1976. Effects of the Santa Cruz Harbor on coastal processes of northern Monterey Bay, California. *Environmental Geology* 1:5: 299–312.

Griggs, G. B., and R. E. Johnson. 1979. Coastline erosion: Santa Cruz County, California. *California Geology* 32:4:67–76.

Griggs, G. B., and Johnson, R. E. 1983. Impact of 1983 Storms on the coastline of northern Monterey Bay. *California Geology* 36:8:163–174.

Griggs, G. B., and Jones, G. D. 1985. Erosion along an "equilibrium" coastline. In *California's Battered Coast*. Conference proc., California Coastal Commission, San Diego, pp. 102–119.

Gordon, B. L. 1979. *Monterey Bay Area: Natural History and Cultural Imprints*. Pacific Grove, Calif.: Boxwood Press.

Hansen, K. L., J. R. Wesling, W. R. Lettis, W. I. Kelson, and L. Mezger. 1994. Correlation, ages, and uplift rates of Quaternary marine terraces: South-central coastal California. In *Seismotectonics of the Central California Coast Ranges*, ed. by I. B. Alterman, R. B. McMullen, L. S. Cluff, and D. B. Slemmons. Geological Society of America Special Paper 292, pp. 45–71.

Jackson, R. A. 1977. *Combing the Coast: San Francisco through Big Sur*. San Francisco: Chronicle Books.

Kennedy, G. L., J. F. Wehmiller, and T. K. Rockwell. 1992. Paleoecology and paleozoogeography of late Pleistocene marine-terrace faunas of south-western Santa Barbara County, California. In *Quaternary Coasts of the United States: Marine and Lacustrine Systems,* ed. by C. H. Fletcher and J. F. Wehmiller. Society for Sedimentary Geology Special Publication 48, pp. 343–361.

LeBoeuf, Bernie, and Stephanie Kaza (Eds.). 1981. *Natural History of Año Nuevo*. Pacific Grove, Calif.: Boxwood Press.

Lettis, W. R., and N. T. Hall. 1994. Los Osos fault zone, San Luis Obispo County, California. In *Seismotectonics of the Central California Coast Ranges*, ed. by I. B. Alterman, R. B. McMullen, L. S. Cluff, and D. B. Slemmons. Geological Society of America Special Paper 292, pp. 73–102.

Moore. L. J., and G. B. Griggs. 2002. Long-term cliff retreat and erosion hot spots along the central shores of the Monterey Bay National Marine Sanctuary. *Marine Geology* 181:265–283.

Noble Consultants. 1989. *Coastal Sand Management Plan, Santa Barbara and Ventura County Coastline*. Main report, prepared for BEACON.

Noda, E. K., and Y. Jen. 1975. *Sand Transport Analysis, Morro Bay*. Report to U.S. Army Corps of Engineers, Los Angeles District. Pasadena: Tetra Tech.

Orme, A. R. 1990. The instability of Holocene coastal dunes: The case of the Morro dunes, California. In *Coastal Dunes: Form and Process*, ed. by K. F. Nordstrom, N. P. Psuty, and R. W. G. Carter. New York: Wiley, pp. 315–336.

Orme, A. R., 1992. Late Quaternary deposits near Point Sal, south-central California: A time frame for coastal-dune emplacement. In *Quaternary Coasts of the United States: Marine and Lacustrine Systems*, ed. by C. H. Fletcher and J. F. Wehmiller. Society for Sedimentary Geology Special Publication 48, pp. 309–315.

Orme, A. R. 1998. Late Quaternary tectonism along the Pacific coast of the Californias: A contrast in style. In *Coastal Tectonics*, ed. by I. S. Stewart and C. Vita-Finzi. Geological Society (London) Special Publication 146, pp. 179–197.

Orme, A. R., and V. P. Tchakerian. 1986. Quaternary dunes of the Pacific coast of the Californias. In *Aeolian Geomorphology*, ed. by W. G. Nickling. London: Allen & Unwin, pp. 149–175.

Rockwell, T. K., J. Nolan, D. L. Johnson, and R. H. Patterson. 1992. Ages and deformation of marine terraces between Point Conception and Gaviota, western Transverse Ranges, California. In *Quaternary Coasts of the United States: Marine and Lacustrine Systems*, ed. by C. H. Fletcher and J. F. Wehmiller. Society for Sedimentary Geology Special Publication 48, pp. 333–341.

U.S. Army Corps of Engineers, San Francisco District, and County of Santa Cruz. 2003. *East Cliff Drive Section* 103 *Shore Protection Project: Draft Detailed Project Report*.

Wieman, Harold. 1980. *Nature Walks on the San Luis Coast*. San Luis Obispo, Calif.: Padre.

Woodring, W. P., and M. N. Bramlette. 1950. *Geology and Paleontology of the Santa Maria District, California*. U.S. Geological Survey Professional Paper 222.

SOUTHERN CALIFORNIA COAST: POINT CONCEPTION TO THE INTERNATIONAL BORDER

Benumof, B. T., and G. B. Griggs. 1999. The relationship between seacliff erosion rates, cliff material properties, and physical processes, San Diego, Calif. *Shore and Beach* 67:4:29–41.

Benumof, B. T., L. J. Moore, and G. B. Griggs. 1999. FEMA and state-of-the-art coastal erosion mapping along the San Diego County, California, coastline. In *Proceedings of California's Coastal Natural Hazards,* ed. by Lesley Ewing and Douglas Sherman. USC Sea Grant Program, pp. 86–97.

Benumof, B. T., C. D. Storlazzi, R. J. Seymour, and G. B. Griggs. 2000. The relationship between incident wave energy and seacliff erosion rates: San Diego County, California. *Journal of Coastal Research* 16:4:1162–1178.

California Coastal Commission, 1999. *Findings and Recommendations for the Santa Monica Mountains/Malibu Regional Cumulative Assessment Project* (ReCAP). San Francisco: California Coastal Commission.

California Department of Boating and Waterways and San Diego Association of Governments. 1994. *Shoreline Erosion and Atlas of the San Diego Region*. State of California, Resources Agency.

California Department of Navigation and Ocean Development. 1977. *Study of Beach Nourishment along the Southern California Coastline*. State of California, Resources Agency.

Campbell, R. H. 1975. *Soil Slips, Debris Flows, and Rainstorms in the Santa Monica Mountains and Vicinity, Southern California*. U.S. Geological Survey Professional Paper 851.

Coastal Frontiers. 1999. *Historical Changes in the Beaches of Los Angeles County, Malaga Cove to Topanga Canyon*, 1935–1990. County of Los Angeles, Department of Beaches and Harbors.

Emery, K. O. 1960. *The Sea off Southern California: A Modern Habitat of Petroleum.* New York: Wiley.

Inman, D. L., and J. D. Frautschy. 1966. Littoral processes and the development of shorelines. In *Santa Barbara Specialty Conference on Coastal Engineering,* pp. 511–536.

Kennedy, M. P. 1973. Sea cliff erosion at Sunset Cliffs, San Diego. *California Geology* 26:27–31.

Knur, R. T. and Y. C. Kim. 1999. Historical sediment budget analysis along the Malibu coastline. In *Sand Rights '99: Bringing Back the Beaches.* Ventura, Calif.: ASCE.

Kuhn, G. G., E. D. Baker, and F. P. Shepard. 1980. Greatly accelerated man-induced coastal erosion and new sources of beach sand, San Onofre State Park and Camp Pendleton, northern San Diego County, California. *Shore and Beach* 48:4:9–13.

Kuhn, G. G., and F. P. Shepard. 1979. Accelerated beach-cliff erosion related to unusual storms in Southern California. *California Geology* 32:3:58–59.

Kuhn, G. G., and F. P. Shepard. 1980. Coastal erosion in San Diego County. In *Coastal Zone '80:* 1899–1918, New York: American Society of Civil Engineers.

Kuhn, G. G., and F. P. Shepard. 1981. Should southern California build defenses against violent storms resulting in lowland flooding as discovered in records of the past century? *Shore and Beach* 49:4:2–10.

Kuhn, G. G., and F. P. Shepard. 1984. *Sea Cliffs, Beaches, and Coastal Valleys of San Diego County: Some Amazing Histories and Some Horrifying Implications.* Berkeley: University of California Press.

Lajoie, K. R., D. J. Ponti, C. L. Powell, S. A. Mathieson, and A. M. Sarna-Wojcicki. 1991. Emergent marine strandlines and associated sediments, coastal California: A record of Quaternary sea-level fluctuations, vertical tectonic movements, climatic changes, and coastal processes. In *Quaternary Nonglacial Geology: Conterminous U.S. (The Geology of North America,* vol. K-2). Boulder, Colo.: Geological Society of America, pp. 190–203.

Leidorsdorf, C. B., R. C. Hollar, and G. Woodell. 1993. Human intervention with the beaches of Santa Monica Bay, California. *Shore and Beach* 62:3:29–38.

McEwen, G. F. 1935. Destructive high waves along the southern California coast. *Shore and Beach* 3:2:61–64.

Moore, L. J., B. T. Benumof, and G. B. Griggs. 1999. Coastal erosion hazards in Santa Cruz and San Diego Counties, California. *Journal of Coastal Research* (special issue) 28:121–139.

Norris, R. M. 1964. Dams and beach sand supply in southern California. In *Papers in Marine Geology,* ed. by R. L. Miller. New York: Macmillan, pp. 154–171.

Norris, Robert M. 1968. Sea cliff retreat near Santa Barbara, California. *Mineral Information Service* 21:6:87–91.

Orme, A. R. 1991. Mass movement and seacliff retreat along the southern California coast. *Southern California Academy of Sciences Bulletin* 90:58–79.

Orme, A. R. 1998. Late Quaternary tectonism along the Pacific coast of the Californias: A contrast in style. In *Coastal Tectonics*, ed. by I. S. Stewart and C. Vita-Finzi. Geological Society (London) Special Publication 146, pp. 179–197.

Orme, A. R. 2000. Evolution and historical development. In *Lower Malibu Creek and Lagoon: Resource Enhancement and Management*, ed. by R. F. Ambrose and A. R. Orme. Oakland: California State Coastal Conservancy, pp. 1–37.

Orme, A. R., and A. J. Brown. 1983. Variable sediment flux and beach management: Ventura County, California. In *Coastal Zone '83*. New York: American Society of Civil Engineers, pp. 2328–2342.

Orme, A. R., and A. J. Orme. 1988. U.S.A.—California. In *Artificial Structures and Shorelines*, ed. by H. J. Walker. Amsterdam: Kluwer Academic, pp. 513–528.

Pipkin, B., and M. Ploessel. 1972. *Coastal Landslides in Southern California*. University of Southern California Sea Grant Publication No. USC-SG-3-72.

Rockwell, T. K., E. A. Keller, M. N. Clark, and D. L. Johnson. 1984. Chronology and rates of faulting of Ventura River terraces, California. *Geological Society of America Bulletin* 95:1466–1474.

Sharp, R. P. 1978. *Coastal Southern California*. Dubuque, Iowa: Kendall/Hunt.

U.S. Army Corps of Engineers. 1986. *Oral History of Coastal Engineering Activities in Southern California, 1930–1981*. Los Angeles: U.S. Army Corps of Engineers.

U.S. Army Corps of Engineers. 1991. *Coast of California Storm and Tidal Wave Study: State of the Coast Report, San Diego Region*. Los Angeles: U.S. Army Corps of Engineers.

U.S. Department of Agriculture. 1995. *Calleguas Creek Watershed Erosion and Sediment Control Plan for Mugu Lagoon, Ventura and Los Angeles Counties, California*. USDA/Natural Resources Conservation Service, Davis, California.

Yeats, R. S., and T. K. Rockwell. 1991. Quaternary geology of the Ventura and Los Angeles basins, California. In *Quaternary Nonglacial Geology: Conterminous U.S. (The Geology of North America*, vol. K-2). Boulder, Colo.: Geological Society of America, pp. 185–189.

EON	ERA	PERIOD	EPOCH	BEGAN (MYA)	ENDED (MYA)	LASTED (MY)	
Phanerozoic	Cenozoic	Quaternary	Holocene	0.01	—	—	1.60
			Pleistocene	1.60	0.01	1.59	
		Tertiary	Pliocene	5.30	1.60	3.70	64.80
			Miocene	23.70	5.30	18.40	
			Oligocene	36.60	23.70	12.90	
			Eocene	54.60	36.60	18.00	
			Paleocene	66.40	54.60	11.80	
	Mesozoic	Cretaceous		144.00	66.40	77.60	178.60
		Jurassic		208.00	144.00	64.00	
		Triassic		245.00	208.00	37.00	
	Paleozoic	Permian		286.00	245.00	41.00	325.00
		Carboniferous		360.00	286.00	74.00	
		Devonian		408.00	360.00	48.00	
		Silurian		438.00	408.00	30.00	
		Ordovician		505.00	438.00	67.00	
		Cambrian		570.00	505.00	65.00	
Proterozoic (Precambrian)				2600.00	570.00	2030.00	4030.00
Archeozoic (Precambrian)				4600.00	2600.00	2000.00	

my = millions of years, mya = millions of years ago

CONTRIBUTORS

REINHARD FLICK

Reinhard Flick is staff oceanographer with the California Department of Boating and Waterways and a research associate at Scripps Institution of Oceanography, UCSD. He received his BS in physics from Cooper Union in 1970, and a Ph.D. in oceanography from Scripps Institution of Oceanography in 1978. Dr. Flick conducts research; manages the California Department of Boating and Waterways Oceanography program at Scripps; lectures; and provides expert consulting in the areas of tides, sea level fluctuations, waves, beach processes and erosion, and coastal storm damage. He was the 2002 recipient of the American Shore and Beach Preservation Association Morrough P. O'Brien Award.

KIM FULTON-BENNETT

Kim Fulton-Bennett is a science writer presently working at the Monterey Bay Aquarium Research Institute (MBARI) in Moss Landing, California. His academic background includes an undergraduate degree in Earth Sciences and Environmental Studies, as well as a master's degree in Earth Science from the University of California at Santa Cruz. He has worked in a wide variety of fields, including geological consulting, land use planning, teaching, and technical writing, and has carried out research in marine geology, marine biology, and aquaculture.

GARY GRIGGS

Dr. Griggs has been a professor of Earth Sciences at the University of California, Santa Cruz, since 1968 and has served as chairman of the

Department of Earth Sciences. He is presently the director of the Institute of Marine Sciences and also the chair of the University of California Marine Council. He was a senior Fulbright scholar in Greece and has also conducted collaborative marine research in Italy and New Zealand. His research over the past 35 years has focused on coastal hazards and land use planning along the California coast, and includes work on coastal erosion, littoral cells and sediment budgets, coastal protection structures and their effectiveness and impacts, and the effects of large coastal engineering structures on the shoreline. He was the 2003 recipient of the American Shore and Beach Preservation Association Joe Johnson Award. Dr. Griggs has written or co-written over 135 articles published in professional journals as well as five books: *Geologic Hazards, Resources, and Environmental Planning; Living with the California Coast; Coastal Protection Structures and Their Effectiveness; California's Coastal Hazards: A Critical Assessment of Existing Land-Use Policies and Practices;* and *Formation, Evolution, and Stability of Coastal Cliffs: Status and Trends.*

KAREN GROVE

Karen Grove received her Ph.D. in geology from Stanford University in 1989 and has been a professor of Geology and Oceanography at San Francisco State University since then. She studies sediments along the coast of California and teaches a variety of geoscience classes.

CHERYL HAPKE

Cheryl Hapke received her undergraduate degree in Geology from the University of Pittsburgh, a master's degree from the University of Maryland in 1992, and a Ph.D. in Geology from the University of California, Santa Cruz, in 2002. From 1997 to 2005, Dr. Hapke's research focused on issues of coastal erosion and coastal hazards along the California coast and in Hawaii while she was employed by the U.S. Geological Survey at their Pacific Science Center in Santa Cruz. She is now working as a coastal geologist with the USGS Patuxent Wildlife Research Center, stationed at their Coastal Field Station at the University of Rhode Island.

CHARLES F. LESTER

Charles F. Lester is the deputy director for the North Central Coast and Central Coast districts of the California Coastal Commission, a state agency that plans for and regulates development along California's coast. Charles received his bachelor's degree in Geochemistry from Columbia College in

New York City. He attended the University of California at Berkeley, receiving a J.D. from Boalt Hall School of Law in 1989 and a Ph.D. in Jurisprudence and Social Policy in 1992. Charles's doctoral dissertation evaluated the implementation of the federal offshore oil and gas program. Before coming to the Coastal Commission, Charles was a professor of Environmental Policy and Law at the University of Colorado at Boulder from 1993 to 1997. His areas of expertise include coastal and environmental policy, law and politics, governance and public law, coastal zone management; and environmental sustainability.

DOROTHY MERRITTS

Dorothy Merritts, a professor in the Department of Earth and Environment at Franklin Marshall College in Lancaster, Pennsylvania, is a geologist with expertise in coastal geomorphology, streams, rivers, and other landforms, and in the impact of geological hazards and human land use on landscapes. Her research in tectonics and topography focuses on active faulting, deformation, and base-level change, and particularly on the landforms of marine terraces (coral and wave-cut), longitudinal stream profiles, fluvial terraces, fault scarps, and alluvial fans. Dr. Merritts has served as a scientific advisor to the South Korean government (Korean Institute of Geology, Mining, and Materials) on coastal tectonics, marine terraces, and earthquake hazards from 1999 to the present. She is the lead author of an introductory textbook titled *Environmental Geology: An Earth Systems Approach.*

ROBERT M. NORRIS

Robert M. Norris is a professor emeritus at the University of California, Santa Barbara, where he taught for 40 years. His undergraduate training was at UCLA and was followed by 2.5 years in the Navy, where he was involved in the Iwo Jima and Okinawa actions. After the war, he earned a master's degree at UCLA and, taking advantage of his sea-going experiences and the newly developed field of marine geology, went to Scripps Institution of Oceanography and earned a Ph.D. under Francis P. Shepard in 1951. His research interests are varied and include numerous studies of coastal erosion, shelf sedimentation, and Quaternary environments. He is co-author of *The Geology of California.*

ANTONY R. ORME

Antony R. Orme (Ph.D. 1961, University of Birmingham, England) is professor of Geography at the University of California, Los Angeles, and editor-in-chief of the journal *Physical Geography.* His main research interests involve geomorphology, Quaternary studies, and environmental management, in which his research along the California coast focuses on relative sea-level change, beach and estuary morphodynamics, barrier-lagoon systems, dunes, wetlands, and management issues. He has also studied the coasts of the British Isles, the Caribbean, Africa, the Middle East, and the Philippines. Apart from research articles, he is the author the book *Ireland* (1970) and editor of and contributor to *The Physical Geography of Africa* (1996), *The Physical Geography of North America* (2002), and *The Physical Geography of South America* (forthcoming), published by Oxford University Press.

KIKI PATSCH

Kiki Patsch received her undergraduate degree in Environmental Science at the University of Virginia and completed her Ph.D. in Earth Sciences at the University of California, Santa Cruz. Her professional interests lie primarily in coastal hazards and shoreline change. Her recent research focuses on the development of sediment budgets for the littoral cells of California and how human activity has influenced those budgets.

BERNARD PIPKIN

Dr. Bernard Pipkin is professor emeritus in the Department of Earth Sciences at the University of Southern California. His specialty is geological hazards and he has authored three books and numerous research papers on the subject. He hosted the 30-segment Emmy-winning educational series *Oceanus,* which was shown nationally on public television.

DEREK RUST

Derek Rust is a geologist at Brunel University in England, having received his B.Sc. from the University of London and his Ph.D. from the University of California. Dr. Rust's research interests center on geological hazards, particularly neotectonics and paleoseismology, volcano edifice instability, and coastal slope failure and retreat. He has carried out published field research in Britain, North and South America, southern Africa, New Zealand, Sicily, Crete, Turkey, the Middle East, Kamchatka, and the Philippines.

LAURET SAVOY

Lauret Savoy received an undergraduate degree in Geology from Princeton University; a master's degree in Earth Sciences from the University of California, Santa Cruz; and a Ph.D. in Geology from Syracuse University.

She writes about the shaping of cultural identity by its relationship with and dislocation from the land. Her goal is to produce multiple narratives of such connections and edges from stories we tell of land and its origin and history, to stories we tell of ourselves in the land and of relational identity. Also a photographer and professor of Geology and Environmental Studies at Mount Holyoke College, Massachusetts, she co-edited *The Colors of Nature: Culture, Identity, and the Natural World* with Alison Deming. Savoy is editing, with Eldridge and Judy Moores, the forthcoming anthology *Reading Earth: The Literature of Geology* (under poet-publisher Barbara Ras). She was also co-author and editor of the first edition of *Living with the California Coast.*

DOUGLAS SHERMAN

Douglas Sherman is a professor of Geography and head of the Department of Geography at Texas A&M University. His academic credentials include a Ph.D. in Geography from the University of Toronto and a postdoctoral scholarship in the Ocean Engineering Department at Woods Hole Oceanographic Institution. He has held visiting positions at the Queen's University of Belfast; the universities of Auckland and Sydney; the Australian Defense Force Academy, University College, Cork; and the University of Amsterdam. He is currently a member of the AAAS Committee for Geology and Geography.

Professor Sherman's research expertise is in near-shore processes and sediment transport in coastal systems. He has produced more than 100 publications, and NSF, NOAA, Sea Grant, National Geographic Society, National Park Service, California Department of Boating and Waterways, and the California Coastal Conservancy have supported his research.

ROBERT WALKER

Robert Walker is a geologist, geochemist, and geochronologist. He has conducted research in east Africa, North America, and Asia and is a leading expert on time scale calibrations and human evolution. Formerly chief geologist for the Institute of Human Origins in Berkeley, California, he is an associate professor in the Department of Earth and Environment at Franklin Marshall College in Lancaster, Pennsylvania. He was elected a Fellow of the California Academy of Science in 1995 and was selected by the American Association for the Advancement of Science to be a Diplomacy Fellow to the U.S. Department of State in 2002–2003. He is the author of over 60 scientific articles and has co-written and co-produced five documentary films.

JERRY WEBER

Jerry Weber is a semi-retired consulting engineering geologist in Santa Cruz, California. He is a geological generalist with wide-ranging interests, but his main research interests involve geomorphology, Quaternary studies, coastal processes, marine terraces, and the neotectonics of the central California coastline. His professional experience includes 6 years as an exploration geologist in the oil industry, 35 years as a consulting engineering geologist, and over 20 years of university and community college teaching, including 19 years of teaching Summer Field Geology at the University of California, Santa Cruz.

INDEX

COMPOSITOR: Publication Services, Inc.

CARTOGRAPHER: Kiki Patsch

NEW LINE ILLUSTRATIONS: Shannon Griggs

TEXT: 10/13.5 Minion

DISPLAY: Franklin Gothic Book and Semibold

PRINTER AND BINDER: Edwards Brothers, Inc.